Built-In Test for VLSI:
Pseudorandom Techniques

Built-In Test for VLSI: Pseudorandom Techniques

PAUL H. BARDELL
WILLIAM H. McANNEY
JACOB SAVIR

International Business Machines Corporation
Poughkeepsie, New York

A WILEY-INTERSCIENCE PUBLICATION

JOHN WILEY & SONS

New York · Chichester · Brisbane · Toronto · Singapore

A NOTE TO THE READER
This book has been electronically reproduced from
digital information stored at John Wiley & Sons, Inc.
We are pleased that the use of this new technology
will enable us to keep works of enduring scholarly
value in print as long as there is a reasonable demand
for them. The content of this book is identical to
previous printings.

Library of Congress Cataloging in Publication Data:

Bardell, Paul H.
  Built-in test for VLSI.

  "A Wiley-Interscience publication."
  Bibliography: p.
  Includes index.
  1. Integrated circuits--Very large scale integration
  Testing.   I. McAnney, William H.   II. Savir, Jacob.
  III. Title.

TK7874.B374   1987   621.381'73   87-23013
ISBN: 0-471-62463-2

10 9 8 7 6 5 4 3

To Adrienne, Patricia, and Bilha

Preface

This book is an outgrowth of the notes that we developed for our tutorial on built-in test at the International Test Conference and owes much to our interaction with and feedback from our audiences. It is a handbook for the experienced professional test engineer and an introduction to built-in testing for graduate students as well as for management and technical leaders. We have collected here all of the relevant topics relating to random pattern (or more properly pseudorandom pattern) built-in test. As with many emerging topics, those who contribute to this field work in several different disciplines, so the notation and the mathematical treatment varies from source to source. A major effort has been made to provide a consistent treatment of the theory and its mathematics.

Chapters 1 and 2 are introductory material used to motivate and set a standard for built-in test. Chapter 1 discusses and describes the problems with conventional testing in light of the growth in complexity of digital circuits, and Chapter 2 reviews the principles of design for testability as a point of departure for the sequel.

From this base, we continue with four chapters on test sequence generation and response data compression. In Chapter 3, on pseudorandom sequence generators, the mathematics of shift-register sequences is developed to show its interesting and useful properties for built-in testing. Chapter 4 introduces various data compression methods, emphasizing a comparison between their information loss characteristics, and Chapter 5 expands upon the most useful of these compression methods—the use of polynomial dividers. The section concludes with Chapter 6, which examines certain unique shift-register sequence generators that have application in special situations requiring the use of built-in testing.

Chapter 7 covers in detail the concept of random pattern testing and the problem of assessing the efficiency of such tests in detecting faults in combinational and sequential digital circuits. The treatment extends to memory testing, the problem of detecting intermittent faults, and random pattern tests for delay faults.

Some of the physical concepts that have been used or proposed for use are described comparatively in Chapter 8 to show the achievable in built-in testing and as a guidebook for future innovation. The penultimate chapter (Chapter 9) discusses the limitations of built-in testing and some of its practical concerns. Finally, Chapter 10 describes the test support systems needed for a successful built-in test. The references cited are included following the text.

It is our hope that by presenting this material in a unified manner, it is more useful to the practicing test engineer as well as easier for the advanced graduate student to grasp the subject and be able to apply the appropriate part of the solution of his problem. Some logic design and statistical background are assumed.

PAUL H. BARDELL
WILLIAM H. MCANNEY
JACOB SAVIR

Poughkeepsie, New York
August 1987

Contents

1

Digital Testing and the Need for Testable Designs

Testing of digital circuits is a major portion of the effort in their design, production, and use. As the electronics industry evolved from discrete components, through the early integrated circuits that contained a single gate or flip-flop per package, to increasingly high levels of integration, the demands on testing methods and their effectiveness have caused test technology to evolve to a high degree of sophistication. As the number of circuits that can be integrated into one piece of silicon approaches one million, it would seem that some small portion of those circuits could be devoted to the testing or assurance of the function the remainder are intended to implement. This concept is called "built-in test" and will be explored in this book.

When digital electronics consisted entirely of discrete components (vacuum tubes, transistors, resistors, capacitors, and diodes), testing was essentially done in three parts. First, each individual part was tested to a specification. Second, these parts were assembled into digital elements—gates, flip-flops, and so forth—and the elements were tested to determine whether they performed their intended function. Finally, functional testing was done at the application or system level, where all of the interactions of timing, loading, temperature, and noise came into play. As computers and other digital systems became increasingly complex, this final test step became quite burdensome. There was no way to determine the adequacy of the test, and the problem of locating and removing faulty components by attempting to operate the system in its intended manner was difficult. R. D. Eldred (1959) showed that there was an effective way to test the *hardware* of a system rather than its function. This was the beginning of structural testing as we know it today. His work spawned an entirely new discipline of test and test generation for digital networks.

1

The first application of these new test generation techniques was to printed circuit boards (or their equivalents). This enabled a manufacturer to eliminate almost all of the assembly-related defects from the subassemblies prior to installation in a system. With this reduced number of defects to contend with, the demonstration of system function was a much easier, but still nontrivial, task.

As integrated circuits (ICs) came on the scene, the test problems were not really different from those of printed circuit boards except for one subtle difference. Whereas the object of circuit board testing is to find defective components and repair the board so that it meets its specification, the object of integrated circuit testing is to determine if the circuit is functioning. In the IC case no repair is possible, so a failed test results in a decision to discard a unit. Initially, the integrated circuits that were tested consisted of single gates, single flip-flops, and the like. This rapidly grew into several gates, multiplexors, small counters, short shift registers, and so forth, but the basic test techniques from the circuit board field were still applicable. As the number of circuits on an integrated circuit chip grew to over 100 (circa 1970), the test problem began to change. The small geometry of integrated circuits permits only a few connections to probe pads or package terminals. The exact relationship varies, but in general the problem of that era was to test hundreds of circuits through tens of terminals. As the advance of technology proceeded into the 1980s a ratio of 1000 circuits behind each pin became the nature of the test problem.

At the level of tens of thousands of circuits per silicon chip, many new applications appear for low cost integrated circuits. As these find their way into the daily lives of everyone, the demand that they be adequately tested becomes intense. The inconvenience, or indeed danger, that can result from a faulty integrated circuit in an application is not something a manufacturer of such circuits can ignore. This pressure for quality translates into a demand for better test techniques that can be applied without appreciable increased cost to ICs designed for these low cost applications.

1.1   THE EVOLUTION OF TEST TECHNOLOGY

The test technology that has evolved along with the digital system technology involves three distinct but interrelated areas. Test hardware, test software, and test theory have developed together into a solid basis for the test technology that supports the digital electronics industry. However, the technology must be extended to include built-in test in order for progress to continue.

Test hardware, or more properly, test systems have evolved from rather simple collections of power supplies and measuring devices interconnected by busses and relays to large electronic systems that have digital-to-analog and analog-to-digital functions at each pin. We also see buffer memory at each pin so that patterns can be applied at speeds approaching the application speed of the devices under test. Some systems have algorithmic pattern generators built into the hardware, and some that are designed to test memory devices have extensive real time results analysis capability.

While such test systems do give excellent performance, they are expensive and they tend to be self-obsolescent. That is they are made from the same generic integrated circuits that they are designed to test, hence by the time they are designed and built by their manufacturers, and purchased and installed by the users, the leading edge of the integrated circuit technology has moved on, creating a test system requirement that the "new" test system can only marginally satisfy. However, owing to their high initial costs (in excess of one million dollars) most firms must keep this equipment in service for many years to justify their investment. This "old iron" problem is one of the driving forces for the development of built-in test of digital circuits. If some or all of the test can be performed by circuitry resident in the integrated circuit itself, the dependence on the latest, most sophisticated test system is lessened to a great extent.

Test software is actually two groups of applications. There is the software that runs the test systems, since nearly every test system has an embedded computer of some variety. The programs that run on this computer control the functions of the tester and apply test programs for specific parts under test. While there is considerable effort to standardize in this area, most testers have a unique language that must be used to communicate with them. This is a problem but not a fundamental limitation.

The other group of application programs has to do with the automation of portions of the design process. These design automation programs have a firm foundation in the underlying test theory. Automatic test generation for combinational networks, fault simulation, and checking of design rules are among the programs that support test. These application programs have been developed to a high degree of sophistication, but as the circuits that they must deal with become ever larger, the computational complexity of tasks like test generation begin to limit progress.

Test theory has developed to the point that much can be said about the testability of a particular circuit. While theory may be an improper word to describe some of the ad hoc procedures that are lumped under the heading "design for testability," the proven effectiveness of these techniques shows that the theory is in reasonably good shape. The test theory aspect of test technology will be extensively explored in this book.

1.2 FAULT MODELS

The object of testing digital networks is to detect faults that are present. Central to this is a model of the faults to be detected. The elements that make up a network are subject to physical defects that can cause them to malfunction. These defects can be manufacturing defects such as open interconnections, shorts between conductors, excess leakage current, and others, or they can be service-related defects such as electromigration or burnout due to electrical overload. Whatever the defect, in order for its effect on the logical behavior of the network to be assessed, it must be modeled in a manner that is consistent with the representation of the network. Unless otherwise noted, networks are represented at the logical gate level in this book. Thus the fault model abstracts the effect of a physical defect into a stuck condition on one of the terminals of the gate that hosts the defect.

The fault model specifies the range of physical defects that can be detected by a given test procedure. The original work of Eldred defined the now classical stuck-at model. By calling a logical level of 0 to be ground (0 V) and a logical level of 1 to be at the positive supply level, he postulated a fault model wherein some signal lines could be shorted to ground (stuck-at-0) or at the potential of the positive supply (stuck-at-1). This has been generalized to apply to any fault condition that causes a logic gate to behave as though one of its inputs or its output is stuck at a logical 1 or 0. This effect can be caused by connections within the gate being open, as well as by shorts as implied by Eldred's definition. This stuck-at model has been the mainstay for the development of test theory ever since. While more complex fault models have been proposed, the stuck-at model is still the model against which all others are compared.

Faults can occur singly or in multiples. A large portion of the literature in this field deals with singly occurring faults. While this simplifies the analysis, there is concern (particularly in manufacturing testing) that multiple faults do occur. When one begins to consider the occurrence of multiple stuck-at faults in the network to be tested, the sheer number of possible multiple fault situations makes the enumeration at best difficult.

Shorts or faults that bridge between two conductors in the network cause two classes of fault behavior. If the two (or more) lines that are shorted together form a feedback path that creates a new state in which the network can exist, the fault is called a *sequential fault*. If the short does not form this type of feedback path, it is called a *combinational fault*. Experience has shown that most (but not all) shorts of the combinational fault type are detected by a high coverage (99%) test set based on the single stuck-at model.

Perhaps the most notable defects that are not adequately abstracted by the stuck-at model are the CMOS (complementary metal oxide silicon) stuck-open and stuck-short defects [first described in Wadsack (1978)]. The stuck-open defect causes a sequential fault whose detection requires the sequential application of a particular pair of test patterns. While this phenomenon is of concern, Zasio [in Zasio (1985)] has shown that by prudent design, the occurrence of stuck-open faults can be avoided in CMOS designs. CMOS stuck-short defects require a form of parametric fault model, where abnormal currents are detected between switching.

Physical defects can cause parametric faults, such as excessive currents, or unacceptable delays. Tests for these faults usually fall into the class of input loading and output driving tests. An example of a test for delay faults is an access-time test of a random-access memory (RAM). Depending on the technology involved and the application, more parametric faults may be subjected to special tests.

Another class of faults is pattern sensitive faults. In this case the circuit under consideration has a fault whose effect is dependent on the state of some other circuit in the network. For example, circuit A has a fault such that a failure is manifested if and only if circuit B is in the 1 state. Circuit B functions properly in both states. Most of the analysis of faults of this class has been directed at the testing of random access memories [for example, Hayes (1975a)], although the phenomena has been observed in combinational logic [Curtin and Waicukauski (1983)].

In the pseudorandom testing covered in this book, it will be shown that such testing yields coverage of faults beyond the single stuck-at fault. However, an important measure of the fault coverage of pseudorandom testing is its effectiveness against the classical stuck-at fault model.

1.3 STRUCTURAL AND FUNCTIONAL TESTING

Prior to the work reported by Eldred, digital networks were tested by demonstrating that they would perform their intended function. That is, an adder unit would be tested to see if it could add; a memory unit would be tested to see if it could be accessed, it could store, and if data could be retrieved from it. This provided a good measure of proper operation, but in the presence of faults, it was extremely difficult to isolate the fault. Another problem with this approach is that exhaustive sets of operands are clearly impractical, so one never knows if a defect is present that will result in a fault for some combination of data that was not in the limited test set used. This problem with the effectiveness measure of functional testing drove most of the work on testing to the structural approach proposed by Eldred. In preparing a structural test for a digital network, the network is first described in terms of logic primitives such as AND, OR, and NOT, and perhaps some macro representations of commonly used more complex functions such as EXCLUSIVE OR, flip-flops, and adders. Each of the logic elements are faulted in turn, and a test is generated for that fault. A practical technique to reduce the search time consumed by test generation algorithms is to fault simulate the test pattern generated for one fault against all other faults in the network. Those additional faults detected by the pattern are marked as tested. Experience shows that initially many faults are detected by the same pattern, while later in the process only one new fault is detected by a test pattern.

Built-in testing can be either structural or functional in nature. Pseudorandom testing is a type of built-in test that performs a structural test of the network involved. A particularly appealing feature of pseudorandom patterns is that they can be generated by simple built-in generators. While there are other built-in test approaches that perform either functional or structural testing, they either work with stored deterministic patterns or execute functional programs that are designed to exercise (test) specifiç parts of the network. Another class of structural built-in test is exhaustive testing, which requires logical partitioning to make the tests practical. The scope of this book is limited to the discussion of pseudorandom testing.

1.4   ON-LINE VERSUS OFF-LINE TESTING

In the initial stages of manufacture, testing is performed to ascertain the proper structure or function of a part, subassembly or system. At some subsequent point in time, the system is operated as intended to perform some assigned task. It is said to be "on-line" in this mode. Often it is desired to test the system after it has been put on-line. This may be to assure readiness for a critical mission or job, it may be required to monitor error recovery procedures, or to measure the fault-tolerant status of the system. If a procedure to test the system is performed as a task in the job stream of the system while other tasks are ongoing, the test is said to be an on-line test. If the system must be shut down and/or dedicated to the test procedure in such a manner that the normal system function stops completely, the test procedure is said to be an off-line test. In other words, the system must be taken off-line for the test to be performed.

Clearly the initial manufacturing test operations are off-line tests. In fact most of the work to date in the field of digital network testing has been focused on off-line testing.

1.5   THE RELATIONSHIP BETWEEN DESIGN AND TEST

The process of taking a requirement for a digital network and implementing that network in hardware begins with design and ends with test. These can be thought of as two separate steps or as two closely integrated tasks. When a designer has completed his design, he first checks it, then releases it to fabrication. Depending on the scale, this may be a prototype card in his own experimental laboratory or an order placed with a silicon foundry. In either event, when he receives the hardware that he designed, he must verify that his design is correct—that it meets some functional specification. In order to do this, the designer must apply some stimuli to the hardware and observe some responses from the hardware. With this information he must decide if the design meets the requirement. To some, this design verification process may be considered "test." Indeed it is a test, but it is not the only (or most effective) process available for verification of a design. In the step called "checking" above, modern simulation tools play an increasingly important role in design verification.

During the simulation of a design, the designer can apply many stimuli to the software model of the digital network and evaluate the responses. In this way he can assure himself that his design meets the requirements that he set out to satisfy. The simulation may be entirely in software, on a general purpose digital computer, or on a special purpose simulation engine designed especially to perform logic simulation. Alternatively, the simulation can be done in a hybrid fashion, where those portions of the system that the designer is designing that are available in hardware are used in conjunction with software models of the portion being evaluated. In either mode, modern simulation tools are available to the designer to verify that his design meets the requirements.

Often the simulation that the designer does is based on a register transfer level of design detail. Since this design must be translated to devices and interconnections at the silicon level, several sources of errors must be dealt with. If a manual translation from register transfer level to gate level description is performed, an equivalence check must be performed. Automatic programs that translate register transfer level descriptions to gate level descriptions (for subsequent test generation) and gate level descriptions to mask level layout descriptions are available in varying degrees of sophistication. These automatic programs preserve the correctness of the design, thus allowing the designer to rely heavily on the software verification done at the higher level.

If sufficient design verification is done in the software stages of implementation, the purpose of test is to determine that the hardware has been properly fabricated. As discussed earlier, a structural test is best for determining proper structure, however a functional test is a more direct step if design verification is still a major objective of the test process. In Chapter 2, we discuss some of the principles of design for testability. These techniques rest on a few fundamental principles: (a) test generation for sequential circuits is "hard," (b) test generation for combinational circuits is "tractable," and (c) by following "simple" design rules to aid testability, fully testable designs are obtained. Most of these design rules implement some version of scan-path-oriented design wherein all internal storage elements are configured as a scan path for testing. In this manner, each storage element becomes a pseudoterminal for the purposes of test. The cost of such a design philosophy is often cited as a reason to avoid design for testability practices. In more and more organizations, the market demands for ever higher quality products are forcing a high testability requirement into the initial specification from which the designer is working. In these cases, the circuit cost of incorporating such structures as scan path into the design is viewed as a necessary part of implementing the requirement. Thus testability is becoming part of the requirement and the disciplines of design and test are becoming inextricably intertwined.

When built-in test is incorporated in a digital network to meet a testability specification the distinction between design and test becomes moot. At that point, test is a part of design.

1.6   THE RELATIONSHIP BETWEEN PACKAGING LEVELS AND TEST

Digital networks are tested in many forms and types of packaging and assembly. The most fundamental test is at the semiconductor wafer level. Unfortunately, today there is a limit to the testing that can be performed at this level due to the need to probe the wafer. Not only must stimuli and response signals be transmitted to and from the wafer under test through probes, but power must also be supplied through these probes. Under some conditions, a high performance semiconductor device (network) cannot be adequately tested in the wafer form because of noise problems associated with the probing environment. These may be problems with inductive power supply leads, mutual inductance coupling between signal probes, or signal reflections from impedance discontinuities produced by the probes. There are other reasons to perform less than complete tests at the wafer level. These are primarily economic trade-offs between the cost of providing an acceptable test at the wafer level in the face of the preceding difficulties, and the extra value that subsequent processing would add to a part that could have been rejected by a complete wafer test.

There is another unpackaged level of test in some manufacturers' flow, that of the diced wafer (chips or die, depending on the jargon). Here all of the electrical problems with probes are still present. The next point in the assembly where the testing problem changes is at the first-level package. This can be a dual-in-line package, a leadless chip carrier, or a complex ceramic multichip carrier. In any case, the package is designed to be assembled to a printed circuit board or some other higher level of packaging. For the test of this device mounted in a first-level package, an adaptation of the next packaging level is usually used as a test socket. In this manner, power can be supplied in a manner consistent with the actual application, signal crosstalk will be at application levels, and actual application impedances will be presented to the part under test. Technically, a complete test can be applied to the device at this point. As the assembly progresses, that is, as the first-level (device) package is assembled to the second-level (printed circuit board) package, testability can degrade if careful attention to testability is not present at all levels of the design.

Another relationship that must be considered when a test process is developed is the increasing value of the device as it moves up the assembly process. While exact costs vary from organization to organization, and are usually rather closely guarded, as a general rule the value of a device increases by 5 to 10 times at each level of assembly. For example, assume a unit cost device. At the first-level package assembly the device (and its package) would be worth 5 to 10 units. At the printed circuit board (PCB) level the value would be 25 to 100 units. Corresponding increases occur in system assembly and in the field. These increases in value come about through labor added, materials added, and inventory carrying costs. Clearly this pyramiding of costs points to testing as thoroughly as one can at the earliest possible stage of the assembly process. One should recall, however, that there were severe limitations at the wafer level which is the first point where complete devices can be tested. By judicious use of built-in test, a superior wafer-level test can be performed, since in principle, only power (and perhaps clocks) need be supplied to the device. The cost equations also must be drastically modified since the "automatic test equipment" required to support built-in test is much less complicated and less costly than standard equipment.

1.7   THE RELATIONSHIP BETWEEN DENSITY AND TEST

As the number of circuits that are designed and fabricated on one silicon chip grows from hundreds, to thousands, and on to hundreds of thousands, the testing problems become more acute and an increasing portion of the total design and fabrication task. It is in easing this test portion of the task where built-in test has its benefit.

As digital circuits grow in size, the ratio of combinational gates to storage elements (flip-flops or scan latches) tends to remain a constant. For networks that cannot be partitioned into disjoint substructures, the number of deterministic tests required to fully test the network tends to grow linearly with the number of gates in the network [Goel (1980)]. In a testable design where scan path techniques are used, since the number of scan latches also grows linearly with the number of gates, Goel predicts that the test application time will grow proportional to the square of the number of gates.

There are two ways in which built-in test can be used to alleviate this rapid polynomial (quadratic) growth of test application time. First, the scan path can be arbitrarily segmented into many paths for applying the tests. Since this is done internal to the device, the problem of interconnections to the outside is minimized. A second improvement is the ability to operate the scan path at its highest shift frequency. In fact, some care should be exercised during the design phase so that the scan path can be shifted at as high a rate as possible. These two techniques should enable the designer to manage test application time such that its growth is contained to an acceptable range.

At the circuit densities that are becoming commonplace in VLSI, a 1–3% circuit overhead for built-in test is enough to implement some quite sophisticated built-in test schemes. These built-in test procedures drastically ease the requirements on supporting automatic test equipment. If lower cost equipment can be used in this area without compromising quality, the pressure on test application time is less. An admittedly simplistic view, but an instructive one nevertheless, says that 1 s on a $1,000,000 tester costs as much as 10 s on a $100,000 tester.

Thus built-in test shows promise in easing the test application time problem that haunts VLSI development and its reduction to manufacturing practice.

1.8   TEST GENERATION TECHNIQUES FOR COMBINATIONAL AND SEQUENTIAL CIRCUITS

Eldred was the first to present a method of generating tests for combinational networks. This concept was extended to various path sensitization techniques. Initially, these were one-dimensional path sensitizations, but Paul Roth [in Roth (1966)] showed that with multiple path sensitization an algorithmic solution to the generation of tests for combinational circuits existed. His d-algorithm will find a test for any fault in a combinational circuit if such a test exists (that is if the fault is not redundant).

Since test generation is an exhaustive search procedure, running a test generator until all the faults are covered can be an extremely lengthy process. It is common to couple test generation with fault simulation to reduce the task. After one or more tests have been generated for specific faults, the tests are applied to the model of the network and the rest of the faults are simulated. Those additional faults which are detected by the test under analysis are marked as tested. In this way the list of faults that the test generator must work on is progressively reduced. This simulation process is quite successful in the early steps of test generation where many faults are detected by each of the early patterns. However, as test coverage approaches 100%, typically only one new fault is detected by a newly generated test.

Many workers observed that the initial high coverage by a single pattern was almost independent of what fault was chosen as the target fault for the test generator or conversely what the pattern generated was. This led some to forgo the test generation step initially and generate 10 to 100 pseudorandom patterns via a random number generation subroutine and fault simulate these patterns. After the fault list was suitably reduced, the conventional test generation/fault simulation procedure was instituted. This procedure is quite widely used today in various test generation programs, but it only shortens the initial portion of the problem. Most programs need more than an exhaustive search algorithm to economically (practically) achieve high (99%) test coverage. This problem arises from the repeated backtracking the d-algorithm performs as it performs the search.

In an attempt to reduce the backtracking of the d-algorithm, an algorithm called PODEM, described in Goel (1981) and Goel and Rosales (1981) was developed. In PODEM, the path from the fault site to a primary input is traced. At each step of the backtrace, branching decisions are made heuristically. Once a primary input is reached, a simulator is called to see if the initial objective function (the target fault sensitized) has been achieved. If not, the algorithm repeats until the target fault is sensitized. At that point, the program drives the effect of the fault to a primary output. In practice, PODEM has been much more effective than the d-algorithm for moderately large circuits (1000 to 20,000 gates). PODEM uses some global information to guide its path tracing. The heuristic for branching decisions uses node controllability values obtained from programs such as CAMELOT [Bennetts et al. (1981)], SCOAP [Goldstein and Thigpen (1980)], and others.

Even with the improvements in PODEM, considerable backtracking occurs in many circuits. In 1983 a test generation algorithm called FAN suited to combinational circuits was reported by Fujiwara and Shimono (1983). In FAN, more information about the circuit topology is used, and different search techniques are used to advantage. In some cases, FAN performs quite a bit faster (25–500%) than PODEM with comparable fault coverages (95–99%).

Despite all the progress in combinational circuit test generation, the problem of generating tests for sequential circuits has remained a largely unsolved problem. Some have given up on the problem and edicted the use of scan-path techniques to convert the problem into a combinational one. Others have tried to develop heuristic methods with which to attack the problem. In general, much more information must be known about a sequential circuit in order for a successful job of test generation to be performed. The state of the circuit must be known at the start of the test. This requires a "homing sequence" or a reset command that determines the state of the circuit. In general the test will be a sequence of stimulus vectors to detect a single fault. Tests can be generated, but much interaction by a knowledgeable test engineer who has detailed knowledge of the circuit and its function is usually required. Many organizations have such test experts, but the general problem of generating tests for sequential circuits is at best difficult. A test generation system utilizing some of the expert system concepts of artificial intelligence has appeared on the scene for use in this arena: It is HITEST[®1] and it appears to have many of the strategies that an expert test engineer would use built into its knowledge base [see Robinson (1983)]. Time will tell if this approach can keep up with the clever designer and the tide of VLSI.

1.9 TEST GENERATION COSTS AND PROJECTIONS

Despite all of the development of efficient test generation programs, the generation of tests for combinational digital networks is expensive. The actual costs of generating tests are not often discussed in the open literature. Many of the using organizations have developed their own programs and they consider them or the circuits for which they are generating tests to be proprietary. Without the algorithmic and/or circuit details, it is hard to compare results. Recently, several companies have appeared whose sole product is test generation programs. It remains to be seen if the marketing claims apply to a broad range of circuits.

The literature does contain some indication of the situation. The 38,182 tests for a 43,000 gate combinational circuit with 165,049 faults took 1000

---

[1] Trademark of GenRad, Inc.

CPU minutes to generate [see Bardell and McAnney (1981)]. The computer used was an IBM S/370 model 168 which had a processing capability of about 3.3 million instructions per second (MIPS). The circuit was considered benign since it did not present the test generator with embedded RAMs or other particularly challenging structures. It was noted earlier that test generation costs in coupled networks grow as the square of the number of gates. (Coupled networks are those that cannot be partitioned into disjoint substructures.) While there are examples of partitioning, they are not particularly effective [see, for example, Bottorff et al. (1977)]. Indeed the example cited earlier from Bardell and McAnney (1981) was partitioned before test generation. There 11 partitions, each of about 30,000 faults, were used to achieve an overall fault coverage of 96.5% with 38,182 tests. One disadvantage of generating tests for a partitioned network is that the partitions overlap and tests for some faults are generated more than once because each partition is considered independently. In the example cited, the overlap was 2.1.

Consider the problem of partitioning in a bit more detail. Initially the network of G gates is partitioned into N subnetworks of size G/N (approximately). This would lead one to believe that the test generation task would be proportional to $N \times (G/N)^2$, or G^2/N, a definite reduction. However, because of the overlap encountered, the overlap factor f ($f > 1$) must be accounted for. The overlap causes each subnetwork to grow by a factor f, thus causing the test generation effort to be $N \times (fG/N)^2$. Thus for any reduction in test generation effort to be realized, f^2/N must be less than or equal to 1. In the example cited, f was 2.1 and N was 11, resulting in a factor of $f^2/N = 0.4$. However, experience has shown that f can easily range as high as 4, and that as N is increased, f also increases. Another practical problem that is encountered is related to the setup overhead that is associated with each partition. The asymptotic G^2 relation was for large networks. As more subnetworks are considered, the setup time for each of them becomes an important factor in the overall time.

Test generation has been shown to be difficult. Goel shows that under some circumstances, test generation as it is done today will increase in cost as the ·square of the number of gates. Even the analysis of Goel assumes a "super" test generator that never has to backtrack. It has been shown that handling backtracking is the key to an efficient test generator.

Test generation has repeatedly been shown to be computationally intensive. In fact the test generation problem has been studied and shown to belong to a class of problems known as NP-complete [for an interesting discussion see Ibarra and Sahni (1975)]. Complexity theory defines this class of problems as being solvable in polynomial time (as opposed to exponential time) by a one-tape Turing machine only if it is augmented with a "guessing" ability, hence the name nondeterministic polynomial (NP) complete. In the worst case, problems of this class are exposed to exponential increases in solution time as the size of the problem increases. In practice, most problems grow in a polynomial fashion for most instances. Thus we can expect the test generation times to grow at least quadratically as circuits increase in complexity, with only incremental reductions as improved heuristic methods are added to the basic algorithms.

While the problem of test generation for combinational circuits of modest size has been solved, a solution applicable to a network of 10,000 gates is not a solution for a network of 1,000,000 gates. The 100^2-fold increase in test generation effort makes it an altogether new problem. It is in this context that material on built-in test is presented in the following portions of this book.

1.10 BUILT-IN TEST AS THE SOLUTION TO TESTING VLSI DESIGNS

In the preceding, we have examined the test problems associated with digital networks of the size that are encountered in working with VLSI. The benefits of the application of increased circuit density will force the levels of integration to continue to grow. This will considerably exacerbate the testing problem as we have shown. Conventional externally applied tests, structural or functional, will become less and less satisfactory. One clear alternative to the externally applied tests is built-in test.

We have shown that providing built-in test can become a part of the design process. As VLSI structures reach commercial applications, quality demands will force some form of design for testability into the designs. The incremental circuit and area overhead taken by the built-in test structures can be a small price to pay for the assured testability and the resultant quality that ensues.

The problems of delivering test patterns to an unpackaged VLSI chip (or wafer) through impedance discontinuities at high speed are considerably eased by built-in test. As will be shown in later chapters, as little as power and a clocking signal need be supplied to a VLSI structure with suitable built-in test in order for an accept/reject decision to be made.

2

Introduction to Testable Design

During the 1970s a quiet revolution occurred in the field of digital circuit testing. Before that time, testing was a manufacturing chore almost completely isolated from design—"we design it, you build and test it." After that time, in many companies the entire responsibility for testing and test pattern generation was in the hands of the designers.

In that 10 year span, as circuit geometries became smaller and more compact the circuit densities available in a single package increased explosively. Pre-1970 circuit packages had perhaps one gate for each two or three package pins. By 1980 the gate-to-pin ratio had not only reversed but approached 20 gates per pin or worse. (At present it is not unusual to find 100 gates or more behind every pin.)

The increasing package density caused dramatic reductions in per circuit costs, but the percentage of those costs consumed by testing stubbornly increased. This effect was forced primarily by the loss of ability to control and observe internal circuit nodes as the gate-to-pin ratio worsened. In 1975, Phillip Writer of the Test Equipment Technical Support Office at the Naval Electronics Laboratory Center in San Diego coined the term "design for testability" and used it as the title of his paper (presented at the Symposium for Advanced Maintainability in October 1975) advocating a set of design constraints meant to simplify at least some of the problems of testing digital circuits. By 1980 the need to design for easier testing was apparent to all, and the idea of placing the burden of test on the designers had taken hold.

Testability is desirable in logic designs for two reasons: it reduces the cost of testing and it improves the quality of the test. The design styles that result in testable designs are collectively grouped under the name "design for testability." While design for testability concepts were originally developed

14

within the framework of conventional test generation, they are equally useful for built-in testing. Built-in test gives better test coverage in less test time when the circuit has been designed to be easy to test.

The choice of design styles for testability depends in large measure upon the control that the designer has over his implementations. Using standard SSI or MSI components forces different approaches than using vendored custom (or semicustom) VLSI or using an in-house semiconductor facility. The former usually relies heavily upon ad hoc techniques while the latter two can take advantage of some of the more structured design techniques. This chapter first covers some of the ad hoc design methods that have been found useful, and later considers structured design for testability techniques. It then considers an assortment of possibilities for nonconcurrent built-in testing. The chapter concludes with a description of some of the techniques that have been developed for measurement and analysis of testability.

Beyond the circular definition of a testable design as one that is easy to test, there are certain general properties of a design that make it easily testable. They are that (1) it contains no logical redundancy, (2) it contains no asynchronous logic, (3) its clocks are isolated from the logic, (4) its sequential circuits are easy to initialize, (5) it is easily diagnosable, and (6) it has a minimal increase in gate or pin count over a "normal" design. These properties, while they are based on practical experiences of conventional test generation for completed designs, are also applicable to built-in testing. A suggestion of the reasons for each of these properties is:

1. *Avoid Logically Redundant Circuits.* A line in the circuit is redundant if a stuck value on the line is undetectable (if the function performed by the circuit is unchanged by the presence of the stuck). Since the stuck fault is undetectable, no test can ever be found for it. This causes two problems. First, a conventional test generator can churn for hours in attempting (and failing) to find a test for such redundancies. Second, the presence of an undetectable fault can cause either a detectable fault to become undetectable or another undetectable fault to become detectable. In some cases, of course, redundancy is deliberately added to a design for sound technical reasons, such as to avoid static hazards in combinational circuits. More frequently, however, redundancy points to nothing more than a design fault.

2. *Avoid Asynchronous Logic.* The absence of clock synchronization in asynchronous sequential circuits creates severe problems for conventional test generation but even worse problems for random pattern built-in testing. Hazards in combinational circuits are glitches or transient pulses caused by the variation in delays through the paths of the circuit. The appearance of such incorrect pulses at the inputs of an unclocked flip-flop can cause indeterminate operation or the actual storage of an incorrect value. Hence a test generator for asynchronous circuits must be capable of dealing with circuit delays in order to create hazard-free tests. Many built-in test techniques use random input patterns. These random, unconstrained, and not necessarily functional

input patterns will usually cause either hazards or multiple-input changes at the inputs to the unclocked bistable elements, making it impossible to predict the state of the elements. A decision to use random pattern testing for asynchronous circuitry creates an almost insuperable circuit design problem.

3. *Isolate Clocks from Logic.* There are several reasons for this rule. From a design viewpoint, if clock signals are combined with logic signals and the resultant is applied to the data input of a clocked latch, a race will occur since both the clock and the data will change simultaneously. From a test viewpoint, it is a necessary rule since isolating clock signals enables the tester to control the unit under test at the speed of the tester rather than the speed of the unit. This also eliminates the need to synchronize the unit to the tester. From a built-in test viewpoint, even if the circuit inputs are randomly stimulated it is likely that the clocks are driven deterministically.

4. *Design the Circuit for Easy Initialization.* When power is applied to a circuit, the sequential circuits may stabilize to either a logical 0 or 1. Before testing can begin they must all be brought to a known state. This requires finding an input sequence that will take the circuit from an unpredictable power-on state to some known starting state. Writer (1975) observed that it is not uncommon to find test programs in which 200 to 300 test vectors are required solely for initializing the unit under test and only 30 to 50 vectors are needed to complete the test. The difficult task of finding initialization sequences can be avoided if the circuit can easily be initialized, either by an initialization circuit that resets all sequential elements at power-on, or by designing the sequential circuits so that their state can be controlled externally.

5. *Make It Easy to Diagnose.* Even at the chip level, diagnosis at least to the net level is critical during production start-up on a new design when zero yield conditions may occur and is always desirable for yield enhancement. Higher levels of packaging are usually not throwaway items and must be repaired. Easy diagnosis facilitates identification of the appropriate repair action. Diagnosis usually involves a probe used in backtracing from a faulty primary output to the faulty component, one whose output signals are incorrect and whose inputs are correct. Designing for diagnosability means avoiding ambiguous fault isolation situations such as high-fan-in wired AND or OR circuits, and unbreakable global feedback loops. The judicious addition of test points at critical locations can supply much useful diagnostic data and reduce the time required to isolate a fault.

6. *Keep It Simple.* Circuit modifications for the purpose of improving testability can create testability problems of their own. A circuit added to improve the controllability of certain faults in the original design must also be tested, and it is not unknown for such a circuit to introduce faults of its own that are as difficult to detect as the target faults the circuit was added to correct. Addition of a test point can increase the observability of those portions of the original circuit that are upstream of the physical location where the net fans out to the test point (at the cost of additional loading on the net),

but can leave unchanged the observability of potential faults on the fan-out branches downstream of the location (Chapter 7 elaborates further on this point).

The goal of any set of guidelines is to create a design that has a small and easily derived test set, regardless of whether that test set is deterministically applied from an external tester or is built into the product under test.

The thrust of design for testability is to improve either our ability to observe values on nets internal to the circuit or to control the values on these nets. Although the relationship of observability and controllability to testability is not unambiguous, the concepts "control" and "observe" usefully describe the goals of most design for testability concepts, and are applicable both to built-in testing and to conventional test generation and application. Some interesting commentary on testability relative to the corresponding notions of observability and controllability are included in Savir (1983).

2.1   AD HOC DESIGN TECHNIQUES

These techniques are just a collection of good design methods that are manually applied with the judgement and skill of the designer. One caveat: None of these ad hoc methods completely solves the problem of testing sequential circuits. For that, one must use the structured designs that are described in Section 2.2.

2.1.1   Logic Partitioning

Empirically it seems that any test-oriented software process involving digital logic structures consumes processor time in proportion to some unfriendly power of the number of gates G in the structure, usually somewhere between G^2 and G^3. Any such process, even on a very fast computer, is not feasible for large structures. While a random pattern test strategy does not require test generation it does require a measure of the effectiveness of the patterns in detecting faults as described more fully in Chapter 7. If the measurement process takes 1 CPU hour for a 1000 gate circuit, and if the costs of measurement grow as G^2, then the process on a 100,000 gate circuit will take 10,000 h. Avoidance of this growth, while still taking advantage of the cost benefits of VLSI, requires methods for dividing the circuit into logically independent parts, each of which can then be processed separately. These methods are collectively called partitioning (or "divide and conquer").

Another strong reason for partitioning is to facilitate a possible exhaustive testing strategy. A small combinational circuit can be tested quite easily by successively applying all possible combinations of the logical values 1 and 0 to the circuit inputs and checking that the output responses match the truth table of the desired function. Such a test on a circuit having scores of inputs is not feasible, even with an extremely fast tester. For a circuit of 64 inputs, the number of possible patterns is 2^{64}. Applying these patterns at a rate of $1/ns$ requires about 585 y of steady testing. Bisecting the circuit into partitions of 32 inputs each will reduce the test time to less than 9 s. Partitioning of sequential circuits is even more compelling, since exhaustively exercising a sequential circuit takes (at a minimum) $s2^n$ patterns, where s is the number of stable states and n is the number of inputs. (An upper bound on s is 2^r, where r is the number of feedback loops in the circuit.)

These two possible reasons for partitioning can be put succinctly as

1. the network is too big for the tools, or
2. the network has too many inputs for exhaustive testing.

In the first case, reduction of the gate count in the partitions is most important and the number of primary input pins (PIs) in each partition is secondary. In the second case, a limitation is placed on the number of PIs in each partition and the goal is to construct partitions within that limit. In both cases, partitioning should not require more than a minimal increase in gate or pin count over the original design.

Partitioning is usually begun at the register transfer level of the design process. Partition boundaries are defined that encompass the desired limits of partition size or number of primary inputs, and stopping points are inserted to control these boundaries. To illustrate the process, however, a completed circuit design is assumed. Each primary output (PO) of the design is an observation point since it is accessible to the tester. Each PO is associated with a cone of logic. The cones are identified by starting at the PO and backtracing through the circuit, collecting all gates encountered, until PIs are reached on all backtrace paths as shown in Figure 2.1. The cones represent the smallest logic sections of the design, and there are as many cones as there are POs. Both primary inputs to a cone and nets internal to a cone may fan out to other cones. Thus a portion of a particular cone can belong to several other cones.

At this point a check is made to see if any cone exceeds the partitioning limits, either because it has too many gates or too many PIs. If so, the offending cones must be manually partitioned using one or more of the techniques shown later. If not, the process stops because the cones are individually small enough to be processed through the test generation tools or to be tested functionally, if that is the goal.

There may be some advantage, at this point, of reversing the partitioning process by combining cones. If, for example, the goal is deterministic test generation, the test generation effort is affected by the number of partitions, the size of each, and the overlap factor between partitions as noted in Chapter 1. The cones that should be combined are those that have the most circuits in common to reduce the amount of circuit overlap between the test partitions.

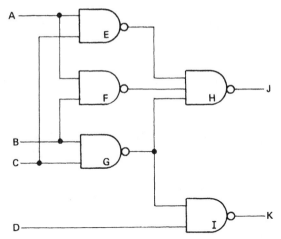

Figure 2.1. Backtracing cones from primary outputs. Backtrace list from PO J: $\{J, H, E, F, G, A, B, C\}$. Backtrace list from PO K: $\{K, I, G, B, C, D\}$.

In general, however, many common circuits are not effectively partition-able. The partitioning process will result in one or more backtrace cones containing more circuits or PIs than can be tolerated. There are a number of ways to deal with these large cones, all of which require manual intervention on the part of the designer. The simplest approach is to break down the large cone into two or more partitions, either mechanically (by a simple division of the cone into two boards that are later rewired by cables or into two portions on one board that may be reconnected by means of jumper wires) or by control gates inserted to isolate various portions of the large cone during test.

A simple example of degating is shown in Figure 2.2. Here one tester controlled gate has been manually inserted to divide the cone into two partitions. The location of the backtrace stop point is chosen to meet the partitioning limits (either number of gates or number of PIs) and to minimize the effect on propagation delay of the added tester-controlled gate.

With the added gate, the two partitions can be forced into logical indepen-dence. Tests can be generated for partition A in isolation by holding off the tester-controlled PI. Test generation for partition B requires holding the tester-controlled PI on and holding off the uppermost PI in partition A.

The use of isolation gates requires that the gates be controllable by the tester at a primary input. The number of added PIs can be reduced by using decoding logic to control the partitions. The number and location of the isolation gates depends upon the complexity of the partitions. In general, it is a good rule to place isolation gates as close as possible to the primary output of the cone to prevent partition expansion due to circuit overlap between cones.

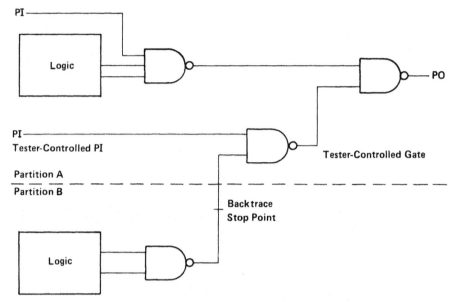

Figure 2.2. Use of degating to partition large cones.

2.1.2 Clock Isolation

The capability of isolating high-speed internal clocks removes the need for synchronizing a high performance tester to the unit under test, and permits the clock to be turned off for stuck-fault testing of the rest of the logic. Figure 2.3(a) shows a simple example of a circuit for isolating a clock generator that is built into the unit it clocks. An additional PO is used to observe the results of analog or timing tests applied to the clock itself.

The same effect can be achieved by a jumper wire as shown in Figure 2.3(b). Removing the jumper wire allows the tester to control the circuit from the test input. Replacing the jumper makes the clock observable on the test input, eliminating the need for the additional clock test PO. The jumper can be implemented conveniently by bringing the clock and the circuit input to card I/O connectors and placing the shorting jumper wire on the board between the corresponding board I/O contacts.

2.1.3 Memory Array Isolation

A major testability problem arises when a random-access memory (RAM) or a read-only memory (ROM) is buried within combinational or sequential logic on a module or card. Faults may exist in the logic driving the memory (the prelogic), in the logic driven by the memory (the postlogic), or in the memory itself. Each of these three aspects of the problem require different test strategies.

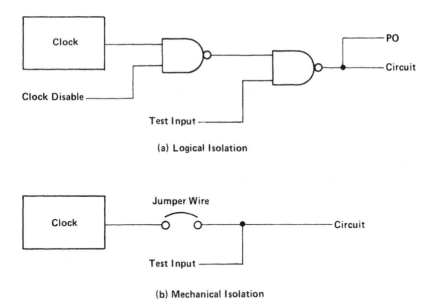

(a) Logical Isolation

(b) Mechanical Isolation

Figure 2.3. Inhibiting internal clocks.

Prelogic Faults. Detection of faults in the prelogic driving the memory requires propagation of the fault effect first through the memory and then through the postlogic driven by the memory.

If the fault occurs in the data prelogic, the effect of the fault will appear at the memory data input port. Propagating the fault effect to the memory outputs requires only a write operation to an arbitrary address followed by a read of the same address. Fault path sensitization through the postlogic may require specific values on the postlogic primary inputs.

On the other hand, when a fault occurs in the address prelogic the effect of the fault is to change the values on the memory address decoder inputs. A memory with M address lines has 2^M accessible addresses, in general. Since in this case all possible decoder input values represent legitimate addresses, the address fault will select an actual, although incorrect, address.

Consider a two-port RAM with one write port and one read port. The detection of read port address faults requires that the word stored at the "good machine" address be different from that at the "failure machine" address. The fault is detected by reading the incorrect word at the failure machine address. This requires that different words be prewritten to the two affected addresses. A write port address fault, however, tends to reduce the size of the memory. A write operation to address A is always transformed by the fault to a write of some faulty address B. Detection of the fault requires recognition of this size change. In a single port RAM, address faults cause this same size change and are similarly detectable.

A ROM address fault is detectable if the word at the good machine address is different than that at the fault machine address. In some cases the words at the two addresses may not be unique. Before the fault can be labeled undetectable, however, a search must be made through all possible tests for the fault since some other test may employ address pairs whose words are unique.

Postlogic Faults. Detection of faults in the logic driven by the memory requires specific binary values on the data output lines of the memory.

For a RAM, this requires a prewrite of the particular data pattern needed. In a ROM, the pattern required may not be one of the words in memory, and again the possible tests for the fault must be searched to see if a particular test pattern is available before the fault is marked undetectable.

Memory Faults. Fully testing the RAM requires verification that every cell is capable of storing a 0 and a 1, that the cell addressing circuitry (or decoders) correctly address every cell, that there is no interaction between cells, that the memory sense amplifiers work correctly, and, for dynamic memories, that the cells can store correctly for the duration of the refresh cycle. ROM testing must ensure that the address decoder functions correctly, and that the contents of each address are correct.

The problem here is to find ways of applying the required memory tests and observing their responses through the embedding logic. Of the tests themselves, testing for interactions between RAM cells is the most difficult problem and is at best a pragmatic business. A multitude of different test sequences have been proposed to detect pattern sensitivity between electrically adjacent cells. The need for many different tests is because the error itself can be a function of the information being written or read, of the particular cells being selected, or of the sequence in which the cells are accessed. A common test for pattern sensitivity is the ripple test, sometimes called "walking ones and zeros." In a memory containing all zeros, a single "target" cell is written to a 1. All other cells are then read to ensure that they still are storing zeros. The 1 in the target cell is then read, rewritten to 0, and the target moves to the next cell. After completion of this "walking a one" through the memory, the patterns are reversed and a zero is walked through a field of ones.

Detection of faults within a ROM requires reading every address and comparing the word read to the correct word for that address.

A further problem arises when a functional memory consists of an array of separate memory chips or modules. Memory testing must be segmented into tests related to a physically replaceable unit for repair. Where the memory chip outputs are joined by wired OR or wired AND connections as they drive a common data bus, considerable diagnostic ambiguity occurs when one chip fails and corrupts the data on the bus.

The most satisfactory solution to the problem of testing embedded memory is to separate the memory, during test, from its embedding logic. Mechanically or logically isolating the memory, by using techniques such as those suggested in Figure 2.3, and providing tester access to the prelogic outputs, the postlogic inputs, and to the memory I/Os dramatically simplifies testing of the prelogic and postlogic, and makes it possible to use a standard set of memory tests.

When memory isolation is unsatisfactory, either due to space constraints or because of unacceptable isolation-added path delay, a pseudo-isolation can be achieved if it is possible to establish a set of primary input states to the circuit that will sensitize paths such that there exists a one-to-one correspondence between each memory input and some circuit primary input, and between each memory output and some circuit primary output [the concept suggested by Eichelberger et al. (1978)]. Under this correspondence condition, memory tests can be applied at the correspondence inputs and their results observed at the correspondence outputs, almost as if the embedding logic were transparent (except, of course, that delay-type tests are ambiguous because of the delays through the correspondence paths). After assuring the memory function, testing the prelogic and postlogic simply requires propagation through the memory by writing and reading a single memory address.

2.1.4 Test Points

Test points are control points (inputs) or observation points (outputs) added to a circuit to reduce the number of tests required for fault detection or to improve the diagnostic fault resolution of the tests or both. The simplest method of adding test points is to use I/O pins, although test points can also be accessed with temporary test probes or with a "bed of nails" test fixture. The number of test points that can be used is limited to the unused package I/O pins or to the probe point configuration of the test probes, so the location of the added test points in the circuit must be carefully chosen. The addition of a parallel-in parallel-out register with serial scan-in and scan-out, as shown in the description of scan/set logic in Section 2.2.4, can increase the number of test points without a large increase in I/O pins. Intuitive candidates as sites for control or observation test points are

1. the outputs of bistable devices,
2. the set or reset inputs of bistables,
3. the junctions of large fan-ins or fan-outs, either for monitoring values or inserting values,
4. the midpoints of long counter chains or linear cascades of logic elements,
5. the data and address bus lines of bus-structured designs,
6. the master clock lines,
7. logically redundant nets.

Very little is known about the theoretical optimum placement of test points in general networks, although some results regarding observation points in fan-out-free and certain limited types of fan-out networks are given in Hayes and

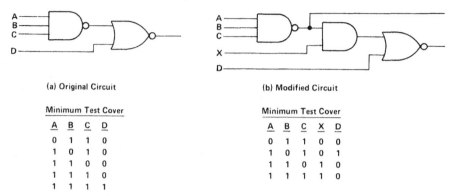

A	B	C	D
0	1	1	0
1	0	1	0
1	1	0	0
1	1	1	0
1	1	1	1

Minimum Test Cover (a)

A	B	C	X	D
0	1	1	0	0
1	0	1	0	1
1	1	0	1	0
1	1	1	1	0

Minimum Test Cover (b)

Figure 2.4. Illustrating the need for both control and test points.

Friedman (1974). There it is suggested that a good heuristic method is to place observation points at locations that tend to subdivide the fan-out-free circuit into partitions, each of which has the same number of input pins.

Some of the testability measures to be discussed later in this chapter may also be of assistance in positioning test points around a circuit. Most of these measures relate testability to two other concepts: controllability (or the ease of controlling a node's logical value from the primary inputs) and observability (or the ease of establishing a sensitive path from a node to a primary output). A test point could be placed first at or near the least observable node in the circuit. Recomputing the testability measure on this modified circuit quantifies the improvement due to the test point and identifies the next least observable node as the next test point candidate.

The location of test points used as control inputs and the type of control logic to be used also seems to be an open problem. There are similarities to the partitioning problem discussed in Section 2.1.1. In some types of circuits the test effectiveness is not changed by using output test points, and control logic has to be added. Consider the circuit in Figure 2.4(a), which implements the function $ABC\overline{D}$. Five tests are required for detection of all single stuck faults. Adding an observation test point to the output of the NAND gate (the only possible location) does not change the number of tests. However, adding both a control and an observation point, as shown in Figure 2.4(b), reduces the test length to four tests. The control input X is held to a logical 1 during normal operation to disable the AND gate, but may be 0 or 1 during test.

Test points can also be usefully added to improve the efficiency of random pattern testing. More on this topic appears in the discussion of random pattern test of Chapter 7.

2.1.5 Bus Access

It was noted in the discussion on test points that the data and address bus lines of bus-structured designs were good candidates for control and observa-

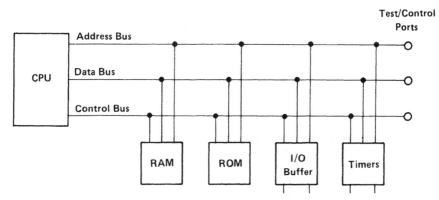

Figure 2.5. Test/control ports added to busses.

tion points. This concept is important enough to warrant a few additional remarks.

Figure 2.5 shows a single board bus-structured computer with added test/control ports on the busses. The added ports are made tester-accessible either by bringing them to primary I/O pins or by probe connections during test.

Given tester access to the busses (and assuming that all bus drivers can be individually controlled to their high impedance state) testing of the computer board becomes a test of the individual board components. An example procedure is outlined as:

1. Set all bus drivers to high impedance and test the busses for stuck-at conditions and inter-bus shorts.
2. Set the CPU bus drivers to high impedance and test the RAM, ROM, I/O buffers, and timers (one at a time, of course).
3. Set all non-CPU bus drivers to high impedance and test the CPU.

A final at-speed functional test of the complete computer using some ad hoc test program completes the test by assuring the individually tested components will work properly together.

2.2 STRUCTURED DESIGN TECHNIQUES

Structured design requires compliance with a set of ground rules centered around a uniform design method for latches. It is quite clear that if the values in all latches can be controlled to any specific value and if they can be observed easily, then test generation is reduced to dealing with combinational logic. There are also some clear advantages that structured designs offer to built-in testing. First, Chapter 8 shows that many built-in test structures are easily constructed with rather minor modifications to structured latch designs. Second, and more importantly, the ground rules for structured designs almost invariably result in synchronous logic having a reduced and well-defined timing dependency. Usually there is no dependence on minimum circuit delay and only a weak dependence on maximum delay. Similar ground rules are usually an imperative for most types of built-in testing, especially random pattern test.

To provide a sampling of structured designs, the following describes four different philosophies: level-sensitive scan design, random-access scan, scan path, and scan/set logic.

2.2.1 Level-Sensitive Scan Design

Level-sensitive scan design (LSSD) is probably the most used and best documented structured design technique [and was first described by Eichelberger and Williams (1978)]. LSSD imposes a clocked structure on all memory elements and forms the elements into shift-register latches, making it possible to shift values into and out of the elements by way of the scan path. The memory elements become "pseudo-inputs" for the purpose of test application and "pseudo-outputs" for test observation.

Figure 2.6 symbolically shows the separation of the combinational logic and the storage elements in an LSSD structure. The primary inputs, the primary outputs, and the combinational circuits are unchanged. The storage elements,

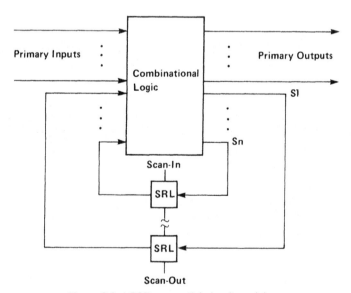

Figure 2.6. LSSD sequential circuit model.

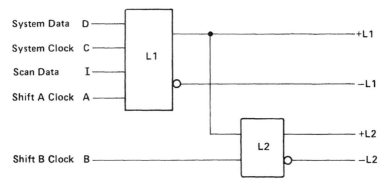

Figure 2.7. General form of a polarity-hold SRL.

however, are implemented as clocked DC latches (latches such that the stored data cannot be changed by any input when the clocks are off). Each latch is augmented to form a shift-register latch (SRL), all of which are chained together into one or more shift-register strings. Testing access to the SRLs is through the scan-in primary input and the scan-out primary output of the string. Separate shifting clocks are used to scan serial data into or out of the string. Test patterns scanned into the string are applied to the combinational logic at the SRL outputs, and response values captured in the SRLs can be scanned out.

Figure 2.7 is the symbolic representation of the LSSD polarity-hold shift-register latch. During normal system operation the shift A and shift B clocks are off and the SRL has just two input signals, system data D and system clock C, that activate the latch memory function. While clock C is off, the $L1$ latch cannot change state. When C is on, the internal state of $L1$ is changed to the value of the system input D.

Conversion of the latch into a shift-register element requires adding a second latch, the $L2$ latch in Figure 2.7. The I input and the $L2$ output of the SRL are the input and output, respectively, of the shift register. When operating as an SRL, the I input value (from the preceding SRL stage) is gated into the $L1$ latch by the shift A clock. After the $L1$ value has stabilized, the shift A clock is turned off and the shift B clock is turned on to capture the $L1$ output value into the $L2$ latch. Each cycle of the shift A and shift B clocks moves the data one step down the shift-register string. The A and B clocks must be nonoverlapping to ensure correct operation of the string.

Figure 2.8 is an implementation of the polarity hold SRL in NAND gates. The memory function of the latch circuit in Figure 2.8 is relatively insensitive to the AC characteristics of clock C, requiring only that the clock be on long enough to stabilize the feedback signal around the latch.

Interconnection of the SRLs into a shift-register string is done by connecting the $+L2$ output of one SRL to the I input of the next, and by connecting all shift A and shift B clock inputs in parallel. Figure 2.9 shows this

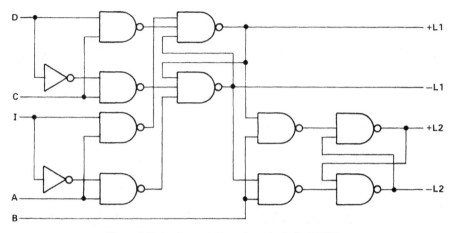

Figure 2.8. Implementation of a polarity-hold SRL.

interconnection for three SRLs. Four additional package pins are needed to implement LSSD. Two pins are used for the scan-in and scan-out ports of the shift-register string, and two are used for the nonoverlapping shifting clocks. Higher packaging levels are interconnected in a fashion similar to that shown in Figure 2.9. (Multiple shift-register strings are sometimes used to reduce the time required to scan test data into or out of the package, in which case additional package pins must be provided.)

There are two global system structures for utilizing LSSD; the single-latch design, and the double-latch design. Figure 2.10 shows a typical LSSD

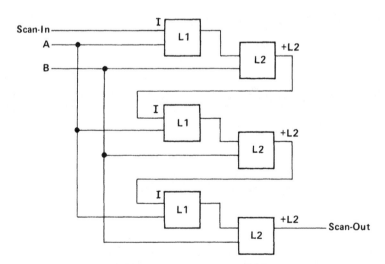

Figure 2.9. Interconnection of SRLs.

single-latch approach. *N*1 and *N*2 are combinational logic networks. Two nonoverlapping system clocks, *C*1 and *C*2, are required to prevent the inputs of a logic network from changing while its outputs are being latched. (Using nonoverlapping clocks also eliminates any dependency on minimum path delay. With overlapping clocks, the amount of the overlap must be less than the time delay through the shortest path through either logic network.) The *L*1

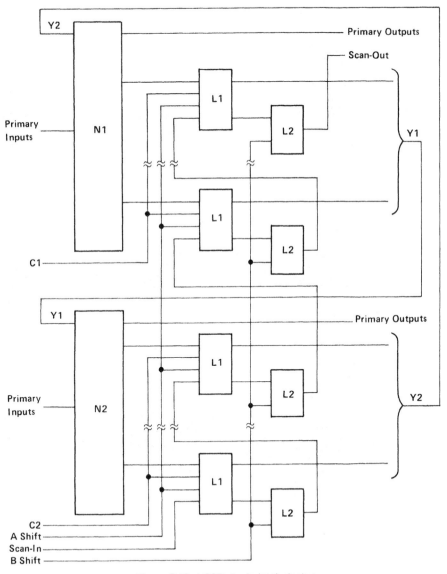

Figure 2.10. LSSD single-latch design.

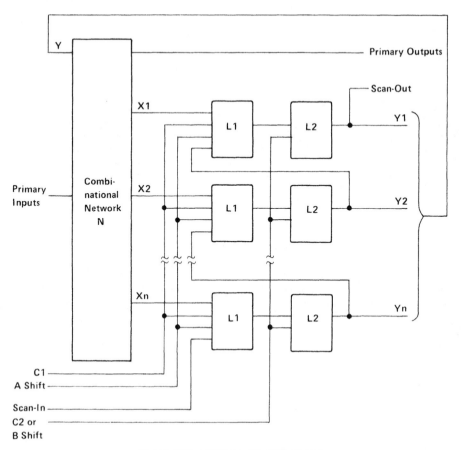

Figure 2.11. LSSD double-latch design.

latch of each SRL is used for system functions, while the $L2$ latch is used only for shifting test data into or out of the system latch.

In contrast to the single-latch design, Figure 2.11 shows a double-latch approach in which both $L1$ and $L2$ are used as system latches, and the system output is taken from $L2$. This design requires that the shift B clock to the $L2$ latch be also a system clock. Again the two system clocks $C1$ and $C2$ (or shift B) must be nonoverlapping to prevent races through the combinational logic.

Any structured design requires that the designer conform with a set of design rules. For LSSD the four most important rules (and a brief reason for each rule) are:

1. All internal storage is in clocked DC latches.
   The latches are controlled by clock signals so that the data in the latches cannot be changed when the clocks are off. This rule insures a synchro-

nous DC design in which the latches can be isolated one from another by simply turning all clocks off.

2. The latches are controlled by two or more nonoverlapping clocks.

Operating the system clocks in a nonoverlapping mode eliminates any system dependency on minimum circuit delay, since a fast circuit cannot create a system malfunction.

3. Latch X can feed latch Y (through combinational logic) if and only if the clock that feeds X is not the same as the clock that feeds Y, and if the clocks are nonoverlapping.

This rule insures that the data at the input to a latch will not change while the latch clock is on.

4. All latches are contained in a shift-register latch, and all shift-register latches are interconnected into one or more shift registers.

This is the central concept in LSSD and reduces the test generation problem to one for the combinational logic alone, since the shift registers can be tested by simply scanning a fixed pattern into the scan-in port and observing the pattern appearance (suitably delayed) at the scan-out port.

Some additional minor rules for LSSD have been added to prevent testing conflicts, to reduce test time, or to minimize the LSSD cost overhead. These rules are also generally useful for built-in test designs that are structured around scan-path methods.

1. Either $L1$ or $L2$ outputs can be used for system functions, but both together may not be used in the same logic structure.

Test data is scanned into the shift register with a series of alternating A and B clock pulses. If the scan ends with a B clock pulse, both $L1$ and $L2$ latches of an SRL contain the same value. When both the $L1$ and the $L2$ are used in the same logic function, there is a test coverage exposure because of the inability to get different values into the two latches. Similarly, if the scan ends with an A clock pulse, there may be an exposure since any two adjacent latches must contain the same value.

2. SRL clock inputs must be controlled from PIs so that when the clock PIs are off, the SRL clock inputs are off.

This enables the tester to keep all clocks off by controlling specific PIs. If clock control were derived from SRLs, the clocks would be toggled inadvertently during any test requiring shift-register scan operations.

3. If multiple shift registers are used, all registers must be capable of being shifted simultaneously and each shift register must have its own scan-in PI and scan-out PO.

The purpose of multiple shift registers is to reduce test socket time by reducing the time required to load and unload the shift-register string. The

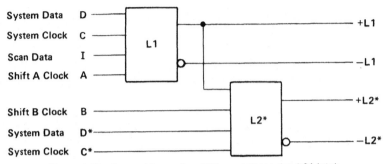

Figure 2.12. General form of an SRL containing the $L2^*$ latch.

rule is necessary to ensure that the multiple strings can be loaded and unloaded in parallel.

4. The $L2$ latch of an SRL may have an independent system data port clocked by a system clock C^* to allow system data storage in the $L2$.

To reduce the silicon cost of implementing LSSD, an SRL whose $L2$ latch (now called an $L2^*$ latch) has two independent data ports is allowed and is shown in Figure 2.12 in block diagram and in Figure 2.13 implemented in NAND gates. Obviously the $L2^*$ implementation is useful only in LSSD single-latch designs, in which the $L2$ latch is mainly used for scan purposes. One disadvantage of the $L2^*$ latch appears during test. It is generally not possible to get different values in both the $L1$ and $L2^*$ latches by scan operations. This gives further emphasis to the rule that restricts the usage of the $L1$ and $L2^*$ outputs. Furthermore, although both $L1$ and $L2^*$ latches

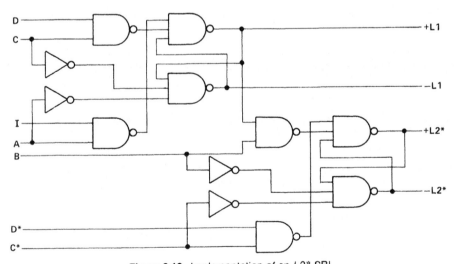

Figure 2.13. Implementation of an $L2^*$ SRL.

contain system-significant data, it is not possible to simultaneously scan out both pieces of data (the data in either the $L1$ or the $L2^*$ is destroyed depending upon whether the A or the B clock is turned on first).

A further test problem arises when a memory array is embedded within the combinational logic. Since it is not economically justifiable to build memory using shift-register latches, testing must access the memory through the embedding logic. Considering the variety of possible embedding logics, accessing the memory to apply its various tests is always difficult and automatic test generation is usually infeasible. In LSSD a special memory embedding rule is used to minimize this difficulty [described in Eichelberger et al. (1978)]. The rule requires that, under a particular set of primary input or SRL values, a one-to-one correspondence must exist between each memory input and a PI or SRL, and between each memory output and a PO or SRL. With this one-to-one correspondence, each memory input is controlled by a specific PI or SRL, and any test pattern can be placed on the memory inputs by placing it on the correspondence inputs. Similarly, each memory output controls a specific PO or SRL, so the test results at the memory outputs can be observed directly (save for possible inversion through the embedding logic). This is not a particularly satisfying solution to the embedded array problem. If the memory input or output correspondence exists to SRLs, an SRL load or unload time has to be added to what is already a long memory test. Furthermore, it is not possible through the correspondence path to apply meaningful timing characteristic tests to the memory.

2.2.2 Random-Access Scan

Random-access scan [described in Ando (1980)] is similar to LSSD in that the states of all storage elements can be controlled from PIs and observed at a PO. It differs in that shift registers are not used. Instead each latch is individually addressable in a fashion similar to that used in a random-access memory. The addressable polarity-hold latch used in random-access scan is shown in Figure 2.14. In normal operation the information on the data line is latched by lowering and raising the clock line -CK. The stored data is available on latch output Q for subsequent system use. During test operations, the test value on the scan data in (SDI) port is gated into the latch by the scan clock SCK. Note that clock SCK affects only that one latch selected by the X-address and Y-address lines.

The scan data outputs (SDO) of all latches are ANDed together to drive a primary output pin. The inverse of the value stored in any latch can be observed at the SDO pin by selecting the X-address and Y-address lines for that latch (holding clock SCK off if the latch value is not to be changed). (Note that latch observability is also available during normal operation, for that latch whose X- and Y-address lines are selected. This is a useful capability for system debug.)

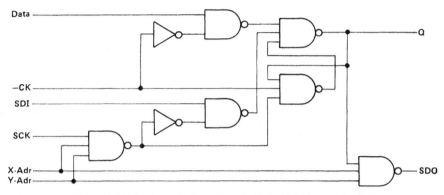

Figure 2.14. Implementation of a polarity-hold addressable latch.

A set/reset addressable latch is also provided and is shown in Figure 2.15. In normal operation, the latch behaves like the polarity-hold latch. Preset line PR is normally off, the clear line -CL is normally on, and the latch accepts the value on its data line whenever system clock -CK is active.

The line -CL is connected in common to all other set/reset latches. Bringing -CL to a logical 0 sets all of the set/reset latches to 0. Those latches that need to be set to 1 are individually selected using their X- and Y-address lines. Turning preset line PR on then sets the selected latch to a 1. The value in the latch can be observed by way of the ANDed SDO nets just as with the polarity-hold latch.

These addressable latches are combined with an X-scan and a Y-scan address decoder into a system structure as illustrated in Figure 2.16. The addressable latch located at the intersection of the selected X- and Y-address lines is accessed, and can be individually written into or observed. There are two control lines into the addressable latches: clear and preset. The clear lines

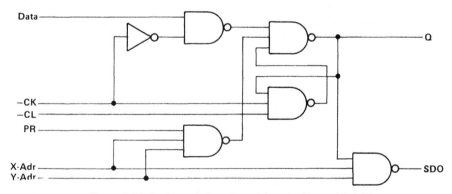

Figure 2.15. Implementation of a set/reset addressable latch.

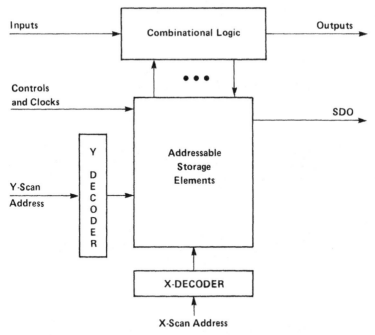

Figure 2.16. General form of the random-access scan design.

(-CL) of the set/reset latches are combined together into one or more groups depending upon the requirements of system initialization. The common preset (PR) input to the set/reset latches can conveniently be derived by ANDing the scan data input line and the scan clock.

The I/O overhead for random-access scan is 4 or 5 pins plus address pins depending upon the number of latches, as follows:

1. scan data in (SDI),
2. scan data out (SDO),
3. scan clock (SCK),
4. clear (-CL),
5. preset (PR), which however can be internally derived if desired as the AND of SDI and SCK,
6. Y-scan address, pins as required,
7. X-scan address, pins as required.

The pin overhead for addresses can be reduced by using a serially loadable shift register as an address counter. This would reduce the pin count to 6 or 7 to accommodate SDI, SDO, SCK, -CL, the address counter scan-in port, and one or two shift-register clocks. ·

2.2.3 Scan Path

The scan-path structure for design is quite similar to LSSD in that it implements storage as stages of a shift register for scanning test data in and test responses out [see Funatsu et al. (1975)]. The scan-path register stages, however, are built with D-type flip-flops. Figure 2.17 shows a typical scan-path flip-flop [from Gutfreund (1983)]. Operation of the device requires two clocks, clock 1 used during normal system operation and clock 2 used during scan operations.

In system operation, clock 2 is held at a logical 1. On the fall of clock 1, the system data value is loaded into latch $FL1$. Clock 1 is held down until the new value propagates around the $FL1$ feedback path, and then is brought up to latch the value into $FL1$. When clock 1 rises, it captures the output of $FL1$ into latch $FL2$. It may happen that, during the rise of clock 1, latch $FL1$ is sensitive to the system data line while $FL2$ is simultaneously sensitive to $FL1$. If the system out line of $FL2$ feeds through a short logic path back to the system data input of $FL1$, a race may occur. This potential race condition is one hazard of using a single system clock.

In similar fashion, the scan operation uses clock 2 (while clock 1 is always at logical 1) to latch the test input into $FL1$ when clock 2 is 0 and into $FL2$ on the clock 2 rise.

The scan-out of one flip-flop is connected to the test input of the next to build the scan path of the card or board. A suggested connection scheme for a integrated circuit card is shown in Figure 2.18. The flip-flops of the card are chained together into a shift-register string. The scan-out point of all IC cards are wired together to a single observable scan-out pin. By selecting the X and Y control lines of a particular card, the card is enabled to drive the scan-out

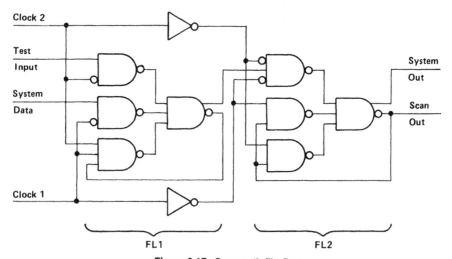

Figure 2.17. Scan-path flip-flop.

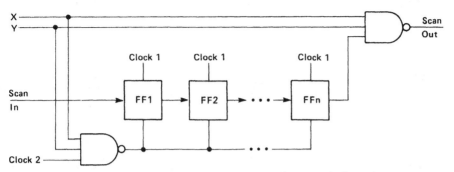

Figure 2.18. Scan-path shift-register connections on a logic card.

pin and the individual clock 2 inputs of its flip-flops are gated to the master clock 2 line.

2.2.4 Scan / Set Logic

Scan/set as described in Stewart (1977, 1978) is a partial scan-structured design, not as rigorous in its design constraints as LSSD or scan path, and should be regarded more as an aid to sequential circuit test generation, rather than a solution for it.

As shown in Figure 2.19, an auxiliary scan/set register is appended to the original combinational or sequential circuit for the purpose of either scanning selected circuit outputs or setting values into circuit storage or both. Since the original circuit remains unchanged except for the added control or observation points, test generation for the circuit is generally more complex than for LSSD or a random-access scan structure. However, the main circuit storage elements are not restricted to a single latch or flip-flop design.

The scan/set register is added overhead for testability. It can capture n bits of observation data in parallel from the main circuit for serial unload from its

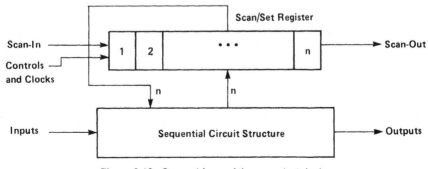

Figure 2.19. General form of the scan/set design.

scan-out port (called a "scan operation"), or can accept n bits of control data serially loaded into its scan-in port for parallel application to test points or storage control points in the main circuit (called a "set operation"). It is recommended in Stewart (1977) that scan points be connected to sense the status of all storage latches.

Since the register is functionally independent of the normal circuit operation, it can be used to take a "snapshot" of the test point status of the operating circuit for subsequent off-loading by way of the scan-out port. This function requires that the register load control be independent of the circuit controls.

2.3 BUILT-IN TEST

The ultimate in testable design is to make the design test itself. Built-in test can be nonconcurrent (an off-line test of either structural or functional integrity) or concurrent (an on-line test using either information redundancy or hardware redundancy or both). Building the test into the design, as might be expected, consumes added circuit and I/O overhead, but at the same time results in visible reductions to the costs of testing when compared with an external test using automatic test equipment. Built-in testing achieves these savings by

1. eliminating (or at least reducing) the costs of test pattern generation and fault simulation,
2. shortening the time duration of tests (by running tests at circuit speeds),
3. simplifying the external test equipment, and
4. easily adopting to engineering changes.

A nonconcurrent built-in test requires a mechanism for supplying test patterns to the unit being tested and a means for comparing the unit responses to the known good responses as suggested in Figure 2.20, and both mechanism and means must be compact enough to reasonably be built into the unit under test. Section 2.3.1 briefly surveys the techniques that have been suggested for comparing output responses to known reference responses, as an introduction for Chapter 4. Section 2.3.2 introduces the four different types of test that are used for built-in testing. Many of the methods included in these sections will be more fully elaborated in later chapters.

Concurrent or on-line testing includes such methods as error detection and correction circuitry, totally self-checking circuits, self-verification, and others. While concurrent testing is beyond the scope of this book, it is worth noting that some of the important techniques [such as triple modular redundancy (TMR)] provide an instantaneous correction of errors caused by either permanent or intermittent faults. An off-line or built-in test of these circuits must

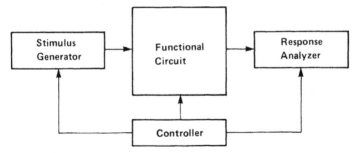

Figure 2.20. General form of a nonconcurrent built-in test structure.

ensure that the redundancy exists and is active (TMR is less reliable than its simplex version if it does not begin operation in a fault-free condition). A complete test of error correction circuitry is a mandatory requirement.

2.3.1 Response Analysis

It is obviously unsatisfactory to build into the circuit a bit-by-bit comparison of its test responses to the expected reference responses because of the large reference storage capacity that would be needed. Instead, it is usual to perform some form of data compression on the responses before making the reference comparison. The compressed response is referred to as the "signature" of the circuit under test, and comparison is made to the precomputed and stored reference signature. Any data compression method tends to lose information (unless the data being compressed is highly redundant). This loss is usually called "masking" and is measured by the probability that a faulty circuit will produce the same compressed signature as the fault-free circuit. When a common estimating framework is used, the relative masking probabilities of various compression techniques can be used as one measure of their efficiency.

Many compression methods have been suggested and some are in actual use. Chapter 4 describes in detail the following forms of data compression:

1. parity checking,
2. transition counting,
3. syndrome generation or ones counting,
4. signature analysis,
5. Walsh spectra.

A brief paragraph on each of these topics is included here as an introduction.

Parity checking of a response bit stream from a circuit under test is the most dramatic data compression scheme of any listed since it reduces a multitude of output data to a signature of length 1 bit. Parity checking by itself, however, is relatively ineffective since a large number of the possible response bit streams from a faulty circuit will result in the same parity as the correct stream. Nonetheless, parity checking in conjunction with another compression method can form a powerful combination with lower masking probability than either.

Transition counting, as its name implies, merely counts the number of times the output bit stream changes from 1 to 0 and vice versa. Curiously enough, the probability of masking in transition count testing is a function of the actual count. Low counts and high counts are less susceptible to masking than are intermediate counts.

Syndrome or ones counting uses as its signature the number of ones in the binary circuit response stream. Syndrome testing differs from ordinary ones counting in that a circuit can be designed for zero masking under syndrome test.

Signature analysis uses a shift register with various stages tapped and fed back to an EXCLUSIVE OR that in turn feeds the register input. The circuit output bit stream is applied to an additional EXCLUSIVE OR input and perturbs the feedback. The signature is the state of the register following completion of the test. Signature analysis is the most widely used data compression method.

Walsh spectral analysis verifies, in a sense, the truth table of the function. Collecting and comparing a subset of the complete Walsh functions has been described as a mechanism for data compression.

Other techniques have been suggested, for example, using a hashing circuit to drive a signature register as in Dervisoglu (1984), but are not as well known as those listed.

2.3.2   Off-Line Built-In Test Methodologies

Just as there are many ways to build in the processing of test responses, so are there many ways to generate the tests. The simplest categorization is in terms of the type of testing used:

1. exhaustive testing,
2. random testing,
3. prestored testing,
4. functional testing.

In exhaustive testing, the test length is 2^n tests, where n is the number of inputs to the circuit. Since all possible test patterns are applied, all possible single and multiple stuck faults are detected (redundancies excepted). The tests are generated with any process that cycles exhaustively through the circuit input space, such as a binary counter, a Gray code generator, or an n-stage nonlinear feedback shift register. Exhaustive testing for high input pin count structures requires relatively long test times, but in Bozorgui-Nesbat and

McCluskey (1980) it is suggested that circuits can be added to partition such structures into subcircuits, each of whose input pin count is low enough to permit exhaustive testing in a reasonable amount of time.

Random testing implies the application of a randomly chosen subset of the 2^n possible input patterns. (Random testing is a misnomer because the tests are actually chosen pseudorandomly so that the test set is repeatable.) A guarantee of the test coverage for the subset can be obtained by running the tests against a fault model of the circuit, or a probabilistic measure of coverage can be obtained by the analytical methods to be described in Chapter 7. The number of applied tests or the size of the subset is constrained by the economically allowable test time. While circuit partitioning is not needed, some logic modification may be necessary to ensure adequate coverage from the limited test set. A linear feedback shift register is the usual choice for a random test generator since its sequential output words, while deterministically generated, appear random and will pass most common tests for randomness.

Prestored testing, on the other hand, requires a preliminary step of test generation. The cost of test generation can be offset by the savings in test time resulting from a much smaller number of applied tests. The certainty of a known test coverage is an added bonus. Given this test set, prestored testing can be achieved in several ways. The conceptually simplest approach is to store the test patterns in an on-board read-only memory (ROM) and to use a counter to cycle through the ROM addresses. For relatively complex circuits, the ROM may be rather large.

Another approach to pre-stored testing uses a much smaller ROM, an address counter, and a linear feedback shift registger (LFSR) in a technique called "store-and-generate" as described in Agarwal and Cerny (1981). The concept is illustrated in Figure 2.21. The ROM contains r words of n bits each. Each of these r words is used sequentially as a starting value for the n-stage LFSR. For each starting value, the LFSR generates s vectors (of n bits each) that are applied to the n-input circuit as test patterns. The counter steps

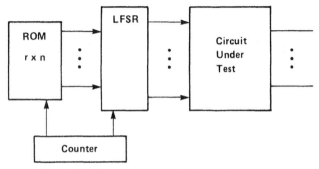

Figure 2.21. Store-and-generate test pattern source.

the LFSR s times for each address of the ROM. This total of $r \times s$ patterns is presumed to be a complete test set for the structural fault model used.

The problem with store-and-generate testing is determining the ROM words and the LFSR feedback configuration that will generate the desired test set. One practical approach is to use any LFSR and to fault simulate the effect of the s LFSR vectors from each ROM word. Given that k words have been chosen, the $(k + 1)$th word should be such as to cause the greatest improvement in test coverage. More exact suggestions for deriving the store-and-generate structure are given in Aboulhamid and Cerny (1983).

In Daehn and Mucha (1981a) it is shown that a nonlinear feedback shift register can be synthesized for use as a pattern source for any arbitrary precomputed test set. The test set is first sorted to order the tests in a way matching the shifting pattern of a feedback shift register. Next a set of "link" vectors is inserted between the tests to link each test to its successor test. Given this set of tests and link vectors, the generation of the shift-register feedback function is simply specifying and minimizing the Boolean function satisfying the transition between vectors on the list. In many cases, however, the feedback function tends to generate the entire exhaustive pattern set.

In a complex digital circuit many different failure models are possible. The assumption that faults can be modeled as logic gate inputs or output fixed to either a logic 0 or logic 1, as in the traditional single stuck-at fault model, admittedly does not cover all these failure modes but has remained successful both because it is computationally tractable (for small circuits) and because it results in a qualitative measure of test coverage. An alternative to the single stuck-fault model is functional testing, an approach that has been suggested to more realistically account for the actual effects of physical failures on logic.

Functional testing is a verification of the intended function of the circuit. Failures are modeled at the register transfer or functional level in terms of variations in expected function. Consider for example a two-input decoder as exemplified by Figure 2.22. Under a physical failure one of the following functional errors will occur:

1. No output line is active.
2. An incorrect output line is active instead of the desired line.
3. An incorrect output line is active in addition to the desired line.

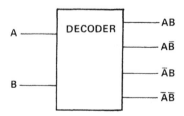

Figure 2.22. Functional level decoder.

Item 3 points out the necessity for assuring that the circuit not only performs its intended function but does not perform some other spurious function. Hence one requirement for successful functional test generation [as given in Abraham (1981)] is

> It is not sufficient for a functional test to determine whether the intended function has been performed correctly; a test should also verify that no unintended function was additionally performed.

While testing the decoder example is trivially simple, functional test techniques for much more complex structures, such as microprocessors, have been developed [for example, see Brahme and Abraham (1984)].

An unusual type of testing that does not fit neatly into any of the categories is called "self-oscillation" [Buehler and Sievers (1982)]. In this scheme, the presumption is that the circuit can be restructured in test mode with feedback such that it will oscillate if fault-free. A further presumption is that oscillation will not occur if faults are present. There is no theoretical basis for either efficiently constructing the self-oscillation test mode or for determining the fault coverage obtainable through self-oscillation. It appears that this mode of testing is generally an ad hoc test mechanism. However, frequency measurements made during self-oscillation offer the possibility of measuring path delays, at least along the oscillation path.

2.4 TESTABILITY MEASURES

It is possible to grossly measure the testability of a circuit without actually generating tests or running a fault simulator against a set of random input patterns. If the goal is deterministic testing, there are testability measures that will give a relative comparison of the costs of test generation and fault simulation, and can help to identify gross circuit areas where test generation will be difficult and where design for testability should be used. For random pattern testing, there are measures that compute the probability of detecting faults in the circuit so that difficult-to-test areas may be redesigned and so that a test quality estimate can be derived. A testability measure is useful only if it requires much less effort than actual test generation or fault simulation. This implies that such a measure can only produce approximate results.

Those testability tools that have been developed specifically for random pattern testing are discussed in Chapter 7. In the remainder of this chapter we describe several testability measures that have been developed for predicting the effort required to do test generation. These measures assign a testability value to each node of a circuit by computing a controllability and an observability value for the node using simplified rules that are in some way related to the actual process of test generation. For example, Figure 2.23 shows two logic blocks Y and Z that are embedded within a much larger logic

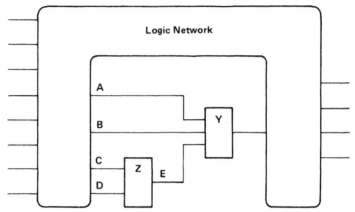

Figure 2.23. Concept of controllability and observability.

structure. The controllability of output E of block Z is a measure of how easy it is to set the value of E to a logical 0 or 1 by assigning values to the primary inputs of the circuit, and is a function of the controllability of lines C and D and the transfer characteristics of block Z. The observability of line E is a measure of how easy it is to propagate the value on line E to a circuit primary output by assigning values to the unassigned primary inputs, and is a function of the controllability of lines A and B and the observability of the output of block Y.

In this section we introduce the concepts and basic procedures of three testability measures. The first was described in 1976 by Jack Stephenson of Bell Laboratories and John Grason of Carnegie-Mellon University, and has since been programmed at Bell Laboratories in a routine called TMEAS. We then describe two contrasting measures called SCOAP and VICTOR. Although all of these measures, save for VICTOR, have mechanisms for handling both combinational and sequential circuits, our descriptions will concentrate on purely combinational functions for simplicity. The details for coping with sequential circuits can be found in the references cited for TMEAS and SCOAP.

While the intuitive concepts of controllability and observability given above could be replaced by more rigorous definitions, such rigor is not justified by the approximations used in the measures described here. An example of each of the measures is given to emphasize the approximate nature of the values obtained.

2.4.1 The Stephenson – Grason Testability Measure

The testability measure described by Stephenson and Grason (1976) and implemented as TMEAS [Grason (1979)] estimates the testability of a circuit from estimates of its controllability and its observability.

Controllability (CY) is normalized to the range 0–1, with a 1 representing perfect controllability, such as for the circuit primary inputs, and 0 representing an absolute lack of control. Controllability values are propagated through circuit components from their inputs to their outputs by means of a controllability transfer factor (CTF) that is developed for each component. The controllability of the component outputs are then related to that of its inputs by

$$CY_{outputs} = (CTF) \times (CY_{inputs}) \qquad (2.1)$$

where CY_{inputs} is the average of the controllability values on the component input lines. Each output of a multiple-output component is assigned the same controllability.

The CTF of a component is a measure of the uniformity of the component's input/output mapping (the number of 1's and 0's in the output Karnaugh map), and varies between 0 (worst) and 1 (best). Components with equal numbers of 1's and 0's have CTF = 1. Components with less uniform I/O maps have lesser CTFs. The argument for this assignment is that controllability is most easily transferred from component inputs to outputs if every possible output value is produced by the same number of input patterns. The CTF for an n-input single-output combinational component is defined as

$$CTF = 1 - \frac{|N(0) - N(1)|}{N(0) + N(1)} \qquad (2.2)$$

where $N(0)$ and $N(1)$ are the number of input patterns for which the component output is 0 and 1, respectively. For example, an INVERTER has $N(0) = N(1) = 1$, so its CTF is 1, which reflects the obvious fact that its output is just as controllable as its input. At the other extreme lies an n-input AND gate whose CTF $= 1/2^{n-1}$ since it has $N(1) = 1$ and $N(0) = 2^n - 1$. The CTF values of a component are related only to the function of the component, and not to its placement or use in the circuit, and thus need be computed just once per component. Figure 2.24 displays CTF values for simple functions. The CTF values for each component in the circuit are normally precomputed and stored in a component library file to support the testability computation.

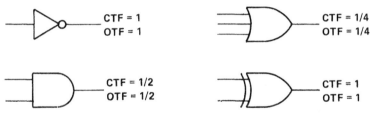

Figure 2.24. CTF and OTF for INVERTER, AND, OR, and EXCLUSIVE OR gates.

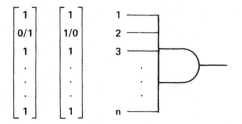

Figure 2.25. Observability patterns for fault on input 2.

The output controllability of a sequential circuit is a function of both the input controllabilities and the current internal state. Its CTF is modeled by adding a feedback net that represents the internal state and by allowing this implicit net to act as both a circuit input and a circuit output. This converted circuit is now specified with an input/output mapping just as is a combinational circuit.

The observability measure (OY) on each net of a circuit also varies between 0 and 1, where a 1 represents perfect observability as would occur on the outputs of the circuit. The measure is computed by backtracing from the circuit outputs, calculating a component input observability from its output observability and an observability transfer factor (OTF) as

$$OY_{inputs} = (OTF) \times (OY_{outputs}) \qquad (2.3)$$

For a component with multiple outputs, $OY_{outputs}$ is taken as the average value of the observability of the individual outputs. The value of input observability computed by Equation (2.3) is assigned to each input of the component.

The OTF of a component ranges from 0 (bad) to 1 (good) and is the probability that a faulty input value (on any component input) is recognized as such on the component output. Let FD_i be the fraction of the input vectors that sensitize a path from input i to the components output. Figure 2.25 shows the observability patterns for the second input of the n-input AND gate. There are two patterns that sensitize a fault on input 2 to the output, and thus $FD_2 = 2/2^n$. Now let p_i be the probability that input i carries the fault value. Since we have no better knowledge of the input carrying the fault, we assume $p_i = 1/n$. Then the OTF for an n-input/single-output combinational component can be computed as

$$OTF = \sum_{i=1}^{n} p_i FD_i \qquad (2.4)$$

Note that since $p_i = 1/n$, OTF is always equal to FD for all components. The AND circuit in Figure 2.25 has $FD_i = 2/2^n$ and $p_i = 1/n$. Hence the OTF $= 1/2^{n-1}$, and applies to any one of the n inputs. An INVERTER has an OTF $= 1$, since $p_i = 1$ and $FD_i = 2/2 = 1$. Figure 2.24 shows the OTF values for some simple functions. The OTF of a sequential circuit is modeled by extracting an implicit feedback net as was suggested for the CTF.

Since testability depends a great deal upon the interconnections of the components, rules are provided for assigning controllability and observability values to stems that fan out to several destinations.

1. Consider controllability first. The problem occurs at points of reconvergence of the fan-out branches where the lack of independence of signal values on the reconvergent paths can cause many remade decisions during test generation. This is effectively a lack of controllability at these points of reconvergence and is modeled by artificially reducing the controllability value at the destinations of such fan-out branches. Let f denote the number of branches (destinations) for a particular fan-out stem. Stephenson and Grason found that good results are produced when the controllability value assigned to each destination is obtained by dividing the controllability of the fanout stem by $(1 + \log_{10} f)$. For example, consider a primary input that fans out to two circuits. The stem controllability is 1. However, at the fan-out branches, the controllability is $1/(1 + \log_{10} 2)$, or 0.7686.

2. For observability, a somewhat different situation exists. The observability of a fan-out stem should somehow be related to the observability at each destination of the fan-out, but should be greater than the observability of any one destination because of the multiple paths along which fault information at the fan-out stem may be observed. Let f be the number of branches for a fan-out stem and let OY_i be the observability value at each destination, $i = 1, 2, \ldots, f$. Then the observability of the fanout stem OY_s is assigned the value

$$OY_s = 1 - \prod_{i=1}^{f} [1 - OY_i] \qquad (2.5)$$

For each node in the circuit we can now set up two equations, a controllability equation in which the component output controllability is a function of its source CTFs and their input controllabilities, and an observability equation in which the component output observability is a function of its destination OTFs and their output observabilities. The set of controllability equations is solved by assigning primary inputs a controllability of 1. Similarly, the observability equations are solved by assigning primary outputs an observability of 1. Notice that the observability values thus obtained are independent of the controllability values.

The testability measure now assumes that any circuit node with either low controllability or low observability is difficult to test. The measure chosen is the geometric mean (the square root of the product) of the controllability and the observability. This measure can be computed for the entire circuit (or for a

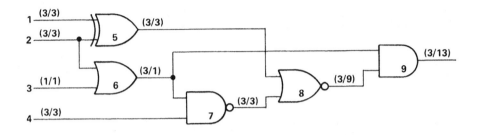

Node	CY	OY	TY
1	1	0.25	0.5
2	1	0.4609	0.6789
3	1	0.2813	0.5304
4	1	0.125	0.3536
5	0.8843	0.25	0.4702
6	0.4422	0.5625	0.4987
7	0.335	0.25	0.2894
8	0.3048	0.5	0.3904
9	0.1612	1	0.4015

Figure 2.26. Circuit example showing Stephenson and Grason measures.

portion of the circuit) by using the average controllability and the average observability of the included components.

An example of the measure is given in the circuit and table of Figure 2.26. The tabulated values are the controllability (CY), observability (OY), and testability (TY) of the nodes in the circuit. The actual number of tests for each fault was computed using fault simulation and in the figure each node is marked with a number of the form (SA0/SA1), where SA0 is the number of tests that exist for that node stuck-at-0 and SA1 is the number of tests for that node stuck-at-1. (The number of possible tests for a node is certainly some measure of the testability of that node.) Notice that there are faults in the circuit that require multiple path sensitization. For example, there is no single sensitized path test for the output of block 6 stuck-at-1. The test for this fault requires a 0 on the first three inputs and a 1 on input 4. The "failure" signal is carried both from output 6 directly to block 9 and also through blocks 7 and 8. One might expect some correlation between the number of tests existing for a node and the testability of that node. However, the best testability occurs on node 2 and the worst on node 7, and both nodes have the same number of possible tests for SA0 and SA1 faults.

Stephenson and Grason compared their testability measure against the actual test generation effort required for three circuits with sizes ranging from 6 to 485 components. They concluded that the relative testability of the three circuits correlated quite well with the actual cost of test generation.

2.4.2 SCOAP

SCOAP (for Sandia Controllability/Observability Analysis Program) is a testability measurement program based on the work by Goldstein [Goldstein (1979); Goldstein and Thigpen (1980)]. SCOAP does not provide an actual testability number for each node, unlike the Stephenson–Grason method. Rather it assumes that the testability of a circuit is related to the difficulty of controlling the logical values on nodes from the primary inputs and of observing nodal values at primary outputs, and provides numeric measures of the difficulty. In many cases the larger the numeric measure, the more difficult that node is to either control or observe.

SCOAP uses six controllability and observability functions, all of which are integer-valued. The *combinational* 0 and 1 controllabilities of a node N are $CC^0(N)$ and $CC^1(N)$. The *sequential* 0 and 1 controllabilities of the node are $SC^0(N)$ and $SC^1(N)$. $CO(N)$ is the *combinational* observability of the node, and $SO(N)$ is its *sequential* observability.

$CC^0(N)$ and $CC^1(N)$ are a function of the number of combinational nodes that must be assigned values to justify a 0 or a 1 on node N. A combinational node is either a primary input pin or an output of a combinational logic cell (AND, OR, NOR, etc.). $SC^0(N)$ and $SC^1(N)$ are a function of the number of sequential nodes that must set to justify a 0 or a 1 on node N. A sequential node is the output of a sequential logic cell (flip-flop, latch, etc.).

The combinational 0 and 1 controllabilities of the output node of a cell are computed by taking either the minimum or the sum of the controllabilities of all input assignments that produce the required output value, and incrementing this by the cell depth. Consider the three-input OR gate shown in Figure 2.27. An output 0 only occurs when all of the inputs are 0. The output 0 controllability, then, is the sum of the 0 controllabilities on the inputs plus the cell depth, or

$$CC^0(Y) = CC^0(X_1) + CC^0(X_2) + CC^0(X_3) + 1 \qquad (2.6)$$

where 1 is the combinational depth of the OR cell. To justify an output 1 requires at least one input be at 1. Then the minimum cost 1 controllability on

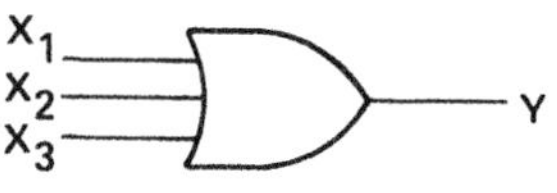

$$CC^0(Y) = CC^0(X_1) + CC^0(X_2) + CC^0(X_3) + 1$$
$$CC^1(Y) = \min[CC^1(X_1), CC^1(X_2), CC^1(X_3)] + 1$$
$$SC^0(Y) = SC^0(X_1) + SC^0(X_2) + SC^0(X_3)$$
$$SC^1(Y) = \min[SC^1(X_1), SC^1(X_2), SC^1(X_3)]$$
$$CO(X_i) = CO(Y) + CC^0(X_j) + CC^0(X_k) + 1$$
$$SO(X_i) = SO(Y) + SC^0(X_j) + SC^0(X_k)$$

Figure 2.27. SCOAP controllability and observability for an OR gate.

the output is the minimum of the 1 controllabilities on the inputs plus the cell depth or

$$CC^1(Y) = \min[CC^1(X_1), CC^1(X_2), CC^1(X_3)] + 1 \qquad (2.7)$$

The OR gate is combinational and has no sequential depth. Thus the sequential 0 and 1 controllabilities are just

$$SC^0(Y) = SC^0(X_1) + SC^0(X_2) + SC^0(X_3) \qquad (2.8)$$

and

$$SC^1(Y) = \min[SC^1(X_1), SC^1(X_2), SC^1(X_3)] \qquad (2.9)$$

The observability of an input to the OR gate is a function of the output observability and of the cost of holding all other inputs at 0, incremented by the cell depth. Thus

$$CO(X_i) = CO(Y) + CC^0(X_j) + CC^0(X_k) + 1 \qquad (2.10)$$

and

$$SO(X_i) = SO(Y) + SC^0(X_j) + SC^0(X_k) \qquad (2.11)$$

SCOAP controllability and observability functions for other simple circuits are shown in Figure 2.28.

In SCOAP, a primary input has $CC^0 = CC^1 = 1$ and $SC^0 = SC^1 = 0$. A primary output pin has $CO = SO = 0$ since it can be observed without the need for controlling any other node values. SCOAP first computes controllability values for each node in the circuit using the cell controllability equations and working breadth first from primary inputs to primary outputs. If there are global feedback paths (external to the cells) this step may need to be repeated two or three times until the controllability values stabilize (as they eventually will). Then, working from primary outputs back to primary inputs, the node observability values are computed using the cell observability equations and the previously computed controllability values. During this backtrace, the observability of a fan-out stem is computed as the minimum observability of any branch of the fan-out. Again it may be necessary to iterate on the observability computation if global feedback exists.

It is useful here to consider how SCOAP differs from the Stephenson and Grason measure described earlier. Considering the SCOAP functions for a three-input OR given in Figure 2.27, it is apparent that SCOAP is rather more complex. Assuming the OR is driven directly by primary inputs, the gate output controllability under the Stephenson–Grason measure is $\frac{1}{4}$. In SCOAP, however, the 0 controllability is 4 and the 1 controllability is 2, reflecting the fact that it is "easier" to get a 1 than a 0. The Stephenson and Grason measures are, in a sense, average values whereas SCOAP measures are mini-

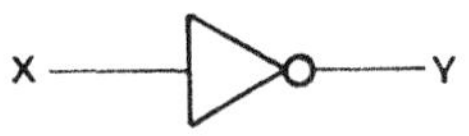

$$cc^0 (Y) = cc^1 (X) + 1$$
$$cc^1 (Y) = cc^0 (X) + 1$$
$$sc^0 (Y) = sc^1 (X)$$
$$sc^1 (Y) = sc^0 (X)$$
$$CO (X) = CO (Y) + 1$$
$$SO (X) = SO (Y)$$

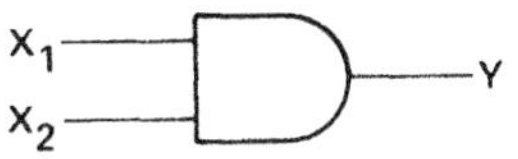

$$cc^0 (Y) = \min[cc^0 (X_1), cc^0 (X_2)] + 1$$
$$cc^1 (Y) = cc^1 (X_1) + cc^1 (X_2) + 1$$
$$sc^0 (Y) = \min[sc^0 (X_1), sc^0 (X_2)]$$
$$sc^1 (Y) = sc^1 (X_1) + sc^1 (X_2)$$
$$CO (X_i) = CO (Y) + cc^1 (X_j) + 1$$
$$SO (X_i) = SO (Y) + sc^1 (X_j)$$

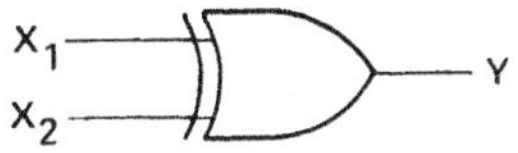

$$cc^0 (Y) = \min[cc^1 (X_1) + cc^1 (X_2), cc^0 (X_1) + cc^0 (X_2)] + 1$$
$$cc^1 (Y) = \min[cc^0 (X_1) + cc^1 (X_2), cc^1 (X_1) + cc^0 (X_2)] + 1$$
$$sc^0 (Y) = \min[sc^1 (X_1) + sc^1 (X_2), sc^0 (X_1) + sc^0 (X_2)]$$
$$sc^1 (Y) = \min[sc^0 (X_1) + sc^1 (X_2), sc^1 (X_1) + sc^0 (X_2)]$$
$$CO (X_i) = CO (Y) + \min[cc^0 (X_j), cc^1 (X_j)] + 1$$
$$SO (X_i) = SO (Y) + \min[sc^0 (X_j), sc^1 (X_j)]$$

Figure 2.28. SCOAP controllability and observability functions.

mum cost values (the cost of controlling a net to 0 as opposed to 1). The question of whether the added complexity of SCOAP is worthwhile can be considered by computing its functions for the combinational circuit of Figure 2.26.

Table 2.1 shows the SCOAP values for combinational 0 and 1 controllability (CC^0 and CC^1) and combinational observability (CO) for the circuit given in Figure 2.26. One way of interpreting these values is to note that a test for a node stuck-at-0 requires that the node be controlled to a logical 1 and simultaneously that the nodal value be observed at a primary output. The test generation effort, then, is defined as the sum of the nodal CC^1 and CO values (assuming their independence). Similarly, a measure of $CC^0 + CO$ is suitable for a SA1 fault on the node. Table 2.2 shows these sums for the SCOAP values of Table 2.1. Also shown for comparison are the number of possible tests for SA0 and SA1 faults on the nodes (these values were originally given in Figure 2.26). Notice that there is a reasonably good correlation between the fault-specific measures and the number of tests. Nodes with low measures are richer in possible tests than those with large measures.

TABLE 2.1 SCOAP Values for the Circuit of Figure 2.26

Node	CC^0	CC^1	CO
1	1	1	10
2	1	1	10
3	1	1	11
4	1	1	10
5	3	3	8
6	3	2	9
7	4	2	7
8	3	8	3
9	4	11	0

TABLE 2.2 Fault-Specific Measures from SCOAP Values

Node	$CC^1 + CO$	SA0 Tests	$CC^0 + CO$	SA1 Tests
1	11	3	11	3
2	11	3	11	3
3	12	1	11	1
4	11	3	11	3
5	11	3	11	3
6	11	3	12	1
7	9	3	11	3
8	11	3	6	9
9	11	3	4	13

A FORTRAN 77 version of SCOAP is available from the Industrial Liaison Program, University of California, Berkeley, CA 94720. An upgraded version of SCOAP called COMET is described in Berg and Hess (1982).

2.4.3 VICTOR

VICTOR (for VLSI identifier of controllability, testability, observability, and redundancy) was developed to generate tests and identify redundant nodes in combinational circuits [described in Ratiu et al. (1982); Ratiu (1983)]. It is a linear algorithm, requiring four passes through the circuit to perform (1) circuit leveling, (2) controllability analysis, (3) observability analysis, and (4) test generation. VICTOR assumes a scan-path design for testability and does not have a mechanism for handling sequential circuits.

Levelization. VICTOR begins by marking and leveling the circuit. Each circuit node is marked with a unique name, where nodes are the primary inputs, the outputs of every functional block, and all fan-out destinations. The circuit is leveled to identify the relative processing order of circuit nodes. All

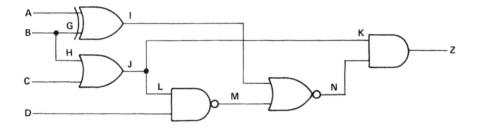

Level 0: A, B, C, D, G, H   Levelized Cell List:
Level 1: I, J, K, L   XOR, OR, NAND, NOR, AND
Level 2: M
Level 3:· N
Level 4: Z

Figure 2.29. Circuit example for testability analysis.

primary inputs and their fan-out nodes are assigned to level 0. A node other than a primary input node takes the level of the functional block that drives the node (a fan-out branch is assigned the level of its stem). The level of a functional block is defined as the level of its highest-level input node plus 1. Markup and leveling is illustrated in Figure 2.29 which will be used as an example in this description of VICTOR.

Controllability. The second pass through the circuit computes the 1-controllability and the 0-controllability of each node, concepts that VICTOR calls "set" and "reset," respectively. Node set and reset measures consist of three values named "pattern," "risk," and "size." The set and reset triplets for a node V are represented as

$$V_{TS} = \{ V_s, \text{risk}(V_s), \text{size}(V_s) \}$$
$$V_{TR} = \{ V_r, \text{risk}(V_r), \text{size}(V_r) \} \tag{2.12}$$

V_s and V_r are the set and reset control patterns, a set of primary input values that are sufficient to force node V to a 1 and 0, respectively (or alternatively to provoke a stuck-at-0 and a stuck-at-1 fault at the node). All patterns in a circuit are the same length, having one symbol for each primary input. VICTOR uses four-valued logic, allowing any of the following four symbols in a pattern:

  0 = logical 0 assignment,
  1 = logical 1 assignment,
  x = no assignment (don't care),
  # = conflicting 0 and 1 assignment (called "clash").

The values risk(V_s) and risk(V_r) in Equation (2.12) are the set and reset risk, and heuristically measure the risk of reconvergence of fan-out nodes and, thus, the possibility of redundant faults. Size(V_s) and size(V_r) are estimates of the number of primary input patterns that will either set or reset the node. The set and reset patterns are the essential control information generated by VICTOR. The risk and size measures in the triplets are used only to guide the selection of patterns.

Computing these controllability measures begins with the primary input nodes (at level 0) and flows forward by level until the primary outputs are reached. A primary input set (reset) pattern is a 1 (0) for the input itself and an x (don't care) for all other inputs.

The reconvergence risk value for a primary input with no fan-out is set at 1. However, where fan-out exists (as with input B of Figure 2.29), the risk value is the product of the number of fan-out nodes and the number of levels to the furthest primary output. Thus the set and reset risks of node B are both $2 \times 4 = 8$.

Finally, the set and reset sizes for primary inputs are both 1 because an input node can be set or reset in only one way. Controllability triplets for the primary inputs and all other nodes of the example circuit will be derived subsequently and are listed for convenience in Table 2.3.

The controllability values for fan-out branches take the values for their driving node. Hence the set and reset triplets for the fan-out nodes G and H in the example circuit are the same as those for node B.

In the propagation of controllability triplets through the circuit, logical elements are handled in level order. Triplets on the outputs of blocks are computed using one of two operations called "pattern selection" and "pattern intersection."

TABLE 2.3   VICTOR Testability Triplets for the Circuit of Figure 2.29

	Set			Reset			Monitor		
Node	ABCD	R	S	ABCD	R	S	ABCD	R	S
A	1xxx	1	1	0xxx	1	1	x111	31	4
B	x1xx	8	1	x0xx	8	1	1x11	22	4
C	xx1x	1	1	xx0x	1	1	####	99999	0
D	xxx1	1	1	xxx0	1	1	001x	37	4
G	x1xx	8	1	x0xx	8	1	1x11	22	4
H	x1xx	8	1	x0xx	8	1	####	99999	0
I	10xx	9	1	00xx	9	1	xx11	21	4
J	xx1x	7	2	x00x	15	1	####	99999	0
K	xx1x	7	2	x00x	15	1	0011	17	9
L	xx1x	7	2	x00x	15	1	0011	23	2
M	xxx0	1	2	xx11	8	2	001x	22	2
N	0011	17	9	xxx0	1	3	xx1x	13	2
Z	0011	24	18	xxx0	1	4	xxxx	0	1

Pattern selection defines an output pattern as the lowest-risk lowest-level input pattern. The risk of the output pattern is the same as the risk of the chosen input pattern. The size of the output pattern is the sum of the sizes of all input patterns.

Pattern intersection defines an output pattern as the symbol by symbol intersection of the input patterns. The intersection operation is governed by the following table:

	0	1	x	#
0	0	#	0	#
1	#	1	1	#
x	0	1	x	#
#	#	#	#	#

The risk of an intersection output pattern is the sum of the risks of the input patterns, and the size is the product of the input sizes. The choice of whether to do selection or intersection is governed by the function of the element. Crudely, one can think of selection as an OR operation and intersection as an AND. The choice will be made clear in the example discussion.

In the circuit of Figure 2.29, the order of the logical elements is EXCLU-SIVE OR (XOR), OR, NAND, NOR, and AND. The first step is to select a set and reset pattern for the XOR (node I). Node I can be set to 1 if either A is set and G is reset or vice versa. These two choices are both intersection operations, thus

$$I_s = \text{intersect}(A_s, G_r) \text{ or intersect}(A_r, G_s)$$

Similarly, there are two choices for resetting I:

$$I_r = \text{intersect}(A_s, G_s) \text{ or intersect}(A_r, G_r)$$

When faced with a choice (either during pattern intersection or pattern selection) VICTOR chooses that pattern with the lowest risk. In the case of a tie, the lowest-level pattern is chosen. If the tie still exists, the choice is arbitrary. As noted earlier, the risk of a pattern resulting from pattern intersection is the sum of the input risks. Consider the choices for setting node I. The resulting patterns and risks are

1. $I_s = \text{intersect}(A_s, G_r) = 10\text{xx}$

$$\text{risk}(I_s) = \text{risk}(A_s) + \text{risk}(G_r) = 1 + 8 = 9$$

2. $I_s = \text{intersect}(A_r, G_s) = 01\text{xx}$

$$\text{risk}(I_s) = \text{risk}(A_r) + \text{risk}(G_s) = 1 + 8 = 9$$

Since both patterns have the same risk and since there is no difference in levels, the set pattern 10xx is arbitrarily chosen. Similarly, the reset pattern choices have the same risk and levels and 00xx is chosen. The size (for both set and reset) is the product of the input sizes or size(I) = size(A) $\times$ size(G) = 1. Thus the {pattern, risk, size} triplets for node I and I_{TS} = {10xx, 9, 1} and I_{TR} = {00xx, 9, 1}.

Node J, the output of the OR gate, is an internal fan-out stem. Whenever a cell output is a fan-out stem, its risk is increased by adding to the computed risk the product of the fan-out count and the number of levels to the furthest primary output. An OR is set if any input is set, so that the set equation for J_s is select(H_s, C_s). Selection chooses the lowest-risk input pattern, which in this case is C_s so that J_s becomes xx1x. The risk for C_s is 1, to which is added the product of the fan-out 2 and the distance 3. The size is the sum of the sizes of H_s and C_s. Thus J_{TS} = {xx1x, 7, 2}.

Resetting node J requires both inputs at 0 or J_r = intersect(H_r, C_r). The intersection J_r = x00x, risk = 9 + 6, and size = 1.

The controllability triplets for the remaining blocks are computed similarly.

Observability. Once the controllability measures are complete on all cells, VICTOR begins the third pass through the circuit computing the observability of the circuit nodes. VICTOR uses the term "monitor" for observability. Monitor measures are computed by starting at the primary outputs and working backward through the levelized cell list. Monitor measures are again a triplet of pattern, risk, and size for each node.

In the example circuit, monitor analysis starts with output Z of the AND gate. An output is observable without any primary input constraints, so the appropriate pattern for Z is xxxx. An output has no risk and its size is 1. Hence the monitor triplet for Z is Z_{TM} = {xxxx, 0, 1}. (Monitor triplets for all nodes are listed in Table 2.3.)

The rules to derive input monitor triplets from output triplets are called the monitor equations, and use the select and intersect concepts described earlier. In the example circuit, the AND gate has monitor equations

$$K_m = \text{intersect}(N_s, Z_m), \qquad N_m = \text{intersect}(K_s, Z_m)$$

which result in

$$K_m = \text{intersect}(0011, \text{xxxx}) = 0011, \qquad N_m = \text{intersect}(\text{xx1x}, \text{xxxx}) = \text{xx1x}$$

Monitor risk and size computation generally follows the rules described earlier. Using those rules, the monitor triplets for K and N would be K_{TM} = {0011, 17, 9} and N_{TM} = {xx1x, 7, 2}. However, when a fan-out branch enters a cell, VICTOR computes a monitor risk adder that is the product of the fan-out and the number of levels from the fan-out stem to the cell output. In the example, the AND gate has an input K that is a fan-out point from stem node J. Since the stem J is at level 1 and the AND gate output is at level 4, the risk for input K is $2 \times (4 - 1) = 6$. This risk is added to the computed risk for the other cell input, node N. Thus the monitor triplets for nodes K and N are

$$K_{TM} = \{0011, 17, 9\}, \qquad N_{TM} = \{xx1x, 13, 2\}$$

The next cell is the NOR with monitor equations

$$I_m = \text{intersect}(M_r, N_m), \qquad M_m = \text{intersect}(I_r, N_m)$$

resulting in the triplets $I_{TM} = \{xx11, 21, 4\}$, $M_{TM} = \{001x, 22, 2\}$.

The NAND gate, processed next, also has a fan-out node as one input. The NAND monitor equations are

$$D_m = \text{intersect}(L_s, M_m) = \text{intersect}(xx1x, 001x) = 001x$$
$$L_m = \text{intersect}(D_s, M_m) = \text{intersect}(xxx1, 001x) = 0011$$

Because L is a fan-out node, the risk adder is the product of the fan-out and the level difference between the fan-out node J and the cell output M, or $6 + 2 \times (2 - 1) = 8$, and is applied to the other cell input D. Without the fan-out, the risk at D is $\text{risk}(L_s) + \text{risk}(M_m)$ or 29. With the adder, $\text{risk}(D_m)$ becomes 37. The risk at L is 23. The size of D_m is 4 and of L_m is 2. Hence $D_{TM} = \{001x, 37, 4\}$ and $L_{TM} = \{0011, 23, 2\}$.

Now to calculate the observability of a fan-out stem, VICTOR first computes the set-monitor and the reset-monitor intersections for each of its branches and then assigns to the stem the triplet of that branch whose set-monitor and reset-monitor patterns are both clash-free. In the example, nodes K and L are the branches from fan-out stem J. The set-monitor and reset-monitor intersections for K are

$$\text{intersect}(K_s, K_m) = \text{intersect}(xx1x, 0011) = 0011$$

and

$$\text{intersect}(K_r, K_m) = \text{intersect}(x00x, 0011) = 00\#1$$

and for L are

$$\text{intersect}(L_s, L_m) = \text{intersect}(xx1x, 0011) = 0011$$

and

$$\text{intersect}(L_r, L_m) = \text{intersect}(x00x, 0011) = 00\#1$$

Note that the reset-monitor intersection for both branches produces a clash in the third symbol. When all patterns have clashes, VICTOR sets the fan-out stem monitor triplet to $\{\#\# \cdots \#, 99999, 0\}$. Hence $J_{TM} =$

{ # # # #, 99999, 0}. In a general network, however, there may be several clash-free patterns, in which case VICTOR takes the pattern that has the most don't cares (x).

Observability of the remaining cells is computed in a similar fashion.

VICTOR Test Generation and Redundancy Identification. It was noted earlier that the set and reset patterns are primary input values that are sufficient to provoke a stuck-at-0 and a stuck-at-1 fault at the node. Similarly, the monitor patterns are sufficient input values to propagate a fault from a node to an observable output. The intersection of the nodal control and monitor patterns, then, can be a test for the node. Two tests are required for each node V:

$V/0$ test: intersect(V_s, V_m).
$V/1$ test: intersect(V_r, V_m).

The set, reset, and monitor patterns from Table 2.3 are reproduced in Table 2.4, where the control and monitor pattern intersections are also shown. Note that eight of the node fault tests contain clashes and are potential redundancies. Clashes occur in the monitor patterns for nodes C, H, and J and carry over to the tests. Nodes K and L are clash-free in the control and monitor patterns, but clashes appear in the $V/1$ test.

$V/0$ and $V/1$ patterns in which no clashes occur are valid tests. From Table 2.4, after test compaction, there are five tests.

A	B	C	D
0	0	1	0
0	0	1	1
0	1	1	1
1	0	1	1
1	1	1	1

These tests, from the VICTOR results, will detect 18 of the 26 possible nodal faults.

For comparison with VICTOR, Table 2.5 shows fault simulation results on the example circuit of Figure 2.29. On the left of the figure are listed the 16 possible input patterns and on the right are the faults detected by each pattern. A 1 means that the pattern detects a s-a-1 fault on the node, a 0 that it detects s-a-0, and no entry signifies no detection. For example, the pattern 0010 detects D s-a-1, M s-a-0, N s-a-1, and Z s-a-1.

Table 2.5 shows that no tests exist for nodes K and L stuck-at-1, and the VICTOR results support this with a clash in the $V/1$ tests (Table 2.4) for these two nodes. However, VICTOR also shows clashes for six other faults that are, in fact, detectable. Table 2.5 shows that the five tests generated by VICTOR also detect two of the potentially redundant faults, namely $C/0$ with test 0011 and $J/0$ with tests 0011 and 1111.

TABLE 2.4   VICTOR Test Generation for the Circuit of Figure 2.29

Node	Set ABCD	Reset ABCD	Monitor ABCD	V / 0 Test ABCD	V / 1 Test ABCD
A	1xxx	0xxx	x111	1111	0111
B	x1xx	x0xx	1x11	1111	1011
C	xx1x	xx0x	# # # #	# # # #	# # # #
D	xxx1	xxx0	001x	0011	0010
G	x1xx	x0xx	1x11	1111	1011
H	x1xx	x0xx	# # # #	# # # #	# # # #
I	10xx	00xx	xx11	1011	0011
J	xx1x	x00x	# # # #	# # # #	# # # #
K	xx1x	x00x	0011	0011	00#1
L	xx1x	x00x	0011	0011	00#1
M	xxx0	xx11	001x	0010	0011
N	0011	xxx0	xx1x	0011	xx10
Z	0011	xxx0	xxxx	0011	xxx0

TABLE 2.5   Fault Simulation Results for the Circuit of Figure 2.29

Test ABCD	A	B	C	D	G	H	I	J	K	L	M	N	Z
0000													1
0001		1				1		1					1
0010			1								0	1	1
0011	1	1	0	0	1		1	0	0	0	1	0	0
0100												1	1
0101	1			0			0					1	1
0110												1	1
0111	1	0		0			0					1	1
1000													1
1001		1											1
1010												1	1
1011	0	1		1			0					1	1
1100				1							0	1	1
1101	0	0		0	0	0	1	0	0	0	1	0	0
1110				1							0	1	1
1111	0	0		0	0		1	0	0	0	1	0	0

A FORTRAN 77 version of VICTOR is available from the Industrial Liaison Program, University of California, Berkeley, CA 94720.

2.4.4   Limitations of Testability Measures

How are the controllability, observability, and testability numbers used? As suggested during the discussion of test points in Section 2.1.4, the controllability or observability of circuit elements can be enhanced by adding observation or control points. We can identify potential locations of test points by running the testability measure against the circuit and then "coloring" a schematic of the circuit to show areas that fall below some threshold of testability. A recomputation of the measure against each of the potential modifications will show the comparative improvement value for each location. Suggestions for automatic design and modification based on controllability and observability measures are given in Chen and Breuer (1985).

It was noted earlier that test generation is an NP-complete process, suggesting that no linear time algorithm can perfectly predict the effort required to do test generation. As a theoretical consequence, the results of testability analysis are only approximate. Faults that are difficult to test will cause problems, but it is precisely those faults for which good measures are needed. For some interesting commentary on testability measures see Agrawal and Mercer (1982) and Savir (1983).

3

Pseudorandom Sequence Generators

A string of binary digits (1's and 0's) is called a pseudorandom binary sequence when the bits appear to be random in the local sense, but they are in some way repeatable, hence only pseudorandom. These sequences have been studied extensively because they find use in communications systems, encipherment, privacy encoding, error-correcting coding, and others. These sequences are of interest here because they form an important class of stimuli for the testing of digital circuits.

While a formal definition of pseudorandomness will be given later in this chapter, the reader can see that if the next output of a sequence generator is equally likely to be a 1 as a 0, independent of what the previous output was, the sequence would appear random on a local scale. However, if it is possible to reset the sequence generator so that exactly the same sequence is produced at another time, the sequence is not random; in fact, it is repeatable. This repeatability gives rise to the *pseudo*random nature of the sequences of interest in this chapter.

3.1 SHIFT-REGISTER IMPLEMENTATION OF A PSEUDORANDOM SEQUENCE

Feedback shift registers have been used by many workers to generate pseudorandom sequences. A shift register is a collection of storage elements (for example, flip-flops) connected so that the state of each element is shifted to the next element in response to a shifting clock signal. A three-stage shift register is shown in Figure 3.1. Here each stage is a pair (master/slave) of bistable latches (for example, $SN7477$) connected to different shift clock signals. The A

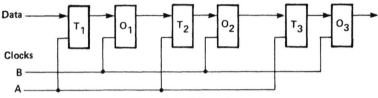

Data

T_1 O_1 T_2 O_2 T_3 O_3

Clocks

B

A

Figure 3.1. Shift register.

clock signal gates the transfer (T) or input latches while the B clock signal gates the output (O) latches. When the A clock goes high, the signal waiting at the input of the register stage is captured in the transfer latches. When the B clock subsequently goes high, this value is set in the output latch. Thus the initial state of the shift register S_0 is changed to the next state S_1 by the application of a single A-B clock signal pair.

Since the sequences that will be treated in this chapter can be generated by feedback shift registers, the case of simple feedback will be examined first. This is shown in Figure 3.2(a), where the output of stage 3 is fed to the input of stage 1. In this figure, the pair of storage elements (bistable latches in Figure 3.1) is represented by a single box, since inner workings of each stage are not important at this point. Note that from

$$\begin{aligned}
&\text{an initial state } S_0: && [101], \\
&\text{we shift through } S_1: && [110], \\
&\text{and } S_2: && [011], \\
&\text{to state } S_3: && [101],
\end{aligned}$$

which is the same as S_0. Thus by this simple feedback, three different states exist in the shift register as it cycles.

If instead, the feedback connection shown in Figure 3.2(b) had been chosen, where the symbol $\oplus$ denotes modulo 2^1 addition, the seven different states shown would have been obtained. In that case, the length of the cycle of states would be $2^3 - 1$ which, as will be shown, is the maximum length cycle that can be obtained from a three-stage shift register with linear feedback. In fact it will be shown later that the number of states that such a linear feedback shift register can cycle through is less than or equal to $2^n - 1$, where n is the number of stages in the shift register. The reader will observe that all possible states except the all 0 state appeared in the cycle of states. To illustrate the use of a feedback shift register as a sequence generator, we take the output of stage 3 shown in Figure 3.2(b) as our sequence. If the initial state of the shift

[1]Modulo 2 means that a is congruent to b if and only if $a - b$ is a multiple of 2. For simple addition this yields $0 + 0 = 1 + 1 = 1 - 1 = 0$ and $0 + 1 = 1 + 0 = 1$. The modulo 2 operator will be denoted $+$, $-$, or $\oplus$ in the remainder of this chapter.

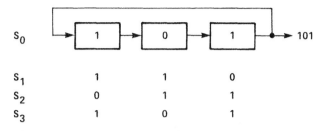

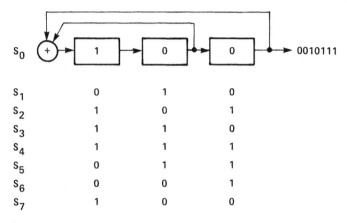

(b) Maximal-Length Shift Register

Figure 3.2. Feedback shift registers. (a) Cyclic shift register. (b) Maximum-length shift register.

register is [100], then the sequence is

$$\{a_m\} = 0010111$$

The sequence $\{a_m\}$ is cyclic in that it will repeat every seven shifts of the shift register.

It has been shown that feedback shift registers can be used as sequence generators and that the length of the sequence that they can generate depends both on the number of stages in the shift register and on the details of the feedback connection. It is this relationship that will be covered in the remainder of this chapter.

3.2 ANALYSIS OF SHIFT-REGISTER SEQUENCES

In this section the formal analysis of sequences will be developed. The concepts of recursion, characteristic polynomials, and generating functions will be introduced.

A sequence of terms where the nth term depends on some combination of the preceding terms obeys a recursion relation. There are many such recursion relations defining everything from Bessel functions to Fibonacci series. As an example, consider the Mersenne series where the nth term is defined as $M_n = 2^n - 1$. With $M_0 = 0$ and $M_1 = 1$, $M_{n+1} = 3M_n - 2M_{n-1}$, where $-$ indicates an arithmetic operation, is a recursion relation that defines the Mersenne series.

Given a sequence of numbers, $a_0, a_1, a_2, \ldots, a_n, \ldots$, a *generating function* $G(x)$ can be associated with the sequence by the rule

$$G(x) = a_0 + a_1 x + a_2 x^2 + \cdots + a_n x^n + \cdots$$

If this series has a radius of convergence $R > 0$, then the properties of the function $G(x)$ may allow the evaluation of the coefficients a_n.

Consider now the binary sequence generated by the shift register shown in Figure 3.2(b). The mth term of the sequence a_m is determined by the modulo 2 sum of the $(m-2)$th and $(m-3)$th terms of the sequence. Thus, the sequence obeys a recurrence relation

$$a_m = a_{m-2} + a_{m-3}$$

This existence of a recurrence relation is a key property of the sequences generated by linear feedback shift registers.

Definition 3.1. A *linear* binary network is one constructed from the following basic components:

Unit delays
Modulo 2 adders
Modulo 2 scalar multipliers

A network or sequential machine constructed of linear elements has the property that its response to a linear combination of inputs will preserve the principle of superposition.[2] As a result, linearity is an important consideration in the analysis of shift-register sequences. A shift register with a linear feedback network is called a linear feedback shift register (LFSR).

Let the sequence $\{a_m\} = \{a_0, a_1, a_2, \ldots\}$ represent the history of the output stage of a shift register where $a_i = 1$ or 0 depending on the state of the output stage at time t_i. The properties of this shift-register sequence can be examined by analysis of the generating function created by the rule

$$G(x) = \sum_{m=0}^{\infty} a_m x^m \tag{3.1}$$

---

[2]*The Principle of Superposition*: The response of a linear network to linear combination of stimuli is the linear combination of the responses of the network to the individual stimuli. (The linear network is initialized such that all storage elements (unit delays) are in the 0 state in each case.)

Because the contents of an n-stage shift register eventually get shifted out to the right to become the sequence under analysis, the *initial* state of the shift register may be thought of as

$$a_{-n}, a_{-n+1}, \ldots, a_{-2}, a_{-1} \qquad (3.2)$$

If the recurrence relation defining $\{a_m\}$ is

$$a_m = \sum_{i=1}^{n} c_i a_{m-i}$$

where $c_i = 0$ or 1 depending on the feedback, then by substituting the formula for a_m into (3.1), $G(x)$ becomes

$$G(x) = \sum_{m=0}^{\infty} \sum_{i=1}^{n} c_i a_{m-i} x^m = \sum_{i=1}^{n} c_i x^i \sum_{m=0}^{\infty} a_{m-i} x^{m-i}$$

$$= \sum_{i=1}^{n} c_i x^i \left[a_{-i} x^{-i} + \cdots + a_{-1} x^{-1} + \sum_{m=0}^{\infty} a_m x^m \right]$$

Thus

$$G(x) = \sum_{i=1}^{n} c_i x^i \left[a_{-i} x^{-i} + \cdots + a_{-1} x^{-1} + G(x) \right]$$

which, upon rearrangement becomes

$$G(x) - \sum_{i=1}^{n} c_i x^i G(x) = \sum_{i=1}^{n} c_i x^i \left(a_{-i} x^{-i} + \cdots + a_{-1} x^{-1} \right)$$

or

$$G(x) = \frac{\sum_{i=1}^{n} c_i x^i \left(a_{-i} x^{-i} + \cdots + a_{-1} x^{-1} \right)}{1 - \sum_{i=1}^{n} c_i x^i} \qquad (3.3)$$

Thus $G(x)$ is expressed entirely in terms of the *initial conditions* $a_{-1}, a_{-2}, \ldots, a_{-n}$ and the *feedback coefficients* $c_1, c_2, \ldots, c_n$. Note that the denominator in (3.3) is independent of the initial conditions. With the initial conditions $a_{-1} = a_{-2} = \cdots = a_{1-n} = 0$, $a_{-n} = 1$, the expression (3.3) reduces to

$$G(x) = \frac{c_n}{1 - \sum_{i=1}^{n} c_i x^i} \qquad (3.4)$$

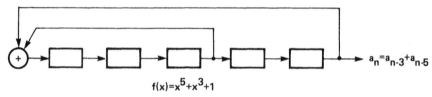

$f(x)=x^5+x^3+1$

Figure 3.3. Linear recurrence relation and characteristic polynomial $f(x) = x^5 + x^3 + 1$.

Note that the denominator of (3.4) is an nth degree polynomial. In the sequel, it will be convenient to refer to

$$f(x) = 1 - \sum_{i=1}^{n} c_i x^i \qquad (3.5)$$

as the *characteristic polynomial* of the sequence $\{a_m\}$ and of the shift register which produced it. Note that $c_n = 1$ for an n-stage feedback shift register with a characteristic polynomial of degree n.

As an example, consider a shift-register sequence with a recursion relation

$$a_m = a_{m-3} + a_{m-5}$$

According to Equation (3.5) the characteristic polynomial of this recursion relation is

$$f(x) = 1 + x^3 + x^5$$

Figure 3.3 illustrates this relationship between the linear recurrence relation and the characteristic polynomial of the shift register and its sequence. Table 3.1 lists the states through which the shift register of Figure 3.3 cycles.

One period of the sequence $\{a_m\}$ that the fifth stage of the LFSR in Figure 3.3 generates can be seen to be

$$\{a_m\} = a_0, a_1, a_2, \ldots, a_{m-1}$$
$$= 00001001011001111100001101110101 \qquad (3.6)$$

It was stated earlier that the feedback network in Figure 3.2(b) resulted in the maximum length sequence for a three-stage LFSR. It also can be seen that the LFSR of Figure 3.3 yields a sequence with period $p = 2^5 - 1$.

The LFSR in Figure 3.3 has $f(x) = x^5 + x^3 + 1$ as its characteristic polynomial. Note that the output of stage 1 represents x or x^1, stage-2's output is x^2, and so forth. The characteristic equation $[f(x) = 0]$ can be written

$$1 = x^5 + x^3$$

since subtraction is the same as addition in modulo 2 arithmetic. Thus the x^0

TABLE 3.1   States of LFSR with Characteristic Polynomial $f(x) = 1 + x^3 + x^5$

State	Stage 1	Stage 2	Stage 3	Stage 4	Stage 5
1	1	0	0	0	0
2	0	1	0	0	0
3	0	0	1	0	0
4	1	0	0	1	0
5	0	1	0	0	1
6	1	0	1	0	0
7	1	1	0	1	0
8	0	1	1	0	1
9	0	0	1	1	0
10	1	0	0	1	1
11	1	1	0	0	1
12	1	1	1	0	0
13	1	1	1	1	0
14	1	1	1	1	1
15	0	1	1	1	1
16	0	0	1	1	1
17	0	0	0	1	1
18	1	0	0	0	1
19	1	1	0	0	0
20	0	1	1	0	0
21	1	0	1	1	0
22	1	1	0	1	1
23	1	1	1	0	1
24	0	1	1	1	0
25	1	0	1	1	1
26	0	1	0	1	1
27	1	0	1	0	1
28	0	1	0	1	0
29	0	0	1	0	1
30	0	0	0	1	0
31	0	0	0	0	1

or 1 term is formed by the summation (modulo 2) of x^5 and x^3. Therefore, to implement an LFSR corresponding to a given characteristic polynomial, number the stages from left to right from 1 to n. Select the outputs of stages corresponding to the exponents of x in the characteristic equation. Connect these taps through a summation modulo 2 to the input of stage 1. This will realize the sequence associated with the given polynomial at any output. (Typically the nth stage output is used.) A characteristic of LFSRs is that if they are initialized with a nonzero state, they cycle through a number of states. The reader is directed to Table 3.1 for the cycle of 31 states this example can have.

Define the *reciprocal polynomial* $f^*(x)$ of $f(x)$ as

$$f^*(x) = x^n f(1/x).$$

(3.7)

In the case at hand,

$$f(x) = x^5 + x^3 + 1$$

Thus,

$$f^*(x) = x^5 \left(\frac{1}{x^5} + \frac{1}{x^3} + 1 \right)$$
$$= x^5 + x^2 + 1$$

If an LFSR is constructed with $f^*(x)$ as its characteristic polynomial, it will generate the sequence

$$\{ b_n \} = 00001010111011000111110011101001 \cdots$$

This sequence is exactly the reverse of the sequence $\{ a_n \}$ shown in Table 3.1. It is a property of a characteristic polynomial that if it is associated with a maximum-length sequence, its reciprocal will also yield a maximum-length sequence.

3.3 PERIODICITY OF SHIFT-REGISTER SEQUENCES

In this section the relationship between the period of a shift-register sequence $\{ a_n \}$ and its characteristic polynomial is analyzed.

A linear feedback shift register is a finite state machine. Each state is uniquely determined from the previous state by the feedback connections. Thus if a state ever repeats, all the following states will repeat, and the sequence of states is periodic. Given an n-stage shift register there are at most 2^n possible states. If the feedback is linear, the successor of the all-zeros state is itself since the output of the feedback circuitry is 0 when all of its inputs are 0. Hence, a linear feedback shift register can have, at most, $2^n - 1$ states. Then the sequence generated by one of the stages of the register is periodic with period no greater than $2^n - 1$.

There are obviously many possible connections between register stage outputs and the linear feedback circuitry, and many corresponding characteristic polynomials. The relation between these characteristic polynomials and the period of the resulting sequence can be determined, for a particular register starting state, by manipulating the generating function of Equation (3.1).

Consider an n-stage linear feedback shift register initialized such that the first $n - 1$ stages contain 0 and the nth stage contains a 1. In terms of Equation (3.2) this corresponds to the initial state

$$a_{-1} = a_{-2} = \cdots = a_{1-n} = 0; \qquad a_{-n} = 1 \qquad (3.8)$$

It is desired to determine the period generated by this initial state.

Theorem 3.1. *Given an LFSR with initial conditions* (3.8), *then the LFSR sequence* $\{a_n\}$ *is periodic with a period which is the smallest integer k for which $f(x)$ divides $1 - x^k$.*

Proof. The theorem is proved in two steps. The first shows that if $\{a_n\}$ is periodic with period k, its characteristic polynomial divides $1 - x^k$. The second shows the inverse, namely that if $f(x)$ divides $1 - x^k$, then $\{a_n\}$ is periodic with period k. Since these two properties hold for any interval for which $\{a_n\}$ repeats, its period must be the smallest such integer k for which $f(x)$ divides $1 - x^k$.

Step 1: Let the generated periodic sequence be represented by the polynomial

$$A(x) = \sum_{i=0}^{k-1} a_i x^i$$

The generating function of $\{a_n\}$ is

$$G(x) = \frac{1}{f(x)} = \sum_{i=0}^{\infty} x^{ki} A(x) = A(x) \sum_{i=0}^{\infty} x^{ki} = \frac{A(x)}{1 - x^k}$$

thus,

$$\frac{1 - x^k}{f(x)} = A(x)$$

proving that if $\{a_n\}$ is periodic with period k its characteristic polynomial evenly divides $1 - x^k$.

Step 2: The division process generates a quotient $B(x)$ and no remainder, or

$$\frac{1 - x^k}{f(x)} = B(x)$$

or

$$\frac{1}{f(x)} = \frac{B(x)}{1 - x^k} = B(x) \sum_{i=0}^{\infty} x^{ki} = B(x)[1 + x^k + x^{2k} + \cdots]$$
$$= B(x) + x^k B(x) + x^{2k} B(x) + \cdots$$

But the generating function is

$$G(x) = \frac{1}{f(x)} = \sum_{i=0}^{\infty} a_i x^i$$

By equating like powers of x it is apparent that $a_i = b_i$ for all i. Thus, if $f(x)$ divides $1 - x^k$, then $\{a_n\}$ is periodic with period k. Taken together the two steps show that the period of the sequence $\{a_n\}$ generated by an LFSR with initial state (3.8) is the smallest k such that $f(x)$ divides $1 - x^k$.   Q.E.D.

It is not possible to generalize Theorem 3.1 to an arbitrary nonzero starting n-tuple, but comments can be made about certain special cases. Recall that, from Equation (3.3), $G(x)$ is defined by

$$G(x) = \frac{\sum_{i=1}^{n} c_i x^i \left(a_{-i} x^{-i} + \cdots + a_{-1} x^{-1} \right)}{f(x)} = \frac{r(x)}{f(x)}$$

The denominator $f(x)$ is of degree n while $r(x)$ is of degree $n - 1$ or less. Suppose first that $r(x)$ has no factors in common with $f(x)$. The proof of Theorem 3.1 still holds. Second, if the characteristic polynomial $f(x)$ is irreducible (cannot be factored), it can have no factors in common with the numerator polynomial $r(x)$, a polynomial of lower degree, unless $r(x) = 0$, which corresponds to the initial condition of all 0's. Thus, when $f(x)$ is irreducible, the period is independent of the initial conditions (except for the all 0's case).

Definition 3.2. If the sequence generated by an n-stage LFSR has period $2^n - 1$ it is called a *maximum-length* sequence or *m*-sequence. The characteristic polynomial of a maximum-length sequence is called a *primitive polynomial*. From Theorem 3.1, the smallest k for which an nth degree primitive polynomial divides an expression of the form $1 - x^k$ is $k = 2^n - 1$.

A primitive polynomial is a special case of an irreducible polynomial. Consider the following:

Theorem 3.2. *If a sequence $\{a_n\}$ is derived from an n-stage LFSR with irreducible characteristic polynomial $f(x)$, then the period of the sequence $\{a_n\}$ is a factor of $2^n - 1$.*

Proof. Every irreducible polynomial of degree $n > 1$ divides the polynomial $1 - x^{2^n - 1}$ [see Van der Waerden (1949)]. In Theorem 3.1 it was shown that the period of $\{a_n\}$ is the smallest integer k such that $f(x)$ divides $1 - x^k$. Since $1 - x^{2^n - 1}$ is of the form $1 - x^k$ (with $k = 2^n - 1$), it is certainly true that $\{a_n\}$ repeats after $2^n - 1$ cycles. Therefore, the period of $\{a_n\}$ is either $2^n - 1$ (in which case the characteristic polynomial is primitive) or a factor thereof.   Q.E.D.

Clearly, if $2^n - 1$ is prime, every irreducible polynomial of degree n corresponds to a shift-register sequence of maximum length.

Next a formula for the number of primitive polynomials of degree n is presented [Equation (3.10)]. This formula is expressed in terms of the Euler ϕ-function which is defined as the number of positive integers less than or equal to n that are relatively prime to n, for any positive integer n, or

$$\phi(n) = n \prod_{p \mid n} \left(1 - \frac{1}{p}\right), \tag{3.9}$$

where p runs through the primes dividing n. In particular, if P denotes a prime, $\phi(P) = P - 1$. If Q is also a prime, $\phi(PQ) = (P - 1)(Q - 1)$. An equivalent definition is: $\phi(n)$ is the number of positive irreducible fractions not greater than 1, with denominator n.

The number of primitive polynomials of degree n is given by

$$\lambda_2(n) = \frac{\phi(2^n - 1)}{n} \tag{3.10}$$

A detailed development of the formula for $\lambda_2(n)$ can be found in Golomb (1982). Values of $\lambda_2(n)$ are tabulated in Table 3.2 for values of n up to 32. The value of 2^n are also tabulated there so that the total number of nth degree polynomials can be compared to the number of primitive polynomials of degree n.

Some general properties of shift-register sequences (including those with nonlinear feedback) are developed in the following.

TABLE 3.2 $\lambda_2(n)$, the Number of Primitive Polynomials of Degree n

n	2^n	$\lambda_2(n)$	n	2^n	$\lambda_2(n)$
1	2	1	17	131,072	7,710
2	4	1	18	262,144	8,064
3	8	2	19	524,288	27,594
4	16	2	20	1,048,576	24,000
5	32	6	21	2,097,152	84,672
6	64	6	22	4,194,304	120,032
7	128	18	23	8,388,608	356,960
8	256	16	24	16,777,216	276,480
9	512	48	25	33,554,432	1,296,000
10	1,024	60	26	67,108,864	1,719,900
11	2,048	176	27	134,217,728	4,202,496
12	4,096	144	28	268,435,456	4,741,632
13	8,192	630	29	536,870,912	18,407,808
14	16,384	756	30	1,073,741,824	17,820,000
15	32,768	1,800	31	2,147,483,648	69,273,666
16	65,536	2,048	32	4,294,977,296	67,108,864

Theorem 3.3. *Distinct states of a feedback shift register have distinct successor states if and only if the feedback relation $a_0 = F(a_1, \ldots, a_n)$ can be decomposed into $F(a_1, \ldots, a_n) = F(a_1, \ldots, a_{n-1}, 0) \oplus a_n$.*

Proof. If two states, represented by the vectors $(a_1, a_2, \ldots, a_n)$ and $(b_1, b_2, \ldots, b_n)$ differ in other than their rightmost component, their successor states are obviously distinct.

Consider the case where the above vectors differ only in their rightmost bit. These vectors can be written as $(a_1, a_2, \ldots, a_n)$ and $(a_1, a_2, \ldots, \bar{a}_n)$. Since it is given that their successors are different, the feedback function F has the property

$$F(a_1, a_2, \ldots, a_{n-1}, a_n) = \bar{F}(a_1, a_2, \ldots, a_{n-1}, \bar{a}_n) \qquad (3.11)$$

For simplicity denote $A = a_1, a_2, \ldots, a_{n-1}$. Equation (3.11) can be written using the Shannon expansion as

$$1 \oplus \bar{a}_n F(A, 1) \oplus a_n F(A, 0) = \bar{a}_n F(A, 0) \oplus a_n F(A, 1)$$

or

$$1 = F(A, 0) \oplus F(A, 1) \qquad (3.12)$$

Equation (3.12) can be written as

$$F(A, a_n) = F(A, 0) \oplus a_n$$

thus proving the if condition. Now it must be shown that given a feedback function of the form

$$F(a_1, \ldots, a_n) = F(a_1, \ldots, a_{n-1}, 0) \oplus a_n$$

distinct states have distinct successor states.

If two vectors are different in a bit that is not the rightmost bit, then their successors are different due to the structure of the feedback. It is still to be proven that if the two vectors differ in their last bit, their successors are also distinct. In other words, it has to be shown that

$$F(A, a_n) = \bar{F}(A, \bar{a}_n)$$

Since

$$F(A, a_n) = F(A, 0) \oplus a_n$$

we have

$$F(A, a_n) = \begin{cases} F(A, 0) & \text{for } a_n = 0 \\ \bar{F}(A, 0) & \text{for } a_n = 1 \end{cases}$$

TABLE 3.3   Truth Table of Shift-Register Feedback Function

$a_1 a_2 \cdots a_{n-1}$	a_n	F
TT_{F_1}	1 $\vdots$ 1	$\bar{F}_1$
TT_{F_1}	0 $\vdots$ 0	F_1

which can be condensed into

$$F(A, a_n) = \bar{F}(A, \bar{a}_n) \qquad \text{Q.E.D.}$$

If F_1 is defined as

$$F_1(a_1, \ldots, a_{n-1}) = F(a_1, \ldots, a_{n-1}, 0)$$

the truth table of the feedback function of Theorem 3.3 can be written as shown in Table 3.3, where TT_{F_1} means the truth table of F_1. Note that the bottom half of the table for the output F is the complement of the top portion of the truth table.

Definition 3.3. A Type 1 feedback shift register has the property that distinct states have distinct successor states.

The simplest feedback shift register is a cycling register with only the output stage fed back to the input ($F_1 = 0$). It is a Type 1 feedback shift register. An interesting property of such a shift register is that it cycles through different groups of states for different initial states. For $n > 2$, such a shift register has an *even* number of cycles [see Golomb (1982)]. This fact will be used in the next theorem.

Theorem 3.4. *Given a Type 1 feedback shift register with $n > 2$ stages, the number of cycles it generates has the same parity as the number of 1's in the truth table of the feedback function F_1.*

Proof. For $F_1 = 0$, the purely cyclic case, the number of cycles is even, thus covering the case of no 1's in the truth table of F_1. Next the effect of adding a single 1 to the truth table of F_1 is analyzed. When a single 1 is added to the truth table of F_1, some state, say S_i, has a new successor state, say S_k. Two cases can occur: either S_k, the new successor, was in the same cycle as S_i, or it was in a different cycle.

For the case that the new successor is in the same cycle, the old successor S_{i+1} has no predecessor. Likewise, S_{k-1}, the previous predecessor of S_k, has no successor. Since Theorem 3.3 must still hold, the single change in the truth table of F_1 will cause a second change in the truth table of F (in the half of F unaffected by the first change). This second change must assign S_{i+1} as the successor to S_{k-1}, thus increasing the number of cycles by one.

For the case where the new successor is in a different cycle, Theorem 3.3 requires that the states S_{i+1} and S_{k-1} be joined, resulting in the two cycles merging into one, thus decreasing the number of cycles by one.

Since the number of cycles is even for the case of no 1's in the truth table of F_1, and the number of cycles is changed by one (increased or decreased) every time a 1 is added to the truth table of F_1, the parity of the number of cycles is the same as the parity of the number of 1's in the truth table of F_1.   Q.E.D.

Several interesting properties of feedback shift registers follow from the preceding theorems. Theorem 3.4 shows that a linear shift register always has an even number of cycles, since F_1 is a linear function in that case and has an even number of 1's in its truth table. Theorem 3.2 shows that if an n-stage LFSR has a feedback network that implements a primitive characteristic polynomial, the resulting sequence will have a period of $2^n - 1$. On some occasions it is desirable to extend the period of a sequence from $2^n - 1$ to 2^n. The properties of these sequences have interested mathematical researchers for some time. Most notable are the studies of Sainte-Marie (1894), de Bruijn (1946), and Good (1946). It will be shown that nonlinear feedback functions are required to implement sequences of period 2^n. These sequences are called de Bruijn sequences.

If the feedback function $F_1(a_1, \ldots, a_{n-1})$ does not make explicit use of all $n - 1$ of its variables, the corresponding shift register has an even number of cycles. To make this clear, consider a single variable, say a_j, not explicitly used by F_1. The truth table of F_1 can be divided into two equal portions, one where $a_j = 1$ and the other where $a_j = 0$. The number of 1's in the output column will be the same in both halves; therefore, the total number of 1's in the truth table of F_1 is even. Hence, by Theorem 3.4, the number of cycles is even. The conclusion of this analysis is that, in order to obtain a sequence of length 2^n from an n-stage shift register, it is necessary to use all n available tap positions in the feedback computations.

Thus, a maximum-length *linear* shift-register sequence can be increased in length from $2^n - 1$ to 2^n by means of the feedback logic

$$F_{new} = F_{old} \oplus \bar{a}_1 \bar{a}_2 \cdots \bar{a}_{n-1} \qquad (3.13)$$

This inserts the all-0 state between $000 \cdots 01$ and $100 \cdots 00$. When all but the rightmost stage of the shift register contain 0's, the 1 in the last stage would normally feed a 1 through the summation to feed the first stage on the next shift. If the complement of the outputs of the first $n - 1$ stages are ANDed (a nonlinear operation) together and the result fed into the summation, the 1 from the nth stage and the 1 from the zero check will sum to a 0 that will be fed into the leftmost stage on the next shift, creating the all 0's state. On the next shift, there will still be a single 1 coming from the zero check fed to the first stage returning to the "linear" portion of the sequence. Often a designer can combine the normal (linear) feedback network with the zero check and apply conventional logic minimization techniques to simplify the resulting total feedback network. Examples of these realizations will be shown in Chapter 6.

3.4 PRIMITIVE POLYNOMIALS

It was shown in Section 3.3 that there are a definable number of primitive polynomials of any given degree. Often the problem is to select such a polynomial as the characteristic polynomial of an LFSR. Several listings of such polynomials exist in the literature. A complete listing of irreducible polynomials up to degree 16 (with primitive polynomials identified) is given in Peterson and Weldon (1972), which also lists many irreducible polynomials up to degree 34.

The Appendix of this book gives a primitive polynomial for each degree up to degree 300. The polynomials are those having the fewest number of terms for each degree.

One of the most practical methods of finding primitive polynomials is a sieve method akin to the Sieve of Eratosthenes used for finding prime integers. A trial polynomial of degree n is selected. A series of tests is applied that determines first irreducibility and finally primitivity.

Test 1. Any polynomial with all even exponents is a square and hence reducible. Therefore, the first test rejects all trial polynomials with all even exponents.

Test 2. An irreducible polynomial has an odd number of terms, one of which is a constant. Thus the second test rejects all polynomials that fail this simple criterion.

This test depends on the fact that unless $f(x)$ has a constant term, $f(0) = 0$, indicating that $x = 0$ is a root and that x is a factor. Likewise, if the number of terms is even, $f(1) = 0$, indicating that $x - 1$ is a factor.

Test 3. Theorem 3.1 states that the maximum-length shift-register sequence has a period of $2^n - 1$. Van der Waerden (1949) shows that if the trial polynomial $g(x)$ is irreducible (a necessary condition for it to be primitive) it divides $1 - x^{2^n - 1}$. In other words, letting $f = 2^n - 1$, if

$$x^f \neq 1 \bmod g(x)$$

then $g(x)$ is reducible and is discarded in favor of another trial polynomial. If, however, $x^f = 1 \bmod g(x)$, the tests continue since the trial polynomial is irreducible and possibly primitive.

Test 4. It must now be determined if there is a prime factor of f, p such that $x^{f/p} = 1 \bmod g(x)$. If no such prime factor can be found, $g(x)$ is primitive.

The test continues to check if $x^{f/p} \neq 1 \bmod g(x)$ for all primes p of f. If no prime is found such that $x^{f/p} = 1 \bmod g(x)$, then $g(x)$ is primitive. Thus the exponent f is divided by each of its prime factors in turn, reduced mod $g(x)$, until the residue is 1 [in which case $g(x)$ is known not to be primitive] or until all prime factors have been used. If in every case, $x^{f/p} \neq 1 \bmod g(x)$, $g(x)$ is primitive.

The key to preparing a computer program to examine binary polynomials for primitivity is a table of the prime factors of $2^n - 1$. Such a table appears in Brillhart et al. (1983).

Suppose a fourth-degree primitive polynomial is needed. The sieve method is applied to each item of a list of several trial polynomials to determine which are primitive:

$$p_1(x) = x^4 + x^2 + 1$$
$$p_2(x) = x^4 + x^2 + x$$
$$p_3(x) = x^4 + x^3 + x + 1$$
$$p_4(x) = x^4 + x^3 + x^2 + x + 1$$
$$p_5(x) = x^4 + x + 1$$
$$p_6(x) = x^4 + x^3 + 1$$

Polynomial $p_1(x)$ is considered first. Test 1 rules $p_1(x)$ out since it is a square. Next, by Test 2, $p_2(x) = x(x^3 + x + 1)$. Hence $p_2(x)$ is discarded. For $p_3(x)$, Test 2 indicates that it contains a factor $x + 1$ since it has an even number of terms. In fact, $p_3(x) = (x + 1)(x^3 + 1)$.

The fourth polynomial $p_4(x)$ passes Test 1 and Test 2. When Test 3 is performed,[3] x^{15} is shown to be congruent to 1, mod $p_4(x)$, indicating that $p_4(x)$ is irreducible. To determine if $p_4(x)$ is primitive, Test 4 is applied. The prime factors of f are 3 and 5. Thus the reduced exponents considered are 3 (15/5) and 5 (15/3). Clearly $x^3 \neq 1 \bmod p_4(x)$, so the only case to test is x^5. Since $x^5 = 1 \bmod p_4(x)$, it is clear that $p_4(x)$ is irreducible but not primitive.

Polynomials $p_5(x)$ and $p_6(x)$ are reciprocals, so what holds for $p_5(x)$ will also hold for $p_6(x)$. Test 1 and Test 2 are clearly satisfied by $p_5(x)$: $x^{15} = 1 \bmod p_5(x)$, so it is irreducible. Continuing to Test 4, as before only x^5 must be reduced mod $p_5(x)$. The result is $x^5 = (x^2 + x) \bmod p_5(x)$. Therefore, $p_5(x)$ and $p_6(x)$ are primitive.

---

[3] The reader may wish to verify the congruences established in the application of Test 3 and Test 4. This can be done by long division, but the reader is again reminded that arithmetic is modulo 2, so addition and subtraction yield the same result.

3.5 CHARACTERISTICS OF MAXIMUM-LENGTH LINEAR SHIFT-REGISTER SEQUENCES

The properties of maximum-length linear shift-register sequences are discussed in this section. These are sequences that have a primitive characteristic polynomial. Some consequences of their periodicity, their randomness properties, and their recursive nature will be examined.

The sequences generated by linear feedback shift registers with primitive characteristic polynomials are variously called pseudorandom sequences, pseudonoise (PN) sequences, maximum-length feedback shift-register sequences, or *m*-sequences. For brevity, the term *m-sequence* will be used for the remainder of this discussion.

Given an *m*-sequence $\{a_n\}$ with characteristic polynomial $f(x)$, a primitive polynomial of degree n, its properties are:

Property I. *The period of* $\{a_n\}$ *is* $p = 2^n - 1$, *that is,* $a_{p+i} = a_i$ *for all* $i \geq 0$.

Proof. This follows from the definition.   Q.E.D.

Property II. *Starting from a nonzero state, the LFSR that generates* $\{a_n\}$ *goes through all* $2^n - 1$ *states before repeating.*

Proof. This is obvious because the period of the sequence is $2^n - 1$.   Q.E.D.

Property III. *The number of* 1's *in an m-sequence differs from the number of* 0's *by one.*

Proof. Since the n-stage shift register that generates the sequence cycles through all $2^n - 1$ nonzero states as it generates the sequence (Property II), it can be thought of as displaying the binary representation of the numbers from 1 to $2^n - 1$. The *m*-sequence can be thought of as the sequence that the units bit (2^0) undergoes as the numbers are enumerated. (Each number between 1 and $2^n - 1$ appears once and only once, but not necessarily in ascending order.) From 1 to $2^n - 1$ there are 2^{n-1} odd numbers and $2^{n-1} - 1$ even numbers. Therefore, in any *m*-sequence, there are 2^{n-1} 1's and $2^{n-1} - 1$ 0's and the disparity is one.   Q.E.D.

Note to Property III. In integrated circuit implementations of shift registers, both the true and the complemented values of a stage are usually available. Consider the sequence generated by the complemented output of the nth stage of an LFSR with a primitive characteristic polynomial. It is an *m*-sequence by definition, but it contains the subsequence of n 0's instead of n 1's. Hence, there are 2^{n-1} 0's and $2^{n-1} - 1$ 1's and the disparity is still one.

Property IV (The Window Property). *If a window of width n is slid along an m-sequence, then each of the 2^{n-1} nonzero binary n-tuples is seen exactly once in a period.*

Proof. By Property I, $\{a_n\}$ has period $2^n - 1$, and by Property II, the LFSR goes though all $2^n - 1$ states before repeating. Since the m-sequence $\{a_n\}$ can be considered the sequence of bits that appears in the rightmost stage of the LFSR, all nonzero n-tuples will appear in the sequence. A window n-bits wide will just reveal one of the $2^n - 1$ states that the LFSR traversed. Sliding the window along the sequence will just reveal the sequence of states that the LFSR traversed. Since there are $2^n - 1$ of these, all $2^n - 1$ of the nonzero n-tuples will appear in the window.   Q.E.D.

The concept of a *run* is now introduced. A maximal contiguous grouping of symbols is called a "run." To avoid ambiguity in periodic sequences, consider the sequence written in a circle to avoid end effects. To count maximal contiguous groupings, the period must start at the transition from a run of 1's to a run of 0's or vice versa.

Property V (The Run Property). *In every period of an m-sequence, one-half the runs have length 1, one-fourth have length 2, one-eighth have length 3, and so forth, as long as the fractions yield integral numbers of runs. The runs of 1's and 0's terminate with runs of length n and n − 1, respectively. Moreover, except for these terminating lengths, there are equally many runs of 1's and of 0's. The total number of runs of 1's equals the total number of runs of 0's. The number of runs in a period is $(m + 1)/2$, where $m = 2^n - 1$.*

Proof. By Property IV, n consecutive 1's occurs exactly once. This run must be preceded by a 0 and followed by a 0, or there would be other runs of n consecutive 1's.

There is no true run of $n - 1$ consecutive 1's. While a run of $n - 1$ consecutive 1's preceded by a 0 occurs exactly once, it is accounted for by the run of n consecutive 1's which is preceded by a 0. Similarly, the run of $n - 1$ consecutive 1's followed by a 0 occurs once, and is accounted for by the fact that the run of n consecutive 1's is followed by a 0.

There is no run of n consecutive 0's since the all-0 n-tuple never appears in the LFSR. Yet a 1 followed by $n - 1$ consecutive 0's must occur, so there is a run of $n - 1$ consecutive 0's.

Now, let $0 < k < n - 1$. To find the number of runs of 1's of length k, consider n consecutive bits beginning with 0, then k 1's, then a 0, and the remaining $n - k - 2$ bits arbitrary. This occurs 2^{n-k-2} ways, since each way of completing the n bits occurs exactly once. Hence, there are 2^{n-k-2} runs of

k consecutive 1's. A similar argument holds for the runs of 0's. The number of runs of 1's is given by

$$\text{runs(1's)} = 1 + \sum_{k=1}^{n-2} 2^{n-k-2}$$

$$= 1 + 2^{n-2} \sum_{k=1}^{n-2} 2^{-k}$$

$$= 1 + 2^{n-2}[1 - 2^{2-n}] = 2^{n-2}$$

Likewise,

$$\text{runs(0's)} = 2^{n-2}$$

so the total number of runs is 2^{n-1}. But $m = 2^n - 1$, so $(m + 1)/2$ is just 2^{n-1}, the number of runs.

The number of runs of 1's of length $k = 1$ is 2^{n-3} and the number of runs of 1's of length $k = 2$ is 2^{n-4}. Thus, one-half the runs of 1's have length 1, one-fourth have length 2, and so on. A similar argument holds for runs of 0's.

Q.E.D.

Property VI (Transitions). *The number of transitions between 1 and 0 that an m-sequence makes in one period is $(m + 1)/2$.*

Proof. Clearly a transition is associated with the beginning of a run. Since the number of runs is $(m + 1)/2$, so is the number of transitions.   Q.E.D.

Property VII (The Shift-and-Add property). *The sum of any sequence A_i and a cyclic shift of itself is another cyclic shift of A_i.*

Proof. Let $A_1 = \{a_1, a_2, a_3, \ldots\}$ be the m-sequence $\{a_n\}$. Let $A_2 = \{a_2, a_3, \ldots\}$, $A_3 = \{a_3, a_4, \ldots\}$, $\ldots$, $A_p = \{a_p, a_{p+1}, \ldots\}$. Also, let $A_0 = \{0, 0, 0, \ldots\}$. Let R be the linear recurrence relation satisfied by A_1. Then R is also satisfied by $A_2, \ldots, A_p$ and trivially by A_0. Since R is linear, it is also satisfied by $A_i + A_j$ whenever it is satisfied by both A_i and A_j, where $A_i + A_j$ denotes the term-by-term sum. Thus $A_i + A_j$ is determined by R from its first n terms. Whatever these n terms are, they are the same as the first n terms of exactly one of the 2^n sequences $A_0, A_1, \ldots, A_p$. Thus the sum of two cyclic shifts of an m-sequence is a cyclic shift of the same m-sequence.   Q.E.D.

The next property (Property VIII) involves the autocorrelation function of a periodic sequence. While this function is not directly involved in any of the built-in test procedures covered in this book, along with Properties III, IV, and V, it justifies the name pseudorandom sequence. These are the properties that one would expect from a sequence obtained by tossing a fair coin $2^n - 1$ times.

To define the autocorrelation function, the sequence $\{a_n\}$ with a_n belonging to $\{0, 1\}$, is changed into an equivalent sequence $\{b_n\}$ by defining $b_i = 1 - 2a_i$, where the minus sign indicates an arithmetic operation. Thus 1's in $\{a_n\}$ are replaced by -1's in $\{b_n\}$ and the 0's by $+1$'s. For a periodic sequence such as a shift-register sequence, the *autocorrelation function* is defined as

$$C(\tau) = \frac{1}{p} \sum_{n=1}^{p} b_n b_{n-\tau}$$

where p is the period of $\{a_n\}$.

Property VIII. *The autocorrelation function of every m-sequence of period $p = 2^n - 1$ is given by*

$$C(0) = 1$$
$$C(\tau) = -1/p \qquad \text{for} \quad 1 \le \tau \le 2^n - 2$$

Proof. The autocorrelation function $C(\tau)$ is defined as

$$C(\tau) = \frac{1}{p} \sum_{n=1}^{p} b_n b_{n-\tau}$$

where $b_n = 1 - 2a_n$. Alternatively, b_n can be written as $b_n = e^{i\pi a_n}$, where $i = \sqrt{-1}$. Using this notation, $C(\tau)$ becomes

$$C(\tau) = \frac{1}{p} \sum_{n=1}^{p} e^{i\pi(a_n + a_{n-\tau})}$$

When $\tau = 0$, all terms have the form $e^{i2\pi a_n}$, clearly all equal to 1. Thus the sum yields p. When $1 \le \tau \le p - 1$, by Property VII (shift-and-add), the sum $a_n + a_{n-\tau}$ in the exponent corresponds to another shifted sequence, say A_j. Thus in this case, $C(\tau)$ amounts to summation of the 1's (represented by -1) and the 0's (represented by $+1$) in the sequence A_j. From Property I it is known that there is one more 1 than 0 in the m-sequence, so the summation is -1. By dividing by the normalizing factor $1/p$, the desired result is obtained.

Q.E.D.

The concept of *proper decimation* of an m-sequence is now introduced. Decimation is the selection of every qth element of the sequence. If q is relatively prime to the period $p = 2^n - 1$, the decimation is called proper.

Property IX (Decimation). *Every proper decimation of an m-sequence is itself an m-sequence.*

The proof of this property is shown in Golomb (1982).

3.6   LINEAR DEPENDENCIES IN *m*-SEQUENCES

One of the primary interests in shift-register sequences is their potential use as stimuli for the built-in test of digital networks. In particular, *m*-sequences are convenient algorithmically generated test stimuli. They are generated by implementation of a recurrence relation and, therefore, the implication of this recurrence relation on their usefulness as test stimuli will be examined. The recurrence relation causes linear dependencies within the sequence that are of importance to their use as test stimuli.

When considered as a source of test stimuli, the *m*-sequence $\{a_n\}$ generated by an *n*-stage LFSR implementing a primitive polynomial $p(x)$ can be thought of as the contents of an infinitely long idler register. Such an idler register arrangement is shown in Figure 3.4. In built-in test schemes the circuit under test can be considered to have its inputs connected to a multiplicity of idler register stages. In general, a test for a particular fault in the circuit under test requires that *k* inputs to that circuit take on certain specified values. Rather than determine what these specific values are, as in conventional test generation, the problem is approached from a different direction. The *m*-sequence is examined to determine if it can exhaustively test (apply all possible *k*-tuples) the *k*-subspace defined by the circuit under test. In some of the analysis to follow, the concept of span is used. The span *S* is the distance from the first stage of the idler register tapped by the circuit under test to the last stage tapped. Thus, *S* is the maximum distance along the idler register that the elements of the *k*-subspace under analysis can be separated.

In a combinational logic network, the probability that a randomly chosen test vector will detect a particular fault is called the detection probability of that fault. If the detection probability for fault *k* is denoted as p_k, T_k as the test set for fault *k*, and *j* as the number of inputs to the network, then

$$p_k = \frac{|T_k|}{2^j}$$

where $|T_k|$ is the cardinality of the test set.

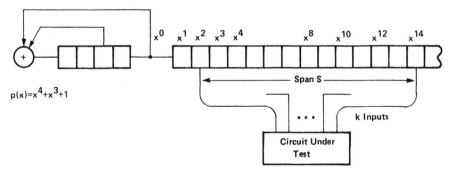

Figure 3.4. LFSR and idler register.

For example, consider a 10-way AND. To detect a stuck-at-1 fault on one input, a ($|T_k| = 1$) pattern with a 0 on that input and a 1 on the other 9 inputs is required. The probability of this happening for a randomly chosen 10 element vector is 2^{-10}. Therefore, for this example, $p_k = 2^{-10}$. Another way to describe this test is to say that 10 inputs to the network must be specified in order to test the fault in question. One should note that $-\log_2(p_k) = 10$. This prompts some workers to use the logarithm (base 2) of the detection probability when describing random pattern testability of a logic network. This quantity is sometimes called "equivalent AND inputs" (EAIs) although it is a measure of the number of network inputs that must be specified for the test of a fault rather than a characterization of the network structure.

In general, logic networks designed for built-in random pattern test will have a constraint that during "test mode" the maximum EAI be bounded, say on the order of 15. Thus, the size of k-subspaces of concern will be bounded. Thus, the question becomes whether the k inputs see all possible binary k-tuples as the m-sequence $\{a_n\}$ is repeatedly shifted through the idler register (each of the m shifted replicas of the sequence appears at least once in the idler register). While all of the k-tuples will not be required to test the circuit and one may choose to truncate the sequence before all the k-tuples appear at the inputs, it is important to understand if some of the k-tuples are precluded a priori. This requirement of k-subspace coverage corresponds to the concept of *possible exhaustion*.[4] In some built-in test arrangements, each shift of the idler register is presented to the circuit under test (bit-pushing) and in others every rth shift (decimation) is presented to the circuit as a test vector.

3.6.1   Cyclic Shift of m-Sequences

Any periodic binary sequence of length $m = 2^n - 1$, $\{a_n\} = a_0 a_1 \cdots a_{m-1}$ can be represented by the polynomial

$$a(x) = a_0 + a_1 x + a_2 x^2 + \cdots + a_{m-1} x^{m-1}$$

A cyclic shift of $\{a_m\}$ by one place to the right,

$$a_{m-1} a_0 \cdots a_{m-2}$$

is represented by the polynomial

$$a_{m-1} + a_0 x + a_1 x^2 + \cdots + a_{m-2} x^{m-1}$$

[4] Possible exhaustion demands a stimulus for the k-subspace such that the k inputs are not precluded a priori from observing all 2^k possible binary patterns. While in practice one may choose not to wait until this event is assured, this condition does not exclude the possibility at the start.

Since the sequence of interest is of length m, x^m is defined as 1 ($x^m = 1$) since m shifts leave the sequence unchanged. Then the shifted sequence is

$$xa(x) = a_0 x + a_1 x^2 + a_2 x^3 + \cdots + a_{m-2} x^{m-1} + a_{m-1} x^m$$
$$= a_{m-1} + a_0 x + a_1 x^2 + \cdots + a_{m-2} x^{m-1}$$

Thus, a cyclic shift by one place to the right corresponds to multiplying by $x \bmod x^m$.

Now consider the contents of the idler register as a polynomial of the form

$$g(x) = c_0 + c_1 x + c_2 x^2 + \cdots + c_r x^r$$

with $r < m$, where the content of the jth stage is c_j, with c_0 being the value at the input to the idler register. The k inputs of the k-subspace under analysis tap the idler register at k stages within a span of S (see Figure 3.4). This can be thought of as a polynomial with k terms of the form

$$h(x) = x^j \left(b_0 + b_1 x + \cdots + b_t x^t \right) \tag{3.14}$$

where $b_0 b_t = 1$, $\Sigma b_i = k$, and $t + 1 = S \le r$. Polynomials such as $h(x)$ are referred to as "sampling polynomials" since they define the k-subspace under analysis by "sampling" the m-sequence.

The *set polynomial* $H(x)$ is now defined as the polynomial formed by the product of $h(x)$ and all of its subpolynomials. Table 3.4 shows the factors of such a polynomial. For this discussion, a subpolynomial is a polynomial formed from a subset of the b_i that defines the polynomial $h(x)$. In these cases $\Sigma b_i < k$. Given an n-stage LFSR, if $k < n$ and $S \le n$, it is clear that all binary k-tuples will appear in the k-subspace. This follows from Property VII

TABLE 3.4 Factors of $H(x)$: $h(x) = x^{18} + x^9 + x + 1$ and its Subpolynomials

$S_1 = x^{18} + x + 1$	$S_7 = x^{18} + 1$
$S_2 = x^{18} + x^9 + 1$	$S_8 = x^9 + x$
$S_3 = x^{18} + x^9 + x$	$\quad = x(x^8 + 1)$
$\quad = x(x^{17} + x^8 + 1)$	$S_9 = x^9 + 1$
$S_4 = x^9 + x + 1$	$S_{10} = x + 1$
$S_5 = x^{18} + x^9$	$S_{11} = x^{18}$
$\quad = x^9(x^9 + 1)$	$S_{12} = x^9$
$S_6 = x^{18} + x$	$S_{13} = x$
$\quad = x(x^{17} + 1)$	$S_{14} = 1$

of m-sequences (Section 3.5). The more interesting case is when $k < n$ and $n < S < 2^n - 1$.

3.6.2 Linear Dependencies

When the span of the sampling polynomial exceeds the length of the LFSR that generates the sequence, that is when $S > n$, the k-subspace is noncontiguous ($k \leq n$). That is, the k taps of the sampling polynomial are not all adjacent stages in the idler register. This case is described by the following lemma:

Lemma 3.1. *The m-sequence generated by $p(x)$ (exhaustively) covers the k-subspace spanned by the sampling polynomial $h(x)$ if and only if its set polynomial $H(x)$ is prime relative to $p(x)$.*

Proofs of Lemma 3.1 can be found in various places including Barzilai et al. (1983), Tang and Chen (1983), and Lempel and Cohn (1985). The elements of the k-subspace are linearly independent if Lemma 3.1 holds. Conversely, if a factor of $H(x) = 0 \bmod p(x)$, it defines a linear dependency on $p(x)$.

Definition 3.4. Given a polynomial $h(x)$ of degree less than m, if $h(x) = 0 \bmod p(x)$, it defines a linear dependency on $p(x)$. Such an $h(x)$ is called a dependency polynomial of $p(x)$.

For example, if $p(x) = x^5 + x^2 + 1$, the subpolynomial s_1 of Table 3.4 is a dependency polynomial of $p(x)$. In other words, the subpolynomial $x^{18} + x + 1 = 0 \bmod p(x)$. If the k-subspace were $[0, 1, 9, 18]$, only eight distinct binary 4-tuples would appear at the tap points. It becomes obvious that these linear dependencies in m-sequences must be understood and care should be exercised in choosing the tap points for the k-subspace. There is no way of avoiding the linear dependencies in an m-sequence, but care can minimize their effect in built-in test applications.

3.6.3 Residues mod $p(x)$

The problem of determining the linear dependencies that appear in an m-sequence is now treated. Consider the following: Given a polynomial $g(x)$ of degree less than r, and a polynomial $p(x) \neq 0$, if $x^r \bmod p(x) + g(x) \bmod p(x) = 0 \bmod p(x)$, then by linear congruence, $x^r + g(x) = 0 \bmod p(x)$. The remainder of x^r divided by $p(x)$ is defined as the residue $x^r \bmod p(x)$.

   A theorem is now presented that can be the basis of an efficient procedure for the enumeration of linear dependencies in an m-sequence. The procedure assumes that the residues $x^r \bmod p(x)$ corresponding to the sequence under analysis are available, either through calculation or via table lookup. Let the ith residue be denoted d_i.

Theorem 3.5. *A trinomial of the form*

$$f(x) = 1 + x^j + x^k$$

will define a dependency polynomial if and only if

$$d_0 + d_j + d_k = 0$$

Proof. The residue d_i is $x^i \bmod p(x)$. We proceed, noting that

$$d_0 + d_j + d_k = 0$$

implies

$$1 + x^j + x^k = 0 \bmod p(x)$$

or

$$f(x) = 0 \bmod p(x).$$

Thus if the residues of the terms of the polynomial sum to zero, the polynomial is a dependency polynomial. Conversely, if $f(x) = 0 \bmod p(x)$,

$$1 + x^j + x^k = 0 \bmod p(x)$$

it implies

$$d_0 + d_j + d_k = 0. \hspace{3cm} \text{Q.E.D.}$$

Corollary to Theorem 3.5. *Given a sampling polynomial $g(x) = \Sigma c_i x^i$ and an m-sequence $\{a_n\}$ with characteristic polynomial $p(x)$ with residue states d_i, $g(x)$ is a dependency polynomial of $p(x)$ if and only if*

$$\Sigma c_i d_i = 0$$

Proof. The proof by induction is obvious. Q.E.D.

Recall that the span S is the maximum distance along the idler register that the elements of a k-subspace under analysis can be separated. In terms of a sampling polynomial like Equation (3.14), the exponent $t + 1 = S$. In a digital network being tested by the m-sequence in question, the span S can be determined. S can be as little as the total number of inputs to the network or, if a scan-path technique is used, the sum of the primary inputs and the scan latches, properly modified for the way in which the sequence is applied to the scan paths. In any event, the span S defines the maximum degree sampling polynomial that must be analyzed for linear dependencies. This materially reduces the task of linear dependency analysis.

3.6.4   Probabilistic Treatment of Linear Dependencies

In this section, the expected number of linear dependencies to be found in an m-sequence $\{a_n\}$ is calculated. To begin, consider an m-sequence with the characteristic polynomial $p(x)$ of degree n, whose length is $m = 2^n - 1$. The residues of $p(x)$, $d_0, d_1, \ldots, d_{m-1}$, form an n-dimensional vector space. Denote these m vectors by $\{d_m\}$. Following an analysis suggested by Chen (1986) and by Dervisoglu (1984), a basis of this vector space can be formed by any n linearly independent members of $\{d_m\}$. Likewise for a k-subspace ($k < n$), any k linearly independent members of $\{d_m\}$ can form a basis for the subspace.

Consider the case of a 3-subspace. Let the first member of $\{d_m\}$ chosen be denoted C_1 and the second C_2. Now C_1 and C_2 form the basis for a two-dimensional space. Of interest is the fact that $C_1 + C_2$ is a member of $\{d_m\}$ also. Since the element $C_1 + C_2$ is dependent on C_1 and C_2, there are only $m - 3$ choices for the third member of $\{d_m\}$ in order to have three independent vectors for the basis of the 3-subspace under analysis. Stated somewhat differently, there are

$$\frac{m(m - 1)(m - 2)}{3!}$$

ways to choose three vectors from m, but only

$$\frac{m(m - 1)(m - 3)}{3!}$$

of them yield independent vectors. In other words, the m-sequence $\{a_n\}$ contains

$$N(n,3) = \frac{m(m - 1)(m - 2)}{3!} - \frac{m(m - 1)(m - 3)}{3!}$$

$$= \frac{m(m - 1)}{3!}$$

three-bit linear dependencies ($N(n,3)$). This result is a function of the degree of the characteristic polynomial and independent of the particular features of the polynomial. The formula for the number of three-bit linear dependencies is deceivingly simple. In general, an i-dimensional space will have $2^i - 1$ linear combinations of its basis vectors that have to be excluded from the choice as the $(i + 1)$th vector. It can be shown that the number of k-bit linear dependencies in an m-sequence of degree n is given by

$$N(n, k) = \prod_{i=0}^{k-1} \frac{(m-i)}{i+1} - \prod_{i=0}^{k-1} \frac{m-(2^i-1)}{i+1}. \tag{3.15}$$

Another way to describe the occurrence of dependencies in m-sequences is to define a dependency fraction $F(n, k)$ as the fraction of linearly dependent choices of k elements divided by the total number of possible choices:

$$F(n, k) = 1 - \prod_{i=0}^{k-1} \frac{m-(2^i-1)}{m-i} \tag{3.16}$$

Table 3.5 shows how the dependency fraction varies with the degree of the m-sequence and the weight of the dependency polynomial.

Stated somewhat differently, the expression $F(n, k)$ is the probability that $h(x)$, a k-weight sampling polynomial with terms chosen at random from the entire $(S = 2^n - 1)$ m-sequence $\{a_n\}$ generated by a characteristic polynomial $p(x)$, will define a set polynomial which is congruent to 0 (mod $p(x)$). Alternatively consider that $F(n, k)$ is the probability that $H(x)$ contains a factor that is a dependency polynomial of $p(x)$. While the formula for $F(n, k)$ is derived assuming that terms of the sampling polynomial were chosen from anywhere in the sequence, in many applications the span of the sampling polynomial is much less than the sequence length. In these cases the formula for $F(n, k)$ holds also. This is intuitive since the choices were made at random. Experiments with random numbers also support this conjecture. These results indicate that the dependency polynomials are in some sense "uniformly distributed" throughout the entire m-sequence. Thus if the placement of the

TABLE 3.5   Dependency Fractions $F(n, k)$ for Various Degree (n) m-Sequences and for Various Weights of the Dependency Polynomials

k	$n = 5$	$n = 10$	$n = 16$	$n = 23$	$n = 32$
3	.0345	9.79×10^{-4}	1.53×10^{-5}	1.19×10^{-7}	2.33×10^{-10}
4	.1724	4.90×10^{-3}	7.63×10^{-5}	5.96×10^{-7}	1.16×10^{-9}
5		.0156	2.44×10^{-4}	1.91×10^{-6}	3.73×10^{-9}
6		.0408	6.41×10^{-4}	5.01×10^{-6}	9.78×10^{-9}
7		.0945	1.51×10^{-3}	1.18×10^{-5}	2.31×10^{-8}
8		.2015	3.33×10^{-3}	2.61×10^{-5}	5.10×10^{-8}
9		.3958	7.09×10^{-3}	5.56×10^{-5}	1.08×10^{-7}
10			.0147	1.15×10^{-4}	2.25×10^{-7}
11			.0299	2.36×10^{-4}	4.61×10^{-7}
12			.0601	4.79×10^{-4}	9.35×10^{-7}
13			.1186	9.65×10^{-4}	1.89×10^{-6}
14			.2287	1.94×10^{-3}	3.79×10^{-6}
15			.4214	3.89×10^{-3}	7.60×10^{-6}

sampling polynomial is not an option for the designer, an engineering alternative is to choose a high enough degree generator polynomial so that the probability of a linear dependency is sufficiently low. For example, if a digital network is to be designed with a maximum EAI of 15, a 32nd-degree sequence would result in a probability of less than 10^{-5} of a dependency being defined by the sampling polynomial.

Another use for the dependency fraction $F(n, k)$ involves catenating m-sequences. Consider the following experiment: Apply each shift of one m-sequence to the digital network under test (Figure 3.4). After the m patterns have been bit-pushed through the idler register the probability that the k-subspace of interest has not been exhausted is $F(n, k)$. If the characteristic polynomial of the LFSR is now changed to a different primitive polynomial and its sequence shifted through the idler register as before, the probability that the second sequence did not exhaustively cover the k-subspace of interest is also $F(n, k)$. However, since the two sequences are independent, the probability that the k-subspace is exhaustively covered (all k-tuples appear at least once) is

$$P(n, k, q) = 1 - F(n, k)^2 \qquad (3.17)$$

Clearly this result can be extended to any number of sequences catenated together.

Again, since the dependency polynomials appear to be distributed throughout the sequence, catenating portions of different m-sequences is of interest. The conjecture of Equation (3.17) is supported by sampling experiments with random numbers.

Equations (3.16) and (3.17) provide a way to determine the probability that a linear dependency will appear in a subspace of a given weight or dimension. These relations are useful guides for the selection of m-sequences for use as stimuli for testing digital networks.

4

Test Response Compression Techniques

Built-in testing requires a method of checking the output responses of the circuit under test that is simpler and less storage-intensive than the conventional bit-by-bit comparison of the actual output values with the previously computed correct values. The usual method is to capture and compare some statistic, called the signature, of the circuit output responses rather than to compare the individual bits themselves. The concept is illustrated in Figure 4.1. Fault detection occurs if the signature realized by a circuit differs from the signature of a fault-free version of the circuit. This chapter provides a brief introduction to six different compression techniques: ones counting, transition counting, parity checking, cyclic codes, syndrome test, and compression using Walsh spectra.

The signature and its collection algorithm should meet four qualitative guidelines:

1. The algorithm must be simple enough to be implemented as part of the built-in test circuitry.
2. The implementation must be fast enough to remove it as a limiting factor in test time.
3. The algorithm must provide approximately logarithmic compression of the output response data to minimize the reference-signature storage volume.
4. The compression method must not lose information. Specifically it must not lose that evidence of a fault indicated by a wrong response from the circuit under test.

There is no algorithm that unambiguously meets all of these criteria. Item 4,

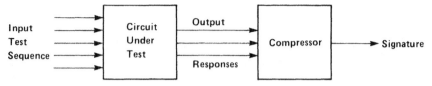

Figure 4.1. Generalized test response compression.

which refers to the possibility that the error pattern from a faulty circuit will be compressed to the same signature as the good circuit, is the greatest problem. Since only a function of the output sequence is being verified rather than the sequence itself, there is a loss of information that can "mask" errors in the sequence. Masking is the compression of an erroneous output sequence from a faulty circuit into the same signature as the fault-free circuit. Masking detracts from the quality of the test, in the sense that the input patterns have detected a fault but the act of compression hides it. An important measure of a good compression technique is how well it minimizes this masking effect.

There are four ways of measuring the masking characteristics of a compression algorithm. The first is to simulate the circuit and the proposed compression method, apply the actual test sequence, and determine those faults that cause masking. Simulation is accurate but expensive (in terms of computer time) because of the large number of faults that have to be simulated against each of the input patterns.

A second method makes decisions about the "typical" faulty output streams that will be seen, and computes masking probabilities against these fault streams. For example, soft failures such as a pattern sensitivity in memories that is induced under certain operating conditions usually cause single-bit errors. A compression algorithm that is strong in detecting single errors may be preferable for soft failures over one that is better at detecting multiple errors but weaker for single-bit errors. In this second method, a list of the possible error patterns is created and analyzed for masking under each potential compression algorithm. The difficulty in this method lies in creating the error pattern list. It could be compiled using fault simulation, but if simulation is to be used it is preferable to also simulate the compression method and thus eliminate the analysis step.

The third method is counting cases. The faults in a circuit are grouped into fault equivalence classes, and the effect of the compression mechanism is assessed for each fault group. As a simple example, suppose an n-input OR circuit is to be exhaustively tested (by applying all 2^n input patterns) with a signature that is the number of ones in the output stream. (The order in which the tests are applied is irrelevant since it cannot affect the count.) In the good circuit, the ones count is $2^n - 1$ since only one input pattern will cause a 0 on the output of the OR. Consider the n faults consisting of a stuck-at-1 on a single input to the OR. These faults are detectable only when the all-0's pattern is applied, and each will change the ones count to 2^n. Each of these n

faults fall into the same category and all are ones count detectable. Counting cases has the difficult problem of determining what output behaviors will occur as a result of faults.

The fourth method of measuring masking is to compute the proportion of all possible error sequences that will mask to the fault-free signature. This proportion can be considered a probability when the assumption is made that all possible error sequences are equally likely to occur from a faulty circuit. For an n-input circuit this assumption means that each of the $2^{2^n} - 1$ faulty versions of the circuit has the same probability of occurrence, a very difficult assumption to justify especially for single faults. Nevertheless, the assumption is widely used as a means of comparing different compression algorithms since there is reason to suppose that a compression algorithm with a low probability of masking (under the equally likely assumption) is in truth less susceptible to masking than another with a higher probability.

Because masking is so difficult to measure in arbitrary circuitry with arbitrary test sets, some compression methods restrict the circuit design or require special test sets or both. This chapter describes compression methods that apply to built-in testing in each of three categories:

1. Those that do not require special test sets and do not require circuit modification.
2. Those that do use special test sets, but do not require circuit modification.
3. Those that require both special test sets and circuit modification.

The descriptions that follow assume a single-output circuit. Any problems that arise in extending the techniques to multiple-output circuits will be noted. To ensure repeatable signatures, circuits containing sequential elements such as flip-flops, counters, and shift registers must be initialized to some fixed state before starting the test since the response of a sequential circuit to a test vector is a function both of the vector and of the internal state of the circuit. If the sequential elements can be externally set or reset, this presetting capability should be used to force the circuit to a known state. Alternatively, one of the uniform latch design methods suggested in Chapter 2 can be used to ensure that the values in all latches are easily controllable. Without a reset or control capability, a deterministic homing sequence must be designed to bring the circuit to a known state prior to test.

4.1 ONES COUNT COMPRESSION

In ones counting, the signature is the sum of the number of 1's appearing on the circuit output during the test. In an output sequence of length N, the ones count $1C(N)$ can range from 0 to N and the required compression circuit is a counter with $\lceil \log_2 (N + 1) \rceil$ stages, where $\lceil x \rceil$ is the smallest integer greater than or equal to x. The counter is clocked as soon after each test vector as the circuit output stabilizes.

Suppose that a single-output circuit is tested with a sequence consisting of N random input vectors, and that r is the expected ones count, $0 \le r \le N$. Masking will occur if a faulty circuit produces r 1's under the same N input vectors. That is to say that the N bits in the faulty output sequence must be composed of r 1's and $N - r$ 0's, in any order. The number of N-bit sequences with $1C(N) = r$ is $\binom{N}{r}$. One of these sequences is the correct response from the circuit, so there are $\binom{N}{r} - 1$ incorrect sequences that will mask to the correct ones count. Since there are 2^N possible sequences, one of which is the fault-free sequence, the ratio of masking sequences to all error sequences (given that $1C(N) = r$) is

$$P(M|r) = \frac{\binom{N}{r} - 1}{2^N - 1} \tag{4.1}$$

$P(M|r)$ can be considered a probability of masking only if it is assumed that all $2^N - 1$ error sequences occur with equal probability. The proportion of masking sequences is a function of the expected ones count r. To show the effect of the value of r in Equation (4.1), for a sequence of length $N = 7$, the ones count must be in the range 0–7. For each possible count the values of $P(M|r)$ are:

r	$P(M|r)$
0	0
1	.047
2	.157
3	.268
4	.268
5	.157
6	.047
7	0

The table displays a characteristic that is common to several compression algorithms: The probability of masking is low when the signature lies at the extremes of its range but increases rapidly toward the midpoint. When the sequence entering the ones count compressor is an N-bit string of all 0's or all 1's, the expected count is either 0 or N, and there is no chance of masking since any error (single or multiple) will change the count. As the expected count moves away from these extremes, the number of possible masking sequences increases to reach a maximum at the midrange of the count.

To minimize masking, then, the test set should be designed to produce a ones count that is near either the minimum or the maximum. Except in very special cases, this is not possible. First of all, it implies a deterministic test set defined upon the circuit function rather than upon the faults in the implementation of the function. Second, there is little or no control over the ones count obtained when using random test patterns. This second point leads to a different analysis of the masking characteristics of 1C testing. Consider the class of all n-input logic functions and assume a random test strategy that applies N test vectors to each function. Then the N-bit output sequence from a randomly selected member of the class is also random. Hence the probability that the N-bit sequence has a ones count of r is

$$P(r) = \frac{\binom{N}{r}}{2^N} \qquad (4.2)$$

Since Equation (4.1) is the masking probability for a N-bit sequence with a ones count of r, the product of $P(r)P(M|r)$ can be summed over all values of r to obtain the masking probability $P(M)$ or

$$P(M) = \sum_{r=0}^{N} P(M|r)P(r) \qquad (4.3)$$

Substitution from Equations (4.1) and (4.2) and doing the summation yields

$$P(M) = \frac{\binom{2N}{N} - 2^N}{2^N(2^N - 1)} \qquad (4.4)$$

Using Stirling's formula for factorials, $n! \simeq (2\pi)^{1/2}n^{n+1/2}e^{-n}$, $P(M)$ asymptotically approaches

$$P(M) \simeq (\pi N)^{-1/2} \qquad (4.5)$$

$P(M)$ is the probability that a 1C test strategy using random test stimuli will fail to detect an error in an N-bit response sequence from an arbitrary logic function.

Another measure of masking mentioned earlier is the ability of the compression method to handle single-bit errors. An error on a single bit of the output sequence will change a 0 to a 1 or vice versa. Either change will affect the ones count. Thus ones counting guarantees detection of all single-bit errors.

There are several methods to handle ones count compression for multiple-output circuits. One approach is to place a counter on each of the k outputs, collect each of the k counts, and compare these with their expected counts. This approach does not increase the test time over the single-output case, but has the disadvantages of a greatly increased hardware overhead (k counters each with $\lceil \log_2(N + 1) \rceil$ stages) and the required storage of k reference counts.

A simpler approach is to do a parallel-to-serial conversion of the multiple-output bits from each test vector prior to entering the counting circuit. This conversion can be accomplished with a parallel-in/serial-out shift register or by an appropriately controlled multiplexer. For a k-output circuit under test with N input vectors, this approach requires a counter with $\lceil \log_2(N + 1)k \rceil$ stages, plus the parallel-to-serial hardware overhead. While this approach requires less hardware and less reference storage than individual counters, it increases the test time because of the delay after each input vector for the parallel-to-serial conversion.

A third approach is to use weights on the different outputs and to collect a weighted ones count as suggested in Barzilai et al. (1981) in connection with weighted syndrome sums.

A different version of ones counting, which may be more useful for sequential circuits, is presented in Losq (1978). There the reference signature is the frequency of logic ones in the circuit output response rather than the actual count. The circuit is judged to be fault-free if the observed signature differs from the reference signature by less than some small value ε. For combinational circuits under test with a pseudorandom pattern generator (such that repeated use of the generator will produce the same test sequence), there is no advantage of this approach over a pure ones count. Sequential circuits, however, can be probabilistically initialized by a long sequence of random input patterns prior to signature collection, so that the state of the circuit is no longer dependent upon its initial state (this assumes that the circuit has no absorbing states). At the beginning of the subsequent signature gathering phase of the test, then, the circuit is in a steady state condition. Under these conditions, Losq shows that the majority of fault-free circuits will have signatures that are extremely close to the reference signature.

Further details on ones counting can be found in Hayes (1976b) and Parker (1976).

4.2 TRANSITION COUNT COMPRESSION

The signature in transition count (TC) testing is the number of logical transitions (from 0 to 1 and vice versa) in the output data stream. The actual bit values are ignored. The observed number of transitions is compared with the expected number for a pass/fail decision. Transition counting is employed in several commercial testers.

If $R = r_1 r_2 \cdots r_N$ is any N-bit binary sequence, its transition count $c(R)$ is

$$c(R) = \sum_{i=1}^{N-1} (r_i + r_{i+1}) \tag{4.6}$$

where Σ denotes ordinary arithmetic summation and $+$ is modulo 2 addition (the EXCLUSIVE OR Boolean operation).

Since an N-bit sequence can have up to $N - 1$ transitions, the data compression circuitry for transition counting is simply a transition detector and a counter with $\lceil \log_2 N \rceil$ stages.

Masking in transition counting depends upon the number of faulty circuits that have the same transition count as the fault-free circuit. Assume that a fault-free circuit has an output sequence R of length N bits and a TC $= i$. What is the probability that a faulty circuit also has TC $= i$? Consider an arbitrary binary sequence R', also of length N. R' has $N - 1$ boundaries between bits where transitions can occur. If R' is to have a transition count of i, there must be i transitions distributed into these $N - 1$ boundaries, and there are $\binom{N-1}{i}$ distinguishable ways of positioning them. Since the sequence $\overline{R'}$ obtained by complementing every bit of R' also has the same transition count, there are $2\binom{N-1}{i}$ possible sequences that have transition count i, one of which is the good circuit response. Thus there are $2\binom{N-1}{i} - 1$ possible error sequences that mask to the same transition count as the fault-free circuit.

In the set of all possible N-bit sequences, one is the correct sequence R, so that there are $2^N - 1$ possible faulty sequences. If all faulty sequences are equally likely to occur as the response of a faulty circuit, the probability of masking (given a transition count of i) is the number of wrong sequences with TC $= i$ divided by the total number of wrong sequences, or

$$P(M|i) = \frac{2\binom{N-1}{i} - 1}{2^N - 1} \tag{4.7}$$

The probability of masking is a function of the transition count i. As an example, let an output sequence R consist of seven bits. Then its transition count i can take on any integer value in the range 0–6. The probability of masking is shown in the following table for each possible value of i.

i	$P(M|i)$
0	.00787
1	.0866
2	.228
3	.307
4	.228
5	.0866
6	.00787

Middle of the range values of the transition count have a much greater probability of masking than values near the extremes. Indeed, for transition counts of either 0 (which means that there are no transitions at all) or $N - 1$, there is only one sequence (other than the good sequence) with the same count, and it is the complete inverse of the good sequence.

TC testing is effective in situations where the order of application of the tests can be controlled to either maximize or minimize the number of transitions. (Transition counting differs in this respect from ones counting where the order of application of the tests has no effect upon the signature.) If, for example, the tests were ordered, with perhaps some repetition of tests, to produce a maximum transition count (that is to produce the alternating 1-0 pattern on the output) then any single-bit error is detectable since it will change the TC. A thorough discussion of transition counting in connection with the ordering of a set of deterministic tests is given in Hayes (1975b, 1976a).

For built-in test schemes using random or pseudorandom test stimuli, the ordering of tests by their response values is impossible. A value for TC masking can still be obtained by noting that the probability of obtaining a transition count of i while testing the set of all possible n-input functions with a random test of N vectors is

$$P(i) = \frac{2\binom{N-1}{i}}{2^N} \tag{4.8}$$

Since $P(M|i)$ from Equation (4.7) gives the conditional probability of masking for TC $= i$, the masking probability for transition count compression can be obtained from

$$P(M) = \sum_{i=0}^{N-1} P(M|i)P(i) \tag{4.9}$$

Substituting from Equations (4.7) and (4.8) produces

$$P(M) = \frac{4\binom{2N-2}{N-1} - 2^N}{2^N(2^N - 1)} \tag{4.10}$$

Using Stirling's approximation for factorials, an asymptotic approximation for masking under transition counting compression is obtained as

$$P(M) \simeq (\pi N)^{-1/2} \tag{4.11}$$

Unlike ones counting, if a single-bit error occurs in an output sequence being compressed by transition counting, there may be masking. Consider an arbitrary binary sequence R of length N bits. Any bit in R can be inverted without changing the transition count if the bit is flanked by a 0 on one side and a 1 on the other. A consequence of this observation is that neither the first nor the last bit of the sequence can be altered without the error being TC detectable.

Consider bit r_j, one of the remaining $(N - 2)$ bits that are candidates for the location of the single error. If an error in bit r_j is to be TC undetectable, bit r_{j-1} must be different from bit r_{j+1}. Since choosing a value for bit r_{j-1} defines the value for bit r_{j+1}, there are 2^{N-1} possible sequences in which an error on bit r_j is TC undetectable. But bit r_j can be any bit in the sequence other than the first or the last. There are $(N - 2)$ possible values for the subscript j, and each is associated with 2^{N-1} possible completing sequences. Hence there is a total of $(N - 2)2^{N-1}$ possibilities for TC undetectable single-bit errors in an arbitrary N-bit sequence. Since in a N-bit sequence there is a total of $N2^N$ possible single-bit errors, the probability of encountering a TC undetectable single-bit error is

$$\frac{(N - 2)2^{N-1}}{N2^N} = \frac{N - 2}{2N} \tag{4.12}$$

or approximately $\frac{1}{2}$ for long sequences. This approximation is for arbitrary sequences. The approximation is important since it allows comparison with equivalent approximations for other compression methods. Compare, for example, the single-bit error masking probability in transition counting with the certainty of detection of single-bit errors in ones counting.

For TC testing of multiple-output circuits, the number of transition detectors is equal to the number of output pins that must be monitored. If each detector drives a counter, the signature is the count on each output. Conversely, a single counter can compress a weighted sum of each of the detectors.

4.3 PARITY CHECK COMPRESSION

Checking parity on the output data stream offers the ultimate in data compression, from the original m-bit stream to one bit signifying odd or even parity.

Figure 4.2 shows the relative simplicity of the compression circuit. It consists of a one-bit shift-register latch (SRL) with feedback through an EXCLUSIVE OR circuit. The SRL shifting clocks, not shown, are clock synchronized with the tests being applied. The SRL initial value must be controlled to ensure repeatable parity. For initialization to 0, the signature S is

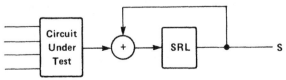

Figure 4.2. Parity check compressor.

the parity of the circuit response bit stream: 1 if the parity is odd and 0 if the parity is even.

Parity checking will detect all single-bit errors and all multiple-bit errors that result in an odd number of error bits in the output stream, but will fail in case of an even number of errors occurring on the circuit output.

The probability of masking for parity checking, assuming all faulty bit streams are equally likely, is not a function of the statistic as it was for ones or transition counting. Over the class of all possible m-bit data streams, exactly half will have even parity and the other half will have odd parity. Regardless of the parity (odd or even) of the good circuit, there are $2^m/2$ possible m-bit sequences with that parity and, therefore, $2^m/2 - 1$ possible error sequences that will mask into the good circuit parity. The probability of masking for parity checking is

$$P(M) = \frac{(2^m/2) - 1}{2^m - 1} \qquad (4.13)$$

or roughly $\frac{1}{2}$ for large m.

Multiple-output circuits can be parity checked in several ways. The simplest is shown in Figure 4.3, where the compressor acts on the EXCLUSIVE OR of the circuit outputs. Since in multiple-output circuits there is a possibility that several output lines are jointly dependent upon one signal or circuit node, a fault on that signal or at that node may cause errors on multiple output lines. An odd number of output lines in error (on a single input vector) will be detected by the output EXCLUSIVE OR, but an even number will fail to be detected. This possibility increases the error-masking probability given in Equation (4.13) for parity compression structures such as Figure 4.3.

The masking probability increase due to multiple output errors can be eliminated by using a parity check compressor on each output, at the expense of additional compressor hardware. Alternatively, the multiple-output circuit can be partitioned such that outputs from common logic structures are grouped together and each output from a group is taken to a different parity checker. A single fault may cause multiple errors in a single group but will not cause multiple-input errors in any of the disjoint groups. Since no parity checker sees more than one input from the faulty group, no multiple output error masking can occur.

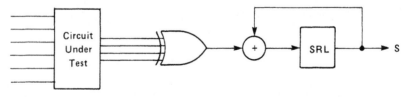

Figure 4.3. Parity check compression of multiple-output circuits.

An interesting parity testing technique requires applying all 2^k logic combinations to a k-input combinational function F which is designed so that under test its output data stream has odd parity [Tzidon et al. (1978) and W. C. Carter (1982b)]. (If the parity requirement is not met by the original function, an unused minterm ANDed with the test mode signal must be added.) The compressor is simply a parity generator, as shown in Figure 4.2, that is initialized to a 0. Since there are an odd number of 1's in the good circuit output stream under test, a good circuit under parity test will have a 1 as its signature at test end.

Case counting can be used to compute bounds on the fault detection probability of the parity testing technique. If input x_j of F is stuck-at-α, then the faulty function F' is

$$F' = F(x_1, \ldots, x_{j-1}, \alpha, x_{j+1}, \ldots, x_k) \qquad (4.14)$$

As the inputs take on all 2^k values, the output values of F' will be repeated twice, once for those input patterns for which $x_j = 0$ and again for those patterns for which $x_j = 1$. Under this fault the output stream will have an even number of 1's, and the test will fail. So any input stuck-at-α will be detected. Similarly, any n-tuple of stuck-at inputs will cause the values of the erroneous function F^n to repeat 2^n times, creating even parity and causing a test failure. A stuck-at on the output of F will also cause a test failure, since the output stream then consists of a string of 2^k 0's or 1's of even parity. It is clear that any combination of input or output faults in F will be detected by the parity testing technique.

W. C. Carter (1982b) also gives a lower bound on the number of multiple faults detectable by the parity testing technique. If the function F has n possible fault locations, there are $3^n - 1$ possible multiple faults (each of the n fault locations can be stuck-at-0, stuck-at-1, or fault-free). Consider a faulty function F' representing the input x_1 of F stuck-at-α along with any of the other $n - 1$ fault locations stuck-at-0, stuck-at-1, or fault-free. The stuck-at input plus whatever other faults exist in F' will cause the values of F' to be repeated at least twice during the application of all 2^k input patterns, so that the error function F' has even parity and will fail the test. There are two choices for the stuck-at value α on x_1 and either can occur in combination with any of the other $n - 1$ fault locations taking on any one of the three values s-a-0, s-a-1, or fault-free. Thus when x_1 is s-a-α, there are $2 \times 3^{n-1}$ possible multiple faults that are detectable by the parity testing technique. Similarly, if x_1 is fault-free, the argument may be repeated for x_2 with another $2 \times 3^{n-2}$ possible multiple fault detections, and so on. In general, there are

$$P = \sum_{j=1}^{k} 2(3^{n-j}) = 3^n - 3^{n-k} \qquad (4.15)$$

possible distinct and detectable multiple faults, where each fault includes at least one input stuck-at fault. The two stuck-at output faults are also detectable, adding $2(3^{n-k-1})$ multiple detectable faults for a total of $3^n - 3^{n-k-1}$ multiple faults. Since there are a total of $3^n - 1$ possible multiple faults, the proportion of detectable multiple faults for an arbitrary combinational function F is

$$P_b \geq \frac{3^n - 3^{n-k-1}}{3^n - 1} > 1 - \frac{1}{3^{k+1}} \tag{4.16}$$

The result is quite astounding. For a combinational function F having as few as six inputs, more than 99.95% of the possible multiple faults are detectable by the parity testing technique (provided that the circuit is modified to have odd parity and provided that an exhaustive test is applied). Notice that this result is obtained without any information about F other than the number of its inputs.

As W. C. Carter points out, this bound on multiple fault detectability using the parity testing technique does not imply correspondingly good detectability of single faults. Without knowledge of the actual circuit, the detectable single faults are limited to stuck-at's on the function inputs and its output.

The simplicity of the compressor for the parity testing technique is its great advantage. One disadvantage is the test time required to apply all 2^n inputs to a circuit with a large number of inputs (say more than 25). A reduction of test length can be obtained, in some cases, by partitioning the circuit into subcircuits and testing each subcircuit separately.

4.4 CYCLIC CODE COMPRESSION

A widely used data compression method is based on a restricted class of data transmission parity codes called cyclic codes. Checking codes of various types have been used for years to detect errors caused by noise in data transmission systems. The codes are a form of redundancy that is built into the message to be transmitted. A block of k message characters is followed by a group of m error-checking bits that encode, in some fashion, the information characters. The probability that a transmission error will go undetected can be made very low by including a large proportion of redundancy bits, but at the cost of severely reducing the information transmission rate of the channel. Most transmission systems compromise with a level of redundancy that will allow some errors, say one in a thousand, to pass undetected.

One of the most popular coding schemes is called cyclic redundancy checking (CRC). Encoding is done using an m-stage shift register with feedback connections as shown in Figure 4.4. The binary message to be transmitted is fed into the shift register (in synchronization with the shifting clock) and simultaneously sent over the channel. After the message has been transmitted, the m-bit residue left in the shift register is shifted out and sent

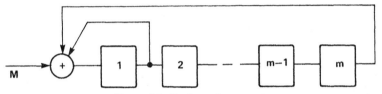

Figure 4.4. Cyclic code generator.

over the channel as a checking code. At the receiving end, the process is reversed. The received block is sent to an identical shift register, and this locally generated residue is compared with the received check code.

CRC coding has been used very successfully for the compression of test response data. The residue left in the feedback shift register after the message (or the responses from the circuit under test) has been compressed is the signature, and the process is usually known by the name signature analysis or signature testing.

The feedback shift register or signature analyzer of Figure 4.4 must have a reproducible initial value to ensure repeatable signatures. It will be shown in Chapter 5, where further results on signature analysis are given, that such a linear feedback structure distributes all possible entering bit streams evenly over all possible signatures.

The proportion of faulty bit streams that mask to the correct signature is independent of the actual signature. Let the incoming bit stream be of length k bits. There are 2^k possible streams, one of which is correct. If the register length is m stages, there are 2^m possible signatures, one of which is correct. Since all streams are evenly distributed over all signatures by the linear feedback, the number of bit streams that will create a particular signature is

$$\frac{2^k}{2^m} = 2^{k-m} \tag{4.17}$$

Hence, given a particular fault-free signature for a circuit under test, there are $2^{k-m} - 1$ wrong bit streams that will produce the same signature. Since there are a total of $2^k - 1$ wrong streams, the proportion of masking error streams is

$$P(M) = \frac{2^{k-m} - 1}{2^k - 1} \simeq 2^{-m} \tag{4.18}$$

where the approximation assumes that $k \gg m$. On the hypothesis that all possible error streams are equally likely, $P(M)$ is the probability that an incorrect bit stream will map into the correct signature, or more positively, the signature analyzer has a probability of $1 - 2^{-m}$ of detecting all errors. This error detection probability is strictly a function of m, the length of the shift register. Increasing the register length by one stage reduces the masking probability by a factor of 2.

It can be seen intuitively, and will be shown in Chapter 5, that single-bit errors are detectable by a linear signature analyzer. Suppose the register has feedback only from its last stage to the EXCLUSIVE OR. As the error bit enters the register, it alters the value of the first stage. This error will shift unchanged down the register until it reaches the last register stage. Since the feedback is linear, the error bit in the last stage will alter the incoming correct bit. Hence the single error will persist, regardless of the location of the error in the incoming bit stream.

Signature analysis is the favored method of test data compression for built-in testing, because it is sensitive to the number of 1's in the data stream as well as to their position in the sequence of 1's and 0's. Chapter 5 describes the mathematical basis for signature analysis of single- and multiple-output circuits, and gives further results on its masking characteristics.

4.5 SYNDROME TESTING

A variation on ones counting is called syndrome testing. Here the syndrome (or signature) is the normalized number of 1's realized by a function under exhaustive application of all possible input patterns. Syndrome testing can be thought of as checking the number of 1's in the output column of the truth table of the function, or equivalently, as checking the number of minterms realized by the function.

Syndrome testing applies all 2^n logical patterns to an n-input combinational function F. The syndrome of F is the normalized ones count of the function or

$$S = \frac{K}{2^n} \tag{4.19}$$

where K is the number of 1's in the function output response stream. Notice that $0 \leq S \leq 1$. Faults are detected by comparing the output syndrome with the previously stored fault-free syndrome.

The advantage that syndrome testing has over ones counting is the concept of syndrome testability, the notion that a circuit can be designed such that no single fault will cause the circuit to have the same syndrome as the fault-free circuit. Not all circuits are syndrome testable. Circuits that may be syndrome untestable are those that contain fan-out branches having unequal inversion parities along their reconverging paths. Consider the function $F = xz + y\bar{z}$. The syndrome of F is $\frac{1}{2}$. But if line z is stuck-at-0 in F, the faulty syndrome is also $\frac{1}{2}$. Thus F is syndrome untestable.

It has been shown, however, that every combinational circuit can be modified to be syndrome testable by the addition of a small amount of control logic and a few control inputs. Consider a modification to F such that

$$F^m = cxz + y\bar{z}$$

where c is a control input held at 1 whenever the objective function F has to be used, and is used as a regular input during the test. If line z is stuck-at-0 as before, the faulty syndrome is still $\frac{1}{2}$ while the fault-free syndrome is $\frac{3}{8}$.

Syndrome testing of sequential circuits would seem practical, but has not achieved the degree of mathematical rigor required to prove syndrome testability. Further details on syndrome testing are included in Savir (1980b) and Barzilai et al. (1981).

4.6 COMPRESSION USING WALSH SPECTRA

A compression method that has been discussed in the literature recently is the verification of one or more of the Walsh spectral coefficients of the function under test [Susskind (1983); Hsiao and Seth (1984)]. The Walsh spectra of a k-variable switching circuit F consists of 2^k coefficients, each corresponding to a distinct Walsh function. The magnitude and sign of the coefficients define the circuit as uniquely as does its truth table, but in a different manner.

Walsh functions are a finite set of continuous square waves with amplitudes of either $+1$ or -1 at the various points within the interval under consideration. Using Walsh functions to represent a switching circuit F of k variables requires replacing the interval of analysis by the 2^k "points" at which the function is defined, interpreting the square waves as discrete values of either $+1$ or -1 at each of the defined points, and associating the arithmetic values $+1$ and -1 with the logical values 1 and 0. For a k-variable circuit there are a total of 2^k distinct Walsh functions, any or all of which may be needed to represent the circuit over its domain. Unfortunately there is no standardized definition for the Walsh functions. Because of the great many notation systems used, much care must be taken in reading the Walsh function literature. This is a field in which any comfortable and consistent definition can be used. The following definition is provided for consideration.

Definition 4.1. Let $F(x)$ be a switching circuit in k binary variables $x_1, \ldots, x_k$. Let n be the decimal value of the binary number defined by the x_i at each of the 2^k points in the domain of $F(x)$, so that $0 \leq n \leq 2^k - 1$. The 2^k Walsh functions $W_i(n)$ assume values of $+1$ or -1 over the range of n as defined in the following:

1. Walsh function $W_0(n)$ is always $+1$ for all n.
2. The k *primary* Walsh functions $W_1(n), W_2(n), \ldots, W_k(n)$ are associated with each of the k input variables $x_1, x_2, \ldots, x_k$ and take on the

arithmetic values $+1$ (-1) as the input variables take on the logical values 1 (0).

3. The *secondary* Walsh functions are subscripted with all possible combinations of the subscripts of the input variables taken two at a time, three at a time, up to k at a time as $W_{12}(n), W_{13}(n), \ldots, W_{1k}(n), W_{123}(n),$ $W_{124}(n), \ldots, W_{12k}(n), \ldots, W_{12\ldots k}(n)$. The arithmetic values of these secondary Walsh functions are $+1$ or -1 as the logical value of the EXCLUSIVE OR of their input variables is 1 or 0.

Thus in Walsh functions of rank 2^k there are k primary (linearly independent) sequences. The remaining secondary sequences are obtained from linear combinations of the primary.

To illustrate these definitions, the eight Walsh functions $W_i(n)$ for a circuit $F(x)$ of three variables are shown in Table 4.1. The plus signs indicate where the function is $+1$ and the minus signs where the function is -1.

Now define $F(n)$ as a function that takes the arithmetic values $+1$ or -1 as the output of switching circuit $F(x)$ takes the logical values 1 or 0, respectively. Any $F(n)$ can be represented uniquely in terms of Walsh functions $W_i(n)$ as

$$F(n) = \sum_i \alpha_i W_i(n) \tag{4.20}$$

where the variable i ranges over all 2^k subscripts of the Walsh functions. The 2^k α_i's are the Walsh coefficients, and can take on values in the range -1 to $+1$.

The coefficient α_i is given by

$$\alpha_i = 2^{-k} \sum_{n=0}^{2^k-1} F(n) W_i(n) \tag{4.21}$$

While the coefficients of a given function can be computed using a fast Walsh transform [see Shanks (1969) or Beauchamp (1975)], a manual compu-

TABLE 4.1   The Walsh Functions $W_i(n)$ of Rank 8

x_1	x_2	x_3	n	W_0	W_1	W_2	W_3	W_{12}	W_{13}	W_{23}	W_{123}
0	0	0	0	+	−	−	−	−	−	−	−
0	0	1	1	+	−	−	+	−	+	+	+
0	1	0	2	+	−	+	−	+	−	+	+
0	1	1	3	+	−	+	+	+	+	−	−
1	0	0	4	+	+	−	−	+	+	−	+
1	0	1	5	+	+	−	+	+	−	+	−
1	1	0	6	+	+	+	−	−	+	+	−
1	1	1	7	+	+	+	+	−	−	−	+

TABLE 4.2   Truth Table for $F(x) = x_1\bar{x}_2 + x_2\bar{x}_3$

x_1	x_2	x_3	$F(x)$	n	$F(n)$
0	0	0	0	0	-1
0	0	1	0	1	-1
0	1	0	1	2	$+1$
0	1	1	0	3	-1
1	0	0	1	4	$+1$
1	0	1	1	5	$+1$
1	1	0	1	6	$+1$
1	1	1	0	7	-1

tation on a small circuit will provide more insight into the meanings of the coefficients. Consider the three-variable function $F(x) = x_1\bar{x}_2 + x_2\bar{x}_3$, whose truth table and equivalent transform $F(n)$ are shown in Table 4.2.

From Equation (4.21) the Walsh coefficients α_i for $F(x)$ are computed by summing the arithmetic product of $W_i(n)$ and $F(n)$ over all values of n. At the end of the summation, the coefficients are normalized by the factor 2^{-k}. Table 4.3 shows the computation of the coefficients of $F(n)$, where the normalization has been omitted for convenience, and again the plus sign indicates $+1$ and the minus sign indicates -1.

Normalizing the coefficients from Table 4.3 by the factor 2^{-k}, the expansion of the transform $F(n)$ of $F(x) = x_1\bar{x}_2 + x_2\bar{x}_3$ into Walsh functions is found to be

$$F(n) = \tfrac{1}{2}W_1 - \tfrac{1}{2}W_3 + \tfrac{1}{2}W_{12} + \tfrac{1}{2}W_{23}.$$

All other Walsh coefficients of $F(n)$ are 0. Given the Walsh coefficients of a function, the inverse transformation to recover the input/output behavior of the original function is simply an application of Equation (4.20). The inverse of the Walsh expansion of $F(n)$ is illustrated in Table 4.4.

The input/output behavior of a switching circuit of k variables is uniquely defined by the values of its 2^k Walsh coefficients. A test procedure that checks

TABLE 4.3   Computation of the Walsh Coefficients for $F(x)$

n	$F(n)$	W_0	α_0	W_1	α_1	W_2	α_2	W_3	α_3	W_{12}	α_{12}	W_{13}	α_{13}	W_{23}	α_{23}	W_{123}	α_{123}
0	$-$	$+$	$-$	$-$	$+$	$-$	$+$	$-$	$-$	$+$	$-$	$+$	$-$	$+$	$-$	$-$	$+$
1	$-$	$+$	$-$	$-$	$+$	$-$	$+$	$+$	$-$	$-$	$+$	$+$	$-$	$+$	$-$	$+$	$-$
2	$+$	$+$	$+$	$+$	$-$	$+$	$+$	$-$	$-$	$+$	$+$	$-$	$-$	$+$	$+$	$+$	$+$
3	$-$	$+$	$-$	$-$	$+$	$+$	$-$	$+$	$-$	$+$	$-$	$+$	$-$	$-$	$+$	$-$	$+$
4	$+$	$+$	$+$	$+$	$+$	$-$	$-$	$-$	$-$	$+$	$+$	$+$	$+$	$-$	$-$	$+$	$+$
5	$+$	$+$	$+$	$+$	$+$	$-$	$-$	$+$	$+$	$+$	$+$	$-$	$-$	$+$	$+$	$-$	$-$
6	$+$	$+$	$+$	$+$	$+$	$+$	$+$	$-$	$-$	$-$	$-$	$+$	$+$	$+$	$+$	$-$	$-$
7	$-$	$+$	$-$	$+$	$-$	$+$	$-$	$+$	$-$	$-$	$+$	$-$	$+$	$-$	$+$	$+$	$-$
			0		$+4$		0		-4		$+4$		0		$+4$		0

TABLE 4.4 The Inverse Transformation of $F(n)$

						$F(n) = \frac{1}{2}W_1 - \frac{1}{2}W_3 + \frac{1}{2}W_{12} + \frac{1}{2}W_{23}$				
n	$F(n)$	W_1	W_3	W_{12}	W_{23}	$+\frac{1}{2}W_1$	$-\frac{1}{2}W_3$	$+\frac{1}{2}W_{12}$	$+\frac{1}{2}W_{23}$	Sum
0	-1	-1	-1	-1	-1	$-\frac{1}{2}$	$\frac{1}{2}$	$-\frac{1}{2}$	$-\frac{1}{2}$	-1
1	-1	-1	$+1$	-1	$+1$	$-\frac{1}{2}$	$-\frac{1}{2}$	$-\frac{1}{2}$	$\frac{1}{2}$	-1
2	$+1$	-1	-1	$+1$	$+1$	$-\frac{1}{2}$	$\frac{1}{2}$	$\frac{1}{2}$	$\frac{1}{2}$	$+1$
3	-1	-1	$+1$	$+1$	-1	$-\frac{1}{2}$	$-\frac{1}{2}$	$\frac{1}{2}$	$-\frac{1}{2}$	-1
4	$+1$	$+1$	-1	$+1$	-1	$\frac{1}{2}$	$\frac{1}{2}$	$\frac{1}{2}$	$-\frac{1}{2}$	$+1$
5	$+1$	$+1$	$+1$	$+1$	$+1$	$\frac{1}{2}$	$-\frac{1}{2}$	$\frac{1}{2}$	$\frac{1}{2}$	$+1$
6	$+1$	$+1$	-1	-1	$+1$	$\frac{1}{2}$	$\frac{1}{2}$	$-\frac{1}{2}$	$\frac{1}{2}$	$+1$
7	-1	$+1$	$+1$	-1	-1	$\frac{1}{2}$	$-\frac{1}{2}$	$-\frac{1}{2}$	$-\frac{1}{2}$	-1

for the correctness of these coefficients, however, requires both an exhaustive test (the application of all 2^k possible input patterns) and the verification of all 2^k coefficients. This procedure would obviously guarantee the correct behavior of the tested circuit, but just as obviously has no advantage over simply checking the response of the circuit to each of the exhaustive input patterns. However, a partial verification of the behavior, by checking the validity of a small subset of the Walsh coefficients, can provide very good fault coverage if the subset is chosen so that any modeled fault changes one or more of the coefficients in the subset.

One coefficient that should be included in the subset is α_0 (derived from Walsh function W_0), which is a strict analogy to the normalized ones count S as used in syndrome testing. The only difference between the two is that S ranges from 0 to $+1$, while the range of α_0 is from -1 to $+1$. Since all faults on unate lines (lines that do not lead to reconvergent fan-out paths with unequal inversion parities) are detectable using syndrome testing, they are equally detectable using α_0.

Faults that are not detectable using α_0 will require adding other coefficients to the subset. The choice of the coefficients to add is not easy. There is a suggestion that the added coefficients should be selected from among those having the largest absolute magnitude. Susskind (1983) shows that adding $\alpha_{12 \ldots k}(n)$, the coefficient depending upon all inputs, to α_0 will provide very good fault coverage, including a significant portion of possible short-circuit faults.

The hardware requirements for measuring the selected subset of Walsh coefficients on a circuit under test can be estimated by noting that an alternative definition of the coefficients is

$$\alpha_i = 2^{-k}(A_i - D_i) \tag{4.22}$$

where A_i is the number of agreements in the value 0 or 1 between W_i (expressed in binary form) and $F(x)$, and D_i is the number of disagreements.

TABLE 4.5   A − D Description of the Coefficients

x_1	x_2	x_3	$F(x)$	W_1	W_3	W_{12}	W_{23}	α_1	α_3	α_{12}	α_{23}
0	0	0	0	0	0	0	0	A	A	A	A
0	0	1	0	0	1	0	1	A	D	A	D
0	1	0	1	0	0	1	1	D	D	A	A
0	1	1	0	0	1	1	0	A	D	D	A
1	0	0	1	1	0	1	0	A	D	A	D
1	0	1	1	1	1	1	1	A	A	A	A
1	1	0	1	1	0	0	1	A	D	D	A
1	1	1	0	1	1	0	0	D	D	A	A
							$A - D =$	$+4$	-4	$+4$	$+4$

This definition is illustrated in Table 4.5 for the example function $F(x) = x_1\bar{x}_2 + x_2\bar{x}_3$, where the normalization by 2^{-k} has been omitted for convenience.

With this alternative definition, the measurement of the (unnormalized) Walsh coefficient α_i is just an up/down counter that increments by 1 whenever the circuit output compares correctly with W_i (in 0/1 form) and decrements by 1 on every miscompare. The comparison values of W_i in binary are easily generated. W_0 is a constant 1. The primary W_i ($W_1, W_2, \ldots, W_k$) are just the values on the input lines $1, 2, \ldots, k$. The secondary W_i are EXCLUSIVE OR combinations of the binary inputs.

4.7   COMPARISONS

Despite the variety of ways in which test response data can be compressed, there is no compression method that will guarantee zero probability of masking. Six different compression techniques were discussed in this chapter: ones counting, transition counting, parity checking, cyclic code compression, syndrome test, and verification of Walsh spectra. It is reasonable to ask which method is best for built-in test. All of the methods satisfy the first three guidelines that were listed at the beginning of the chapter. The choice between the methods depends upon the masking characteristics needed, which in turn depends upon the size of the circuit, the willingness of the designer to modify his design, and the allowable test time.

The six methods have quite different masking propensities that may be categorized into these groups:

Nonuniform masking	Ones counting
	Transition counting
Uniform masking	Parity check
	Cyclic coding
Guaranteed masking	Syndrome testing
	Walsh spectra

Nonuniform masking refers to methods in which the probability of masking depends upon the numeric value of the signature statistic. In the discussion of ones counting it was noted that the number of error sequences with the correct ones count was a function of the actual count. Similarly the number of transition count masking sequences depends upon the transition count of the fault-free circuit.

In uniform masking, there is no relationship between the value of the signature and the propensity for masking. Half of all possible error sequences will mask when using a standard parity check on the circuit output, regardless of the signature parity. Likewise, there is no best fault-free signature with signature analysis.

A guarantee of zero masking for a restricted class of faults (single stuck-at faults) can be obtained from syndrome test compression if the circuit is designed to be syndrome testable. Verifying the complete Walsh spectra, while it does not provide compression, will guarantee the absence of all faults (except possibly for those that cause the circuit to become sequential).

A simple example of the differences between these masking characteristics may be illustrative. Suppose a compressor is desired for a binary sequence of length 128 bits, and our choices are between ones counting, signature analysis, and syndrome testing. Ones count compression or syndrome testing would require an eight-bit counter (to accommodate possible ones counts in the range 0–128). An eight-stage signature analyzer could be used with about the same hardware requirements. The ratio of masking sequences to all error sequences is given in Equation (4.1) for the ones counter, and in Equation (4.18) for signature analysis. Comparing these equations for the example shows that ones counting has a lower proportion of masking sequences than signature analysis when the expected ones count is less than 51 or more than 77. Inside that range, signature analysis performs better. Syndrome testing, assuming that the input test patterns creating the 128-bit sequence are exhaustive, will guarantee no masking sequences, but requires analysis of each circuit for syndrome testability and may require added circuit hardware to modify syndrome untestable areas.

Each of the compression methods discussed here has its own masking characteristics. Suggestions have been made for using two or more methods such that the weak points of one are compensated for by the strong points of another. References of interest in this vein are:

1. Parity with syndrome testing, and with signature testing [W. C. Carter (1982a)].
2. Output modification and ones counting [Zorian and Agarwal (1984)].
3. Collection of multiple signatures [Hassan and McCluskey (1984)].

A more in-depth discussion of the most important compression method, signature analysis, is given in Chapter 5.

5

Shift Register Polynomial Division

Polynomial division with a linear feedback shift register uses the remainder left in the register after completion of the test as the retained statistic for comparison with the known good remainder. It is an extension of the well known CRC code and is the most popular data compression technique because it is sensitive to the sequence of ones and zeros in the output bit stream and because it is easily modified for use with multiple-output circuits. The remainder is usually called a signature, and the technique is called signature analysis, a name coined by Hewlett-Packard and first used in Frohwerk (1977). This chapter addresses the techniques of signature analysis for single- and multiple-output circuits, the error masking probabilities of the method, and some suggested techniques for improving its efficiency.

In the first parts of this chapter, attention will be directed towards single-output combinational circuits, postponing a discussion of multiple-output circuits and sequential circuits to later portions.

5.1 POLYNOMIAL REPRESENTATION OF BINARY DATA

Coding theory treats binary streams as polynomials in a dummy variable, usually represented by x. Let

$$R = r_{m-1}r_{m-2} \cdots r_1 r_0$$

be any m-bit binary sequence. Using the dummy variable x, the sequence can be converted into a polynomial in x by letting each bit in the sequence be the coefficient of a unique power of x:

$$R(x) = r_{m-1}x^{m-1} + r_{m-2}x^{m-2} + \cdots + r_1x + r_0$$

This transformation of a bit stream into a polynomial is only for convenience. It associates each bit of the stream with the coefficient of a unique power of the dummy variable and permits mathematical manipulation of the bit streams in polynomial form.

The "degree" of a polynomial is the largest power of x in a term of the polynomial with a nonzero coefficient.

As an example, the sequence $R = 11001$ may be represented by the fourth-degree polynomial

$$R(x) = x^4 + x^3 + 1$$

Only the coefficients are of interest, not the actual value of x. [However, for $x = 2$, $R(x)$ is the decimal value of the bit stream.] The arithmetic that must be used in the manipulation of these polynomials is the arithmetic of the coefficients, a linear algebra over the field of two elements 0 and 1. This arithmetic follows the laws of ordinary algebra with one exception: Addition is done modulo 2. Modulo 2 addition uses the following rules for handling coefficients: $0 + 0 = 0$, $0 + 1 = 1$, $1 + 0 = 1$, $1 + 1 = 0$. An example of a modulo 2 adder is the sum output of a binary full adder or an EXCLUSIVE OR (XOR) circuit. In the terms of the equivalent polynomials, then,

$$(x^a + x^a) = 0 = (x^a - x^a), \quad \text{and hence} \quad x^a = -x^a$$

so that addition is the same as subtraction modulo 2.

As examples of modulo 2 arithmetic, the sum of two polynomials $x^4 + x^3 + x + 1$ and $x^4 + x^2 + x$ is

$$
\begin{array}{l}
x^4 + x^3 + x + 1 \\
+ x^4 + x^2 + x \\
\hline
x^3 + x^2 + 1
\end{array}
$$

The product of $x^4 + x^3 + x + 1$ and $x + 1$ is

$$
\begin{array}{l}
x^4 + x^3 + x + 1 \\
\times x + 1 \\
\hline
x^5 + x^4 + x^2 + x \\
x^4 + x^3 + x + 1 \\
\hline
x^5 + x^3 + x^2 + 1
\end{array}
$$

Division of one polynomial by another produces a quotient polynomial and, if the division is not exact, a remainder polynomial:

$$\frac{P(x)}{G(x)} = Q(x) + \frac{R(x)}{G(x)}$$

For example, let $P(x) = x^7 + x^3 + x$, and $G(x) = x^5 + x^3 + x + 1$. The longhand division is (remembering that subtraction is the same as addition, modulo 2):

$$
\begin{array}{r}
x^2 + 1 \\
x^5 + x^3 + x + 1 \overline{\smash{\big)}\ x^7 + x^3 + x} \\
\underline{x^7 + x^5 + x^3 + x^2} \\
x^5 + x^2 + x \\
\underline{x^5 + x^3 + x + 1} \\
x^3 + x^2 + 1
\end{array}
$$

The quotient $Q(x) = x^2 + 1$ and the remainder $R(x) = x^3 + x^2 + 1$.

In a basic CRC coding scheme, the remainder $R(x)$ from the division of the message word $P(x)$ by a divisor polynomial $G(x)$ is used as a check word. The transmitted code consists of the unaltered message word followed by the check word. Upon receipt, the reverse process occurs: the message word is divided by the known divisor polynomial, and a mismatch between the received check word and the remainder from the division process indicates an error in transmission.

In this chapter, signature testing means the use of CRC encoding as the data compressor and the use of the remainder $R(x)$ as the *signature* of the output data from the circuit under test.

5.2 IMPLEMENTATION OF A CRC GENERATOR

The modulo 2 division example given earlier had a dividend $P(x)$ and a divisor $G(x)$:

$$\frac{P(x)}{G(x)} = \frac{x^7 + x^3 + x}{x^5 + x^3 + x + 1}$$

To make the relationship with the mathematics of the coefficients clearer, the division can be written with the coefficients of the powers of x explicitly shown:

$$\frac{P(x)}{G(x)} = \frac{1x^7 + 0x^6 + 0x^5 + 0x^4 + 1x^3 + 0x^2 + 1x + 0}{1x^5 + 0x^4 + 1x^3 + 0x^2 + 1x + 1}$$

The longhand division can be conducted in terms of the detached coefficients only:

$$101 = Q \quad Q(x) = x^2 + 1$$

$$101011 \overline{)\,10001010}$$
$$\underline{101011}$$
$$00100110$$
$$\underline{101011}$$
$$\overline{001101} = R \quad R(x) = x^3 + x^2 + 1$$

Only the remainder is of interest, since it is the CRC code word.

This division process can be mechanized using a shift register with interspersed modulo 2 adders, a structure that is usually called a linear feedback shift register or LFSR. Figure 5.1 shows a five-stage LFSR implementing the divisor polynomial $G(x) = x^5 + x^3 + x + 1$. The blocks marked SRL are shift-register latches or master–slave flip-flops and each stores the coefficient of a unique power of x.

The divisor polynomial is defined by the feedback connections. In Figure 5.1, when a shift occurs that would create x^5 in an endless shift register, x^5 is replaced by $x^3 + x^1 + x^0$. For example, if the register contents are 00001 and the input is 0, after the next shift cycle the register contents are 11010 or $1 + x + x^3$. Hence the feedback connections are

$$x^5 = x^3 + x + 1$$

Thus whenever a quotient coefficient (the x^5 term) is shifted off the rightmost stage of the LFSR, $x^3 + x + 1$ is added to the register (or subtracted from the register since addition is the same as subtraction modulo 2). Effectively the dividend has been divided by $x^5 + x^3 + x + 1$, or

$$G(x) = x^5 + x^3 + x + 1$$

If the LFSR is initialized to zero, and the message word (or dividend) $P(x)$ is serially streamed to the LFSR input, high-order coefficient first, the content of the LFSR after the last message bit is the remainder from the division of the message polynomial by the divisor $G(x)$. Notice that $P(x)$ enters the

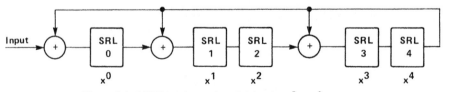

Figure 5.1. LFSR implementing division by $x^5 + x^3 + x + 1$.

TABLE 5.1   LFSR States During Division of P = 10001010

Clock	Input	x^0	x^1	x^2	x^3	x^4	
0		0	0	0	0	0	← Initial State
1	1	1	0	0	0	0	
2	0	0	1	0	0	0	
3	0	0	0	1	0	0	
4	0	0	0	0	1	0	
5	1	1	0	0	0	1	
6	0	1	0	0	1	0	
7	1	1	1	0	0	1	
8	0	1	0	1	1	0	← Remainder R

register high-order bit first because in division the high-order terms of the dividend must be processed first.

Consider the states of the Figure 5.1 LFSR during the division of $P(x) = x^7 + x^3 + x$ that was shown in longhand form earlier. The input sequence is

$$P = 10001010$$

The LFSR is initially zero. P is serially fed to the LFSR input, high-order bit first, in synchronization with the LFSR shifting clock. The various states of the LFSR are shown in Table 5.1.

The first four shifts of the LFSR have no effect because the 0 in the last stage inhibits the feedback. On the fifth shift, the high-order bit of P [the coefficient of x^7 in $P(x)$] enters the fifth stage and the coefficient of x^3 enters the first stage. At this point the register contains 10001 or $1 + x^4$.

On the sixth shift clock, the first subtraction occurs. In an endless register, the $1 + x^4$ would become $x + x^5$. In the actual register, the x^5 term is discarded (by shifting it out) and is simultaneously replaced by $1 + x + x^3$. The register then effectively subtracts $1 + x + x^3 + x^5$ from $x + x^5$ to end up with 10010 or $1 + x^3$.

Subtracting the leftmost terms of the divisor and the dividend is automatic as the coefficient in the fifth stage of the LFSR is shifted off the end of the register (and becomes the first term of the quotient).

The second subtraction (shown in the long division) is similar.

The LFSR emits the quotient of the division process (101 or $x^2 + 1$) by shifting it out from the right-hand end of the register. The remainder is left in the register after all bits of P have been shifted in.

This LFSR structure, then, creates a "code word" as the remainder after dividing the message polynomial by the feedback polynomial. Similarly, any data, such as the test response results from a circuit, can be compressed into a "code word" by an LFSR. This code word, the remainder from the division process, is called the *signature* of the input data stream, and the LFSR itself is called a *signature analyzer*.

Using a signature analyzer to compress a long stream of test output data into a short signature that is then used to make a pass/fail decision on the circuit obviously raises the question as to how effective the compressor is in reflecting any test errors into the final signature.

5.3 ERROR POLYNOMIALS

One concern with the use of CRC check words as the signature of a circuit output sequence is the possibility that an erroneous sequence from a faulty circuit will be compressed into the same signature as the fault-free circuit. This phenomenon is called "masking," since the effect of the fault is masked by the compression process. Masking is a loss of information caused by the compression of the output sequence. The probability of masking is addressed in this section.

A fault in a circuit can cause an error on its output during one or more tests. The possible error patterns can be treated as error polynomials. Each nonzero coefficient in the error polynomial $E(x)$ represents an error in the corresponding bit position of the circuit output sequence. For example,

$$P = 10001010, \qquad P(x) = x^7 + x^3 + x \qquad = \text{correct data}$$

$$P' = 00001101, \qquad P'(x) = x^3 + x^2 + 1 \qquad = \text{incorrect data}$$

$$E = 10000111, \qquad E(x) = x^7 + x^2 + x + 1 = \text{error polynomial}$$

so that $P'(x) = P(x) + E(x)$.

Starting from an initial state of all zeros, the signature of the correct data $P(x)$ is the remainder $R(x)$ in

$$P(x) = Q(x)G(x) + R(x), \quad \text{or} \quad R(x) = P(x) \bmod G(x)$$

An undetectable error will occur if

$$P'(x) = P(x) + E(x) = Q'(x)G(x) + R(x)$$

since then both $P(x)$ and $P'(x)$ have the same signature. Hence any undetectable error polynomial $E(x)$ must be evenly divisible by the divisor polynomial $G(x)$. In other words, an undetectable error is represented by an error polynomial that is a nonzero multiple of the divisor.

For example, with $P(x) = x^7 + x^3 + x$ as the data stream polynomial and with $G(x) = x^5 + x^3 + x + 1$ as the divisor, it was shown earlier that the remainder left in the LFSR after the division was $R(x) = x^3 + x^2 + 1$. If a fault exists such that the error polynomial is $x^7 + x^2 + x + 1$, as in the earlier example, the actual data entering the signature analyzer is 00001101 or $x^3 + x^2 + 1$. The remainder left in the LFSR of Figure 5.1 is $x^3 + x^2 + 1$, exactly the same as the signature from the fault-free circuit.

An alternative image of the masking process can be obtained by ignoring the data stream $P(x)$ and assuming that only the error polynomial is being compressed (this is another example of the principle of superposition as described in Chapter 3). R is the signature of the fault-free sequence P. Let R' be the signature of some faulty sequence P'. Let R_e be the modulo 2 sum of R and R'. Now

$$R(x) = P(x) \bmod G(x)$$

and

$$R'(x) = P'(x) \bmod G(x)$$

Then, since the signature analyzer is linear,

$$R_e(x) = R(x) + R'(x) = [P(x) + P'(x)] \bmod G(x)$$

Since $E(x) = P(x) + P'(x)$, $R_e(x) = E(x) \bmod G(x)$. That is to say that $R_e(x)$ is the signature of the error polynomial $E(x)$. In the masking example, $E(x) = x^7 + x^2 + x + 1$. By setting $P(x) = E(x)$, and streaming this $P(x)$ into the LFSR [using $G(x)$ as the divisor], the remainder $R(x) = 0$. [There is a nonzero quotient $Q(x) = x^2 + 1$, but only the remainder is of interest since it is the retained signature.] An alternative description of an undetectable error polynomial, then, is an error polynomial that has a remainder of 0 when divided by $G(x)$.

Since a correct m-bit data stream is represented by an $m - 1$ degree polynomial, any error can be represented as a polynomial of degree $m - 1$ or less. If the correct data were

$$P = 10111, \qquad P(x) = x^4 + x^2 + x + 1$$

the greatest error polynomial is $E(x) = x^4 + x^3 + x^2 + x + 1$ ($E = 11111$), since then every bit in P is wrong. The incorrect data with this error polynomial is $P' = P + E = 01000$.

There are $2^m - 1$ possible error polynomials for an m-bit data stream, since there are m coefficients in the error polynomial and each coefficient can be either 0 or 1. [The 1 is subtracted from 2^m to eliminate the error-free case $E(x) = 0$.]

Since all undetectable error polynomials are nonzero multiples of the divisor polynomial, it is natural to ask how many undetectable error polynomials there are.

For example, the eight-bit data stream $P = 10001010$ is represented by the polynomial $P(x) = x^7 + x^3 + x$ which has $2^8 - 1$ possible error polynomials. The undetectable error polynomials for a divisor $G(x) = x^5 + x^3 + x + 1$ are shown in Table 5.2.

For this example, the largest multiplier is $x^2 + x + 1$ since any multiplier of higher degree results in an error polynomial of a degree greater than $P(x)$.

TABLE 5.2   Undetectable Error Polynomials for $P(x) = x^7 + x^3 + x$ and Divisor $G(x) = x^5 + x^3 + x + 1$

Multiplier	$E(x)$	$P'(x) = E(x) + P(x)$
1	$x^5 + x^3 + x + 1$	$x^7 + x^5 + 1$
x	$x^6 + x^4 + x^2 + x$	$x^7 + x^6 + x^4 + x^3 + x^2$
$x + 1$	$x^6 + x^5 + x^4 + x^3 + x^2 + 1$	$x^7 + x^6 + x^5 + x^4 + x^2 + x + 1$
x^2	$x^7 + x^5 + x^3 + x^2$	$x^5 + x^2 + x$
$x^2 + 1$	$x^7 + x^2 + x + 1$	$x^3 + x^2 + 1$
$x^2 + x$	$x^7 + x^6 + x^5 + x^4 + x^3 + x$	$x^6 + x^5 + x^4$
$x^2 + x + 1$	$x^7 + x^6 + x^4 + 1$	$x^6 + x^4 + x^3 + x + 1$

Thus there are 7 undetectable error polynomials for this example, out of a total of 255 possible error polynomials.

If the divisor polynomial $G(x)$ is of degree r, then it has $2^{m-r} - 1$ nonzero multiples that result in a polynomial of degree less than m. That is to say that there are $2^{m-r} - 1$ wrong m-bit streams that can map into the same signature as the correct bit stream. Since there are $2^m - 1$ possible wrong bit streams, if all possible error polynomials are equally likely the probability that the degree r divisor will not detect an error is

$$P(M) = \frac{2^{m-r} - 1}{2^m - 1}$$

For large m, $P(M) \simeq 1/2^r$. Hence the probability of not detecting an error (or the probability of masking) can be made arbitrarily small by increasing the degree r of the divisor polynomial (which is the same as increasing the length of the signature register).

In this analysis, the masking probability is a function only of the length of the signature register and not of the feedback polynomial. Even a simple LFSR with a single feedback from the last stage could be used and would achieve the same masking probability.

The basic assumption used in the masking probability analysis is that all possible error polynomials are equally likely. This implies that a fault is just as likely to cause errors in every bit as it is to cause a single-bit error. Typically, however, the error patterns of faulty logic circuits are not independent and not equally likely. It is therefore useful to examine the masking probabilities for a few specific types of error polynomials.

Single-Bit Errors.   Any divisor polynomial with two or more nonzero coefficients will detect all single-bit errors.

It was shown earlier that all undetectable errors are represented by an error polynomial $E(x)$ that is a nonzero multiple of the divisor polynomial $G(x)$. If the divisor $G(x)$ has two or more nonzero coefficients, all nonzero multiples of $G(x)$ must also have at least two nonzero coefficients. Hence, any error polynomial with only one nonzero coefficient cannot be a multiple of $G(x)$ and must be detectable. A single-bit error corresponds to an error polynomial $E(x) = x^i$, and detection requires only that $G(x)$ not evenly divide x^i. Any $G(x)$ of more than one term will not divide x^i evenly, and hence will detect all single errors. The simplest divisor polynomial with two nonzero coefficients is $x + 1$. This polynomial defines a single-stage LFSR with feedback from the stage output to one leg of the XOR input, the other input leg driven by the data stream. The resulting LFSR is nothing more than a parity check on the input bit stream, and parity is well known for its ability to detect all single-bit errors.

Burst Errors. Define a burst error of length b as one where the erroneous bits are within b consecutive bit positions and at most b bits are in error. A divisor polynomial $G(x) = x^b + 1$ will detect all burst errors of length b.

All undetectable errors must be nonzero multiples of $G(x) = x^b + 1$. Any error polynomial that is a nonzero multiple of $G(x) = x^b + 1$ must have two nonzero coefficients appearing farther apart than b. But the error polynomial representing the b length burst error does not have two nonzero coefficients farther apart than b and therefore cannot represent an undetectable error.

Repeated Use Errors. A repeated use error is one where the erroneous bits are spaced at intervals that are powers of 2. These types of errors can occur in certain operations on a byte-organized computer. It has been shown in Smith (1980) that masking is improbable under repeated use errors. This result may not be too relevant for built-in testing unless the pattern source for the circuit is functional in nature so that repeated use errors do occur.

Some further analytical results on the probability of masking with LFSR signature analysis are given in Carter (1982) and David (1978). Practically, however, the question must be "how has signature analysis performed as a test response data compression tool in actual practice?" Some experimental results on relatively complex structures are given in Segers (1981) and Koenemann et al. (1979). It is difficult to relate such experiments to different circuits, and in the final analysis a simulation of the signature analyzer against various faulty versions of the circuit is the best analysis of masking. The problem, of course, is to obtain the output responses for all possible faulty circuits. The method may be useful as an adjunct to analytical study of a particular LFSR length or feedback polynomial. Some interesting results on the use of one such simulator, called the SIGLYZER (for SIGnature anaLYZER), are given in Sridhar et al. (1982).

Obviously LFSR simulation is computationally expensive and furthermore cannot be generalized to circuits other than the one simulated. A possibly fruitful area of analysis lies in understanding the characteristics that can be exhibited by faulty circuits, that is, the types of error polynomials that can be produced. Such an analytical approach is suggested by the work in Bhattacharya and Gupta (1983).

5.4   ALTERNATIVE LFSR IMPLEMENTATION OF POLYNOMIAL DIVISION

The LFSR divider structure of Figure 5.1 may be inconvenient for implementation because of the need for EXCLUSIVE OR gates between stages of the shift register. An alternative structure (and the dual of Figure 5.1) in which the feedback logic is grouped at one end of the register is shown in Figure 5.2. This alternative divider generates the correct quotient but the remainder is not always correct. The choice of the two implementations for signature analysis depends upon the circuits available.

The divisor polynomial for an LFSR structured like Figure 5.2 is determined by the feedback connections. Feedback from the ith stage of an n-stage register adds x^{n-i} to the divisor polynomial x^n. In Figure 5.2, feedback exists from register stages 2, 4, and 5. Hence the divisor polynomial is $x^5 + x^{5-2} + x^{5-4} + x^{5-5}$ or $x^5 + x^3 + x + 1$.

The behavior of these alternative dividers is easiest to visualize by assuming an input sequence and observing the effect of the feedback. Let the polynomial

$$R(x) = r_5x^5 + r_4x^4 + r_3x^3 + r_2x^2 + r_1x + r_0$$

enter the LFSR of Figure 5.2, high-order bit first. The LFSR is initially zero. As the r_5 term (the coefficient of x^5) moves through the register it affects the modulo 2 adder through the feedback connections. Let $R(x)$ be

$$R(x) = x^5 + x^3 + x + 1 \quad \text{or} \quad R = 101011$$

As the high-order term of $R(x)$ moves through the LFSR it cancels the x^3, the x, and the 1 terms, and is finally shifted off the register at the last shift clock. For this input sequence, then, the quotient is 1 and the remainder is 0, and in a divider this could only happen if it were implementing division by $x^5 + x^3 + x + 1$.

However, it was noted earlier that the alternative LFSR does not always produce the correct remainder. This can be seen by comparing the behavior of the two implementations against the same input sequences. Figure 5.3 shows two different LFSRs that divide the input data stream by $x^4 + x + 1$. LFSR1

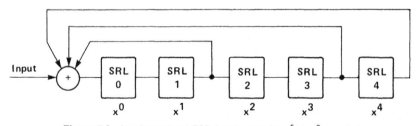

Figure 5.2. An alternative LFSR for division by $x^5 + x^3 + x + 1$.

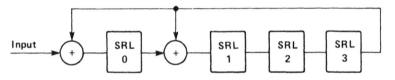

LFSR1: A True Polynomial Divider

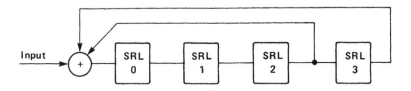

LFSR2: The Dual to LFSR1

Figure 5.3. Two configurations for division by $x^4 + x + 1$.

is a true polynomial divider that generates the correct quotient and the correct remainder. LFSR2 is the dual of LFSR1 and generates the correct quotient but not always the correct remainder.

Assume the same input sequence to both LFSRs and let the sequence be of degree 4 (the lowest degree sequence that will create a quotient).

Table 5.3 lists the 16 possible input polynomials of degree 4, beginning with x^4 or the sequence 10000 and ending with $x^4 + x^3 + x^2 + x + 1$ or 11111.

TABLE 5.3   Remainder in LFSR1 and LFSR2 for Various Dividends

Dividend	Quotient	Remainder	LFSR1	LFSR2
x^4	1	$1 + x$	1100	1100
$x^4 + 1$	1	x	0100	0100
$x^4 + x$	1	1	1000	1000
$x^4 + x + 1$	1	0	0000	0000
$x^4 + x^2$	1	$1 + x + x^2$	1110	1110
$x^4 + x^2 + 1$	1	$x + x^2$	0110	0110
$x^4 + x^2 + x$	1	$1 + x^2$	1010	1010
$x^4 + x^2 + x + 1$	1	x^2	0010	0010
$x^4 + x^3$	1	$1 + x + x^3$	1101	0101
$x^4 + x^3 + 1$	1	$x + x^3$	0101	1101
$x^4 + x^3 + x$	1	$1 + x^3$	1001	0001
$x^4 + x^3 + x + 1$	1	x^3	0001	1001
$x^4 + x^3 + x^2$	1	$1 + x + x^2 + x^3$	1111	0111
$x^4 + x^3 + x^2 + 1$	1	$x + x^2 + x^3$	0111	1111
$x^4 + x^3 + x^2 + x$	1	$1 + x^2 + x^3$	1011	0011
$x^4 + x^3 + x^2 + x + 1$	1	$x^2 + x^3$	0011	1011

Also shown are the true polynomial remainders and the actual binary remainders in both registers. For all of these input polynomials the quotient from either register is 1. Notice that there is an exact match between the true remainder and the binary representation of that remainder for the LFSR1 implementation. LFSR2, on the other hand only creates the correct remainder for half of the input polynomials.

Although in the example given exactly half of the input polynomials map through LFSR2 into the correct remainder, this is not a general truth. In the more usual case only a few polynomials will generate the correct remainder. However, it is always true that the divisor polynomial (or multiples of the divisor) maps to a remainder of 0 in either implementation (note that in the example the input $x^4 + x + 1$ results in a remainder of 0). This is important because it means that both LFSR structures have the same masking characteristics. Specifically, let $R(x)$ be the signature remaining in an LFSR of either type shown in Figure 5.3 when $G(x)$ is the divisor and $P(x)$ is the input data stream. For an error polynomial $E(x)$, $P(x)$ and $P(x) + E(x)$ will have the same signature if and only if $E(x)$ is a multiple of $G(x)$.

One further point to be made regarding Table 5.3. Both LFSRs have 4 stages, and hence there are 16 distinct signatures. Note that a different signature results from each of the 16 possible degree 4 input polynomials. That is to say that the LFSRs distribute the incoming bit streams evenly over all possible signatures. To stress the point, there are 32 possible degree 5 input polynomials and they will be distributed over the 16 possible signatures (by either register) such that two of the polynomials share each signature.

5.5 VARIOUS FEEDBACK STRUCTURES

Obviously very many choices exist for the feedback configurations. Consider, for example, the simple linear feedback shift register shown in Figure 5.4. This is the parity-checking compressor that was shown in Figure 4.2 of Chapter 4 and it is a true polynomial divider, implementing division by the polynomial $x + 1$. Suppose the register is initially 0. As the input sequence is clocked in, the first 1 bit causes the LFSR to flip to 1. A second 1 bit flips it back to 0, and so on. Thus, as a signature analyzer the LFSR is computing the parity of the incoming bit stream, and the final signature is the parity over the entire stream. Any odd number of error bits (an error polynomial having odd parity) will be detected by this LFSR, while an even number of error bits will mask to

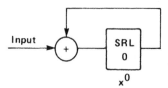

Figure 5.4. An LFSR that computes parity on its input.

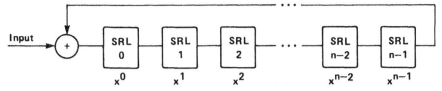

Figure 5.5. An LFSR mechanizing division by $x^n + 1$.

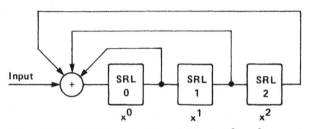

Figure 5.6. An LFSR mechanizing division by $x^3 + x^2 + x + 1$.

the correct signature (will have the correct parity). If all possible error polynomials are equally likely, the masking probability of this LFSR is approximately $\frac{1}{2}$.

The parity-generating LFSR of Figure 5.4 can be generalized to the LFSR shown in Figure 5.5. This is again a true polynomial divider that implements division by $x^n + 1$. It is usually called a *pure cycling register* because in the absence of an input, the register merely cycles its contents. When used as a signature analyzer, each bit in the signature is just the parity over every nth bit of the input sequence. David (1980, 1984) presents a strong argument for using such an LFSR in signature analysis.

Figure 5.6 shows a register having feedback from every stage. It is not a true divider. The divisor polynomial is $x^3 + x^2 + x + 1$, and again only error polynomials that are multiples of the divisor will mask to the correct signature.

As a curiosity, it is interesting to consider the ultimate case of a register with no feedback at all. The divisor polynomial is x^r, where r is the number of stages of the register. An example for $r = 5$ is shown in Figure 5.7.

To understand the effect of this degeneracy, assume $m > r$ and consider an m-bit data stream $P(x)$ entering an r-stage register with no feedback (such as shown in Figure 5.7), high-order bit first. $P(x)$ can be represented by the

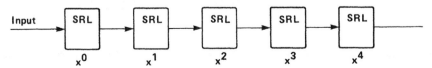

Figure 5.7. A degenerate LFSR with divisor x^5.

polynomial

$$P(x) = a_{m-1}x^{m-1} + \cdots + a_r x^r + a_{r-1}x^{r-1} + \cdots + a_1 x + a_0$$

Dividing $P(x)$ by $G(x) = x^r$,

$$\frac{P(x)}{x^r} = \frac{a_{m-1}x^{m-1} + \cdots + a_r x^r + a_{r-1}x^{r-1} + \cdots + a_1 x + a_0}{x^r}$$

$$= Q(x) + \frac{a_{r-1}x^{r-1} + \cdots + a_1 x + a_0}{x^r}$$

For the degenerate LFSR, the remainder is just the last r bits of the data stream. The first $m - r$ data bits affect the quotient $Q(x)$ but not the remainder. Masking will occur no matter what the first $m - r$ bits are as long as the trailing r bits are correct. There are 2^{m-r} possibilities for the leading $m - r$ bits, one of which is correct, so that there are $2^{m-r} - 1$ wrong m-bit data streams that will map into the same signature as the correct stream. Since there are $2^m - 1$ possible wrong m-bit streams, the probability of masking is

$$P(M) = \frac{2^{m-r} - 1}{2^m - 1} \simeq \frac{1}{2^r}$$

again on the assumption that all possible error streams are equally likely. This result can also be obtained by noting that when the degenerate LFSR is used for signature analysis, only the last r tests need to be applied. The fault may have been detected within the first $m - r$ tests, but the evidence of the fault is lost in the LFSR.

5.6 ALTERNATIVE DESCRIPTIONS OF SIGNATURE REGISTERS

There are several other methods for analyzing linear systems, and specifically for describing the feedback structure of an LFSR. All of these methods result in the same relationship between the LFSR and its divisor polynomial. These alternatives are most useful when considering the LFSR as an autonomous machine, that is, with a constant 0 on the input so that the feedback connections to the modulo 2 adders are the only activating force.

One description is in terms of the "recurrence relationship" or the difference equation generated by the feedback structure, as discussed in Chapter 3. Consider the n-stage LFSR of Figure 5.8. The input to the first stage of the LFSR (the output of the modulo 2 adder) is a_k. The outputs of the stages are $a_{k-1}, a_{k-2}, \ldots, a_{k-n}$, the values of a_k at the preceding n clock periods. As the LFSR clock advances from k to $k + 1$, the stages take on the values of their inputs. Then the value of a_k can be described by the recurrence relation

$$a_k = c_1 a_{k-1} + c_2 a_{k-2} + \cdots + c_{n-1} a_{k-n+1} + a_{k-n}$$

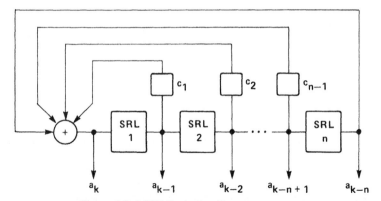

Figure 5.8. LFSR illustrating the recurrence a_k.

The recurrence satisfied by the LFSR is just a rule for determining the value of a_k from the values stored in the LFSR stages. As an example, for the LFSR of Figure 5.2, with input equal to 0, the recurrence is

$$a_k = a_{k-2} + a_{k-4} + a_{k-5}$$

where, as usual, addition is modulo 2. Calculating the behavior of an LFSR from its recurrence requires that its initial state be known, that is, $a_{-1}, a_{-2}, \ldots, a_{-n}$.

The recurrence relation defines an infinite sequence of binary values

$$\{a_k\} = \{a_0, a_1, a_2, \ldots\}$$

From this sequence a generating function $g(x)$ can be defined as

$$g(x) = \sum_{k=0}^{\infty} a_k x^k$$

In Chapter 3 it was shown that

$$g(x) = \frac{\Sigma c_i x^i \left(a_{-i} x^{-i} + \cdots + a_{-1} x^{-1} \right)}{1 - \Sigma c_i x^i}$$

where the summations are over i. The n-degree polynomial in the denominator

$$f(x) = 1 - \sum_{i=1}^{n} c_i x^i$$

is called the characteristic polynomial of the LFSR. Notice $f(x)$ is also equivalent to

$$f(x) = 1 - \Sigma x^j$$

where the sum is over those LFSR stages j whose outputs feed back to the modulo 2 adder. Since subtraction is equivalent to addition (modulo 2), it is convenient to rewrite $f(x)$ as

$$f(x) = 1 + \Sigma x^j$$

For the LFSR of Figure 5.2, the characteristic polynomial is

$$f(x) = 1 + x^2 + x^4 + x^5$$

Finally, the divisor polynomial is the reciprocal of the characteristic polynomial, where the reciprocal of an m-degree polynomial $G(x)$ is $x^m G(1/x)$. For the LFSR of Figure 5.2, the reciprocal of the characteristic polynomial is the divisor:

$$G(x) = x^5(1 + x^{-2} + x^{-4} + x^{-5}) = x^5 + x^3 + x + 1$$

5.7 MULTIPLE-INPUT SIGNATURE REGISTERS

For built-in testing of multiple-output circuits, the overhead of a single-input signature analyzer on every output would be high. Of course the single-input analyzer could be multiplexed to the various circuit outputs, one at a time, and the test sequence repeated for each output. However, a parallel signature testing method using a structure called a multiple-input signature register (MISR) is the preferred technique for multiple-output circuits. Such a MISR is shown in Figure 5.9.

Usually the MISR has as many stages as there are circuit outputs. A modulo 2 adder placed between each of the MISR stages is driven by one of the circuit outputs. Note that the divisor feedback also feeds the adders at the appropriate stages. The MISR in Figure 5.9 implements the divisor polynomial $x^5 + x^3 + x + 1$. (If inputs 1 through 4 are set to zero, the MISR becomes the LFSR shown earlier in Figure 5.1.)

Consider an m-output circuit whose test responses are being compressed by an m-stage MISR. Let test cycle i be an arbitrary test that lies part way

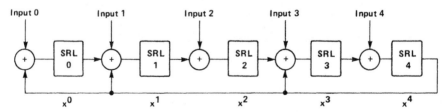

Figure 5.9. Multiple-input signature register (MISR).

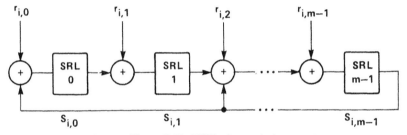

Figure 5.10. MISR after cycle i.

through the complete test sequence. Define the phrase "after test cycle i" as meaning that test i has just been applied to the circuit, that the test responses are stable on the circuit outputs, and that the shifting clock has not yet been applied to the MISR. Let $R_i(x)$ be the $(m - 1)$th degree polynomial representing the test responses on the m outputs of the circuit after test cycle i. Then

$$R_i(x) = r_{i,m-1}x^{m-1} + r_{i,m-2}x^{m-2} + \cdots + r_{i,1}x + r_{i,0}$$

Similarly, let $S_i(x)$ be the polynomial defining the state of the m outputs of the MISR after test cycle cycle i. Then

$$S_i(x) = S_{i,m-1}x^{m-1} + S_{i,m-2}x^{m-2} + \cdots + S_{i,1}x + S_{i,0}$$

The status of the MISR and the circuit outputs after test cycle i is shown in Figure 5.10.

The next state of the MISR (after test cycle $i + 1$) is

$$S_{i+1}(x) = [R_i(x) + xS_i(x)]\bmod G(x)$$

To give an example, suppose that $G(x) = x^5 + x^2 + x + 1$.

If $S_i(x) = x^4 + x^2 + x + 1,$ $S_i = 10111$

and $R_i(x) = x^2 + x,$ $R_i = 00110$

then, $S_{i+1}(x) = x^3 + x^2 + x + 1,$ $S_{i+1} = 01111$

This result can be seen from the following computation:

$$[R_i(x) + xS_i(x)]\bmod G(x) = [x^2 + x + x(x^4 + x^2 + x + 1)]\bmod G(x)$$
$$= (x^5 + x^3)\bmod G(x)$$
$$= x^3 + x^2 + x + 1$$

or by observing the existing and next state of the MISR, shown in Figure 5.11, that implements this transition.

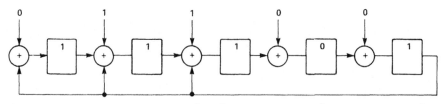

Figure 5.11. MISR at cycle i: $G(x) = x^5 + x^2 + x + 1$, $R_i(x) = x^2 + x$, and $S_i(x) = x^4 + x^2 + x + 1$.

The initial state S_0 of the MISR must be known. If the initial state is zero, the sequence of MISR states will be

$$
\begin{aligned}
S_1(x) &= \left[R_0(x) + xS_0(x)\right] \bmod G(x) \\
&= R_0(x) \bmod G(x) \\
&= R_0(x) \\
S_2(x) &= \left[R_1(x) + xS_1(x)\right] \bmod G(x) \\
&= \left[R_1(x) + xR_0(x)\right] \bmod G(x) \\
S_3(x) &= \left[R_2(x) + xS_2(x)\right] \bmod G(x) \\
&= \left[R_2(x) + xR_1(x) + x^2R_0(x)\right] \bmod G(x) \\
&\;\;\vdots \\
S_n(x) &= \left[x^{n-1}R_0(x) + x^{n-2}R_1(x) + \cdots \right. \\
&\qquad \left. + xR_{n-2}(x) + R_{n-1}(x)\right] \bmod G(x)
\end{aligned}
$$

This last expression, then, is the remainder "signature" of the fault-free circuit (the final state of the MISR) after a sequence of n tests has been applied. The signature can be computed by knowing the circuit responses for each test of the test sequence.

Now to account for the possibility that a fault in the circuit will cause one or more of its outputs to be in error on one or more tests, the concept of error polynomials is used again. Let $E_i(x)$ be the error polynomial indicating the incorrect outputs of the circuit after test cycle i. Then

$$
E_i(x) = e_{i,m-1}x^{m-1} + e_{i,m-2}x^{m-2} + \cdots + e_{i,1}x + e_{i,0} \qquad (5.1)
$$

Since all of the operations in the MISR are linear, the error polynomials can be considered as being compressed independently. By analogy with the state sequence given earlier, the final error signature (after n tests) is

$$
S_{en}(x) = \left[x^{n-1}E_0(x) + x^{n-2}E_1(x) + \cdots + xE_{n-2}(x) \right.
$$
$$
\left. + E_{n-1}(x)\right] \bmod G(x) \quad (5.2)
$$

$S_{en}(x)$ is the cumulative error signature polynomial after n tests have been applied to the circuit or is the "signature" of the errors caused by the faulty circuit. Obviously masking will occur if $S_{en}(x) = 0$.

5.8 MASKING IN MULTIPLE-INPUT SIGNATURE REGISTERS

For an error on one or more of the circuit outputs at one or more test cycles to go undetected by the MISR it is necessary that $S_{en}(x) = 0$ or

$$S_{en}(x) = \left[x^{n-1}E_0(x) + x^{n-2}E_1(x) + \cdots \right.$$
$$\left. + xE_{n-2}(x) + E_{n-1}(x) \right] \bmod G(x) = 0$$

Thus a MISR has masking characteristics that are quite similar to those of a single-input signature analyzer in that all undetectable error polynomials are represented by a nonzero multiple of $G(x)$. Now let

$$E(x) = x^{n-1}E_0(x) + x^{n-2}E_1(x) + \cdots + xE_{n-2}(x) + E_{n-1}(x)$$

so that

$$S_{en}(x) = E(x) \bmod G(x)$$

Since $E(x)$ is of degree $m + n - 2$, it has $2^{m+n-1} - 1$ possible nonzero values. Since $G(x)$ is of degree m, it has $2^{n-1} - 1$ nonzero multiples that result in polynomials of degree less than or equal to $m + n - 2$. Hence there are $2^{n-1} - 1$ cumulative error polynomials that map into the same signature as the fault-free circuit.

Then the probability that a MISR using $G(x)$ as the divisor will not detect an error is

$$P(M) = \frac{2^{n-1} - 1}{2^{m+n-1} - 1} \simeq \frac{1}{2^m}$$

again on the assumption that all possible error polynomials are equally likely.

There are some special types of error polynomials about which additional statements can be made.

Single Cycle Errors. Recall the error polynomial that represents the erroneous outputs of an m-output circuit at an arbitrary test cycle i:

$$E_i(x) = e_{i, m-1}x^{m-1} + e_{i, m-2}x^{m-2} + \cdots + e_{i,1}x + e_{i,0}$$

$E_i(x)$ is a "single cycle error polynomial." The greatest degree of $E_i(x)$ is $m - 1$ (if $e_{i, m-1} = 1$). But the divisor $G(x)$ is of degree m. Since the degree of $G(x)$ is greater than the degree of the largest single cycle error polynomial, no single cycle error can be a multiple of $G(x)$. Hence all errors that occur only in a single cycle will be detected. (Hence, also, all single-bit errors will be detected.)

Single-Output Errors. As might be expected, if only one circuit output is affected by a particular fault, the MISR masking characteristics are the same as the single-input signature analyzer. To show this, assume a fault such that only one output of the circuit is ever in error, and let k be that output. Under this assumption, the single cycle error polynomial is

$$E_i(x) = e_{i,k}x^k, \qquad 0 \le k \le m - 1 \tag{5.3}$$

The cumulative error signature polynomial $S_{en}(x)$ defined in Equation (5.2) is

$$S_{en}(x) = \left[x^{n-1}E_0(x) + x^{n-2}E_1(x) + \cdots \right.$$
$$\left. + xE_{n-2}(x) + E_{n-1}(x)\right] \bmod G(x)$$

Substituting the single cycle error polynomial from Equation (5.3):

$$S_{en}(x) = \left[x^{k+n-1}e_{0,k} + x^{k+n-2}e_{1,k} + \cdots + x^k e_{n-1,k}\right] \bmod G(x)$$

Now define $E(x)$ as

$$E(x) = x^{k+n-1}e_{0,k} + x^{k+n-2}e_{1,k} + \cdots + x^k e_{n-1,k}$$

so that

$$S_{en}(x) = E(x)\bmod G(x)$$

Since $E(x)$ has n terms it must have $2^n - 1$ possible nonzero values. Since there are $2^{n-m} - 1$ nonzero multiples of $G(x)$ that yield a polynomial of degree less than or equal to $n - 1$ that will map into the same MISR signature as the fault-free circuit, the probability of masking when a single output is in error is the same as for a single-input signature analyzer or approximately $1/2^m$.

Error Cancellation. There is one "new" type of undetectable error that was not seen in the discussion of masking in single-input signature analyzers. Substituting the values of $E_i(x)$ from Equation (5.1) into the final error signature polynomial $S_{en}(x)$ in Equation (5.2) results in

$$S_{en}(x) = \left[x^{n-1}\left(e_{0,m-1}x^{m-1} + e_{0,m-2}x^{m-2} + \cdots + e_{0,1}x + e_{0,0}\right)\right.$$
$$+ x^{n-2}\left(e_{1,m-1}x^{m-1} + e_{1,m-2}x^{m-2} + \cdots + e_{1,1}x + e_{1,0}\right)$$
$$+ \cdots$$
$$+ x\left(e_{n-2,m-1}x^{m-1} + e_{n-2,m-2}x^{m-2} + \cdots + e_{n-2,1}x + e_{n-2,0}\right)$$
$$+ e_{n-1,m-1}x^{m-1} + e_{n-1,m-1}x^{m-2} + \cdots$$
$$\left. + e_{n-1,1}x + e_{n-1,0}\right] \bmod G(x)$$

The error polynomial $E(x)$ can be written as

$$E(x) = e_{0, m-1}x^{m+n-2} + (e_{0, m-2} + e_{1, m-1})x^{m+n-3}$$
$$+ (e_{0, m-3} + e_{1, m-2} + e_{2, m-1})x^{m+n-4}$$
$$+ (e_{0, m-4} + e_{1, m-3} + e_{2, m-2} + e_{3, m-1})x^{m+n-5}$$
$$+ \cdots$$
$$+ (e_{n-3,0} + e_{n-2,1} + e_{n-1,2})x^2$$
$$+ (e_{n-2,0} + e_{n-1,1})x + e_{n-1,0}$$

Notice that a new type of masking, called "error cancellation," can occur. In the expression for $E(x)$, consider the term

$$(e_{0, m-2} + e_{1, m-1})x^{m+n-3}$$

The coefficient $(e_{0, m-2} + e_{1, m-1})$ represents a possible error occurring on output $m - 2$ at test cycle 0 and a possible error on output $m - 1$ at test cycle 1. If both of these errors occur, the MISR will capture the error at cycle 0 from output $m - 2$, but the error will be cancelled by $e_{1, m-1}$ occurring on the next output at the next test cycle. This type of cancellation occurs before the division by the MISR, and thus is not dependent on the feedback polynomial. It is seen from the structure of $E(x)$ that error cancellation can occur at many different combinations of circuit outputs and test cycles.

The probability of masking due to error cancellation can be computed on the assumption that all error polynomials are equally likely. Let the number of applied tests n be greater than the number of circuit outputs m (this is the usual case). In the expression for $E(x)$ the highest degree term is x^{m+n-2} and the lowest is x^0 and there are $m + n - 1$ unique exponents. The exponents can be collected by the number of summed e terms with which they are associated:

two exponents $(x^{m+n-2}$ and $x^0)$ have one e term
two exponents $(x^{m+n-3}$ and $x^1)$ have two e terms
two exponents $(x^{m+n-4}$ and $x^2)$ have three e terms
two exponents $(x^{m+n-5}$ and $x^3)$ have four e terms
$\cdots$
two exponents $(x^n$ and $x^{m-2})$ have $m - 1$ e terms

The rest of the exponents have m e terms, and there are $n + m - 1 - 2(m - 1) = 1 + n - m$ of them.

As a simple example, let the responses of $n = 5$ tests on a circuit with $m = 4$ outputs be compressed in a four-stage MISR. The final error polynomial is

$$E(x) = e_{0,3}x^7 + (e_{0,2} + e_{1,3})x^6 + (e_{0,1} + e_{1,2} + e_{2,3})x^5$$
$$+ (e_{0,0} + e_{1,1} + e_{2,2} + e_{3,3})x^4 + (e_{1,0} + e_{2,1} + e_{3,2} + e_{4,3})x^3$$
$$+ (e_{2,0} + e_{3,1} + e_{4,2})x^2 + (e_{3,0} + e_{4,1})x + e_{4,0}$$

For masking to occur as a result of error cancellation, the coefficient of every unique exponent of x in $E(x)$ must be equal to 0, modulo 2. Consider the term

$$(e_{0, m-3} + e_{1, m-2} + e_{2, m-1})x^{m+n-4}$$

in $E(x)$. There are 2^3 or eight possible values for the three terms in the coefficient, and four of the values will cause error cancellation (they are 000, 011, 101, and 110). Specifically, a term having k e terms will have 2^{k-1} cancellation patterns. Table 5.4 lists the cancellation patterns for each term of a general error polynomial.

The product over the error cancellation patterns in Table 5.4 is $2^{(m-1)(n-1)}$ and is the number of possible sequences in which error cancellation can occur. One of these sequences is the all-0 sequence, in which no errors at all occur, and this represents the good machine pattern. Thus the number of error polynomials in which all errors cancel is $2^{(m-1)(n-1)} - 1$.

Since n tests are applied to the m-output circuit, there are 2^{mn} possible sequences of output responses, one of which is the expected response. Then

TABLE 5.4   Number of Error Cancellation Patterns per Term

x Value	e Terms	Cancellation Patterns
x^{m+n-2}	1	1
x^{m+n-3}	2	2
x^{m+n-4}	3	4
x^{m+n-5}	4	8
...	...	...
x^n	$m-1$	2^{m-2}
x^{n-1}	m	2^{m-1}
x^{n-2}	m	2^{m-1}
...	...	...
x^{m-1}	m	2^{m-1}
x^{m-2}	$m-1$	2^{m-2}
...	...	...
x^3	4	8
x^2	3	4
x^1	2	2
x^0	1	1

the proportion of error sequences in which error cancellation can occur is

$$\frac{2^{(m-1)(n-1)} - 1}{2^{mn} - 1}$$

or approximately 2^{1-m-n}.

Hence the proportion of error polynomials that lead to error cancellation is 2^{1-m-n}. This can be considered a probability only if all error polynomials are equally likely.

Since error cancellation is highly unlikely, the masking characteristics of multiple-input signature analyzers are very nearly as good as those of single-input analyzers.

5.9   MISR AS A SINGLE-INPUT SIGNATURE ANALYZER

A MISR can also be analyzed as if it were a single-input signature analyzer. Earlier it was shown that the remainder (or signature) in the MISR after all n tests were compressed is

$$S_n(x) = \left[x^{n-1}R_0(x) + x^{n-2}R_1(x) + \cdots \right.$$
$$\left. + xR_{n-2}(x) + R_{n-1}(x) \right] \bmod G(x)$$

if the initial state of the MISR were all zeros. $R_i(x)$ is the $(m-1)$th-degree polynomial representing the test responses on the m outputs of the circuit after test cycle i or

$$R_i(x) = r_{i,m-1}x^{m-1} + r_{i,m-2}x^{m-2} + \cdots + r_{i,1}x + r_{i,0}.$$

Substituting $R_i(x)$ into $S_n(x)$ results in

$$S_n(x) = \left[x^{n-1}\left(r_{0,m-1}x^{m-1} + r_{0,m-2}x^{m-2} + \cdots + r_{0,1}x + r_{0,0} \right) \right.$$
$$+ x^{n-2}\left(r_{1,m-1}x^{m-1} + r_{1,m-2}x^{m-2} + \cdots + r_{1,1}x + r_{1,0} \right)$$
$$+ \cdots$$
$$+ x\left(r_{n-2,m-1}x^{m-1} + r_{n-2,m-2}x^{m-2} + \cdots + r_{n-2,1}x + r_{n-2,0} \right)$$
$$+ r_{n-1,m-1}x^{m-1} + r_{n-1,m-2}x^{m-2} + \cdots$$
$$\left. + r_{n-1,1}x + r_{n-1,0} \right] \bmod G(x).$$

By rearranging terms,

$$S_n(x) = \left[x^{m-1}\left(r_{0,m-1}x^{n-1} + r_{1,m-1}x^{n-2} + \cdots + r_{n-2,m-1}x + r_{n-1,m-1}\right) \right.$$
$$+ x^{m-2}\left(r_{0,m-2}x^{n-1} + r_{1,m-2}x^{n-2} + \cdots + r_{n-2,m-2}x + r_{n-1,m-2}\right)$$
$$+ \cdots$$
$$+ x\left(r_{0,1}x^{n-1} + r_{1,1}x^{n-2} + r_{2,1}x^{n-3} + \cdots + r_{n-2,1}x + r_{n-1,1}\right)$$
$$+ r_{0,0}x^{n-1} + r_{1,0}x^{n-2} + r_{2,0}x^{n-3} + \cdots$$
$$\left. + r_{n-2,0}x + r_{n-1,0}\right] \bmod G(x).$$

Now let F_j be the sequence

$$F_j = r_{0,j}, r_{1,j}, r_{2,j}, \ldots, r_{n-2,j}, r_{n-1,j}$$

and let $F_j(x)$ be the polynomial

$$F_j(x) = r_{0,j}x^{n-1} + r_{1,j}x^{n-2} + r_{2,j}x^{n-3} + \cdots + r_{n-2,j}x + r_{n-1,j}$$

Then $S_n(x)$ becomes

$$S_n(x) = \left[x^{n-1}F_{m-1}(x) + x^{m-2}F_{m-2}(x) + \cdots \right.$$
$$\left. + xF_1(x) + F_0(x)\right] \bmod G(x)$$

Each of the F_j represents the bit sequence entering a single input of the MISR, and the subscript j indicates the particular input. F_0 is the sequence entering stage zero of the MISR (see Figure 5.10), F_1 is the sequence entering stage one, and so on. Now in $S_n(x)$ consider the sum of the last two terms $xF_1(x) + F_0(x)$. It was noted in Chapter 3 that multiplying a vector by x was equivalent to a shift of the vector by one place to the right. The sum then represents the modulo 2 addition of F_0 and a right shift of F_1. For example, let $F_0 = 1010101$ and let $F_1 = 1110001$. Then $xF_1(x) + F_0(x)$ represents the operation

$$xF_1(x) \leftrightarrow 1110001$$
$$F_0(x) \leftrightarrow 1010101$$
$$\overline{xF_1(x) + F_0(x) \leftrightarrow 11011011}$$

Similarly the sum $x^2F_2(x) + xF_1(x) + F_0(x)$ represents the modulo 2 addition of a two-shifted F_2, a one-shifted F_1, and F_0. Letting F_2 be 0011001, the sum is represented by

$$x^2F_2(x) \leftrightarrow 0011001$$
$$xF_1(x) \leftrightarrow 1110001$$
$$F_0(x) \leftrightarrow 1010101$$
$$\overline{x^2F_2(x) + xF_1(x) + F_0(x) \leftrightarrow 110101111}$$

It is apparent that the sum

$$F(x) = x^{m-1}F_{m-1}(x) + x^{m-2}F_{m-2}(x) + \cdots + xF_1(x) + F_0(x)$$

represents the modulo 2 summation of time-shifted copies of the individual F_j and that the MISR signature is

$$S_n(x) = [F(x)]\bmod G(x).$$

That is to say that the MISR signature $S_n(x)$ is exactly the same signature that would be obtained from a single-input signature analyzer having the same feedback as the MISR and having $F(x)$ as its stage zero input. Notice that this conclusion requires first that the MISR implement true polynomial division (has feedback similar to that shown in Figure 5.9) and second that the MISR stages are all initially zero.

5.10 RECIPROCAL POLYNOMIALS

In Chapter 3 it was noted that the reciprocal of a polynomial $G(x)$ is $G^*(x) = x^m G(1/x)$, where m is the degree of $G(x)$, and that the sequence generated by an LFSR constructed with $G^*(x)$ is exactly the reverse of that obtained from an LFSR using $G(x)$. This point is emphasized in Figure 5.12 in which the LFSR implementing $x^3 + x + 1$ (register A) generates the repeating sequence 0010111 as the output of stage A3, while the reciprocal (register B) generates the sequence 1110100 as the output of stage B3.

Along with this reversal of the output there is a corresponding reversal of the state sequences of the two registers. Reversing each state of register B but

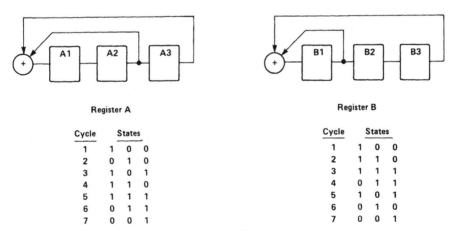

Register A

Cycle	States		
1	1	0	0
2	0	1	0
3	1	0	1
4	1	1	0
5	1	1	1
6	0	1	1
7	0	0	1

Register B

Cycle	States		
1	1	0	0
2	1	1	0
3	1	1	1
4	0	1	1
5	1	0	1
6	0	1	0
7	0	0	1

Figure 5.12. Implementing $x^3 + x + 1$ and its inverse.

listing them in their original order produces

Cycle	Register B Reversed States
1	001
2	011
3	111
4	110
5	101
6	010
7	100

The reversed states of register B are exactly the states of register A in inverse order. When register B is being clocked forward, it is generating the same state sequence as A but in time-reversed order. This section describes two useful applications of this characteristic of reciprocal polynomials.

Zero Signature. A typical self-test process involves loading an arbitrary but repeatable initial value (which may be all zeros) into the signature analyzer, applying a sequence of pseudorandom test stimuli, unloading the signature left in the analyzer at the end of the test, and checking it against the correct signature as previously computed.

The unload and checking portions of the test process can be eliminated by choosing the initial analyzer value such that the final test signature is a constant regardless of the circuitry being tested. This final constant test signature can easily be checked by a built-in checking circuit wired to the stages of the analyzer. The use of the all-zeros word as the final test signature is emphasized since checking its correctness requires only an OR tree driven by the outputs of each of the analyzer stages; however, the method described can be used to obtain any desired final signature.

Let S_f be the desired final test signature. The procedure for obtaining S_f is:

1. Initialize the analyzer to zero and compute the final fault-free test signature S using the actual test patterns.
2. Compute S_1 as the sum modulo 2 of S and S_f.
3. Set all analyzer inputs to zero so that it behaves in its autonomous manner. Compute the initial analyzer value S_2 that will result in a final analyzer state equal to S_1 after L cycles, where L is the number of test patterns applied to the circuit during the test procedure. Then using S_2 as the initial analyzer value (at the start of the test), the final fault-free signature in the analyzer after test completion will be S_f.

If a final test signature S_f of zero is chosen, then $S_1 = S$. When S_f is zero, it is

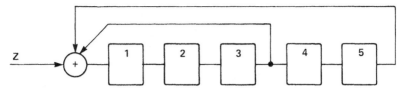

Figure 5.13. Signature register with a primitive feedback polynomial.

only necessary to form the OR function of the analyzer stage outputs to determine if a fault has been detected during the test. The analyzer hardware would be modified to include an OR tree driven by each of its stages. The circuit passes the test if the output of the OR tree is zero at the end of the test. If the output of the OR tree is 1, a fault has been detected and diagnosis may proceed from that point.

The procedure depends only on the linearity of the analyzer and is independent of its actual feedback. A proof of the procedure is given in McAnney and Savir (1986). An example will be given here to show the use of the reciprocal polynomial.

The example uses the single-input five-stage signature analyzer given in Figure 5.13. The analyzer is first cleared (initialized to zero) and the fault-free input sequence Z is applied. The final state of the register is S.

S_1 is computed as the sum modulo 2 of S and S_f. S_2 is the Lth predecessor of S_1 in the autonomous behavior of the analyzer of Figure 5.13. To avoid a backward search for S_2, the simplest approach is to simulate the reciprocal autonomous register, seed it with S_1, and run it forward for L cycles. The analyzer implements $G(x) = 1 + x^2 + x^5$, and its reciprocal $G^*(x) = 1 + x^3 + x^5$. The reciprocal register is shown in Figure 5.14. Note that the coordinate system has been changed in Figure 5.14 in accordance with the earlier observation that the time-inverse state sequence is obtained by the reverse of the individual states of the reciprocal register. By loading the reciprocal register with S_1 and running it for L cycles, the final state of the register will be the desired S_2.

Finally, initializing the analyzer of Figure 5.13 with S_2 and applying Z, the final signature will be S_f.

As an example, suppose $Z = 01001011100$. Initialize the analyzer of Figure 5.13 to the all-zeros state and compress the sequence Z (starting with its

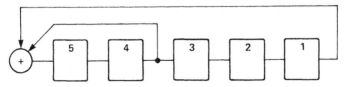

Figure 5.14. Reciprocal autonomous LFSR for generating S_2.

leftmost bit). After all 11 bits of Z have been entered, the signature $S = 10110$ (reading the register contents beginning with stage 1). If the desired S_f is zero, $S_1 = S$. Seeding the autonomous LFSR of Figure 5.14 with S_1 results in state $S_2 = 10011$ after 11 clock cycles. (Care must be taken in seeding the LFSR and reading its final state because of the coordinate reversal. The stage marked 1 in Figure 5.14 is initialized with the S_1 value from the stage marked 1 in Figure 5.13, and so on for the other stages.) Now if the original analyzer (Figure 5.13) is initialized to S_2, the final signature S_f from input sequence Z is all zeros.

Suppose now that the desired S_f is something other than zero, perhaps 11111 so that an AND gate can be used to check the correct signature. If the input sequence Z remains the same, S does not change (it is still 10110), but S_1 becomes 01001, the modulo 2 sum of S and S_f. Using S_1 to seed the reciprocal register and applying 11 clock cycles, the appropriate S_2 is 10111.

Notice that if the analyzer implements a primitive polynomial, when $L > 2^m$ the reciprocal register need only be clocked $L \bmod (2^m - 1)$ times instead of L times, since the register cycle length is $2^m - 1$.

Identifying the Failing Test. If a faulty signature is observed after a test and if the circuit must be repaired, there are two major diagnostic methods: a moveable signature probe, used to backtrace to the failing component, and fault simulation on the failing tests. Identification of the failing tests is necessary in the second method and would be useful in the first. Assuming that only a single test in the sequence has failed, there is a method using the reciprocal of the signature analyzer polynomial for identifying the failing test. The method operates on the signature of the fault-free circuit S and the observed incorrect signature S'. It begins with the error signature $S_e = S + S'$ and tries to identify an error polynomial E that will fully explain S_e.

Let m be the length of the input data stream and let n be the degree of the divisor polynomial. Suppose now that only a single test has failed. The error polynomial E consists of a single 1 somewhere in a field of $m - 1$ 0's. Then $E(x) = x^i$, an error in the ith position of E (counting from the left and numbering the leftmost position zero). The method tries to find i.

Now if the signature analyzer is initialized to all 0's and the error polynomial E is applied, then the final state of the analyzer will be the error signature S_e. The essence of the method is to note that in the analyzer under these circumstances the $(m - i)$th predecessor of the error signature S_e is the n-bit pattern $100 \cdots 00$, since this pattern can only occur at the test during which the single 1 in E entered the analyzer.

The simplest procedure for computing i is to implement (in hardware or software) the inverse of the characteristic polynomial used in the signature analyzer, to seed it with the error signature S_e, and to run it forward until the pattern $100 \cdots 00$ is encountered. The number of cycles required until the pattern is encountered is $(m - i)$.

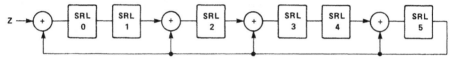

Figure 5.15. Divider implementing $G(x) = 1 + x^2 + x^3 + x^5 + x^6$.

The length m of the input stream must be restricted to be less than the cycle length of the autonomous analyzer. If the analyzer feedback is primitive, the cycle repeats itself after $2^n - 1$ patterns and if $m > 2^n - 1$ there is a possibility that the pattern $100 \cdots 00$ may occur on several cycles. The restriction is to avoid this possible diagnostic ambiguity. The length of the analyzer (and hence the length of its cycle) must be matched to the test length to meet this restriction.

For an example of the method, Figure 5.15 shows a single-input signature analyzer implementing the primitive polynomial $G(x) = 1 + x^2 + x^3 + x^5 + x^6$. Suppose that under test a fault-free single-output circuit has the 12-bit output response

$$Z = 110010100101$$

which, on entering the signature analyzer of Figure 5.15, leftmost bit first, produces the signature

$$S = 011000$$

Now let the fifth bit of Z be in error so that

$$Z' = 110000100101$$

which on entering the Figure 5.15 analyzer produces the signature

$$S' = 100011$$

Then S_e (the sum of S and S') is 111011.

The reciprocal polynomial for the analyzer of Figure 5.15 is $G^*(x) = 1 + x + x^3 + x^4 + x^6$, which is shown implemented in Figure 5.16. Notice that the register of Figure 5.16 still shifts from left to right, as does the analyzer of Figure 5.15, but that its coordinate system has been reversed as before.

The error signature S_e is used to seed the reciprocal register of Figure 5.16 (remembering that the coordinate system has been reversed and so S_e must be reversed). The inverse register is counting backwards through the test sequence, starting with the final test. For the example, the states of the inverse register are also shown in Figure 5.16. Note from Figure 5.16 that the pattern

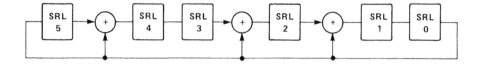

Test	States					
No.	0	1	2	3	4	5
12	1	1	1	0	1	1
11	1	0	1	1	0	1
10	0	0	0	0	0	1
9	0	0	0	0	1	0
8	0	0	0	1	0	0
7	0	0	1	0	0	0
6	0	1	0	0	0	0
5	1	0	0	0	0	0
4	0	1	1	0	1	1
3	1	1	0	1	1	0
2	1	1	0	1	1	1
1	1	1	0	1	0	1
0	1	1	0	0	0	1

Figure 5.16. Reciprocal LFSR of Figure 5.15 and the example states.

being searched for is reached at test 5, the test on which the single error occurred.

The following list summarizes the procedure for identifying one failing test out of m applied tests for single-input signature analyzers:

1. Compute the sum S_e of the fault-free signature S and the faulty signature S'.

2. Use S_e to seed the reciprocal register of the analyzer.

3. Starting at m, count down in increments of one and cycle the reciprocal register once for each count. When the reciprocal register contains a 1 in stage 0 and 0's elsewhere, the count indicates the failing test.

The preceding method applies to multiple-input signature registers (MISRs) exactly the same as described except for uncertainty as to which channel carried the fault into the MISR. Because there are multiple channels entering the MISR, the method cannot narrow down to a single failing test. A window of uncertainty, equal to the number of MISR inputs, is unavoidable. For a MISR of k stages, the method will only say that the failing test occurred somewhere between test j and test $j + k - 1$ inclusive, thus bounding the failing test. This window of uncertainty for MISRs is unfortunate but the method still (on the assumption of a single test failure) reduces dramatically the search space for the actual failing test.

5.11 WHAT POLYNOMIAL SHOULD BE USED FOR SIGNATURE ANALYSIS?

It was noted earlier that a MISR of n stages has a masking probability approximately equal to $1/2^n$ for equally likely error patterns and long data streams. The simplest degree n divisor polynomial is x^n. It was also shown that a divisor polynomial with two or more nonzero coefficients will not mask single-bit errors. To satisfy this requirement, the simplest polynomial is $x + 1$. Considering this, why not use $x^n + 1$ as the divisor? The MISR would then be the cycling register shown in Figure 5.17 in which the n outputs from the circuit under test are labeled "In 0" to "In $n - 1$." (Incidentally, this register dramatically illustrates the effects of error cancellation. Suppose that every one of the n circuit outputs is incorrect on two successive tests k and $k + 1$. After test k each SRL is incorrect. Then test $k + 1$ adds an incorrect input value to each incorrect SRL value, and the two test errors cancel.)

In addition to error cancellation, there is still the certainty that a divisor polynomial $G(x)$ will mask an error polynomial $E(x)$ that is a multiple of $G(x)$. The current trend in signature analysis is to incorporate a rather complex feedback structure into the LFSR, on the assumption that a multi-term $G(x)$ will prevent masking by anything but an equally complex $E(x)$. For example, the Hewlett-Packard 5004A signature analyzer uses a 16-stage LFSR with $G(x) = x^{16} + x^9 + x^7 + x^4 + 1$ [as described in Frohwerk (1977)].

There is also a trend toward using, as the feedback connection, a structure that results in a maximum-length shift-register sequence. An n-stage signature analyzer with such a feedback structure, a constant 0 on its input, and a nonzero initial state, will cycle through all possible $2^n - 1$ nonzero states before repeating. These maximum-length LFSRs were discussed in Chapter 3.

The very unsatisfying conclusion is that the $G(x)$ chosen should be such that no expected fault in the circuit will create, as its error polynomial, a multiple of $G(x)$. Since this recommendation is difficult to satisfy except through exhaustive simulation, the accepted truth in the field is to make $G(x)$ reasonably complex.

As an example, a nonprimitive feedback polynomial may be used to guarantee detection of any odd number of errors (any error polynomial having odd parity). A single- or multiple-input signature register can be constructed

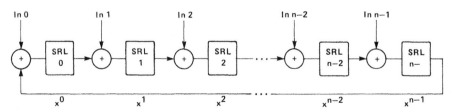

Figure 5.17. MISR implementing $x^n + 1$.

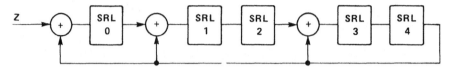

Figure 5.18. Signature analyzer implementing $f(x) = 1 + x + x^3 + x^5$.

so that the parity of its signature is the same as the parity over the entering data streams. This equivalence can be used to protect against the possibility of masking when the error polynomial has odd parity.

When a divisor polynomial $f(x)$ is the product of the parity generator polynomial $g(x) = 1 + x$ and a primitive polynomial $p(x)$ of degree $n - 1$, the n-stage MISR implementing $f(x)$ as a true divider has the property that the parity of the signature after compression of the input data streams is also the parity over all the bits in the input streams. [The idea of implementing the product of a generator and a primitive polynomial is also used in Hsiao et al. (1977) for a different purpose.]

As an example, let $p(x) = 1 + x^3 + x^4$. Then $f(x) = (1 + x)p(x) = 1 + x + x^3 + x^5$. A signature analyzer implementing $f(x)$ is shown in Figure 5.18. An example of its use may be illustrative. Let the analyzer be initialized to zero and assume a sequence of six tests that causes the input values Z listed in Table 5.5. The analyzer states are shown after clocking in each bit of Z. Notice that at any test the parity of the analyzer contents reflects the parity history of the entire sequence that entered the analyzer up to that test. The important point is that the signature of the input sequence has the same parity as the sequence itself.

This assumes that the analyzer was initialized to zero. This result, that the analyzer parity reflects the parity history of all inputs, holds whenever the initial state of the analyzer is any even value (as for example zero). If the initial state is odd, on the other hand, the parity relationship is reversed. With an odd starting state, if the signature parity is even, the inputs had odd parity and vice versa.

Now if a faulty circuit causes an odd number of errors in the input sequence Z to create a faulty sequence Z', the parity of Z' will differ from the

TABLE 5.5   An Example Input Sequence and Its Register States

Test	Z	Register	Parity
1	1	10000	Odd
2	1	11000	Even
3	0	01100	Even
4	1	10110	Odd
5	0	01011	Odd
6	1	01111	Even

parity of Z. Because of the parity correspondence, then, the parity of the faulty signature will differ from the parity of the fault-free signature. Hence an analyzer constructed as shown will guarantee that masking will not occur for odd numbers of errors. Any even number of failing tests, however, will result in the same parity of Z and Z' and masking may occur, depending on whether or not $E(x)$ is a multiple of $f(x)$.

5.12 IMPROVING THE EFFECTIVENESS OF SIGNATURE ANALYSIS

The effectiveness of signature analysis is conditional upon its probability of masking. The masking probability can never be reduced to zero except by abandoning data compression and using the entire test response sequence as the signature. Masking is discouraging because it implies the loss of hard won information. The test sequences are good enough to be able to detect a particular fault, but the erroneous output sequence is compressed into the same signature as the correct one and vanishes. Furthermore, there is little to say about the probability of masking without some (usually unpalatable) assumptions about the error polynomials.

It would be fruitful to suggest means by which the masking probability of LFSR signature analysis can be reduced, independent of the types of errors exhibited by the circuit. Necessarily this is not possible, but a few possibly useful enhancements can be suggested, without much formal consideration of their qualitative advantages.

Lengthen the LFSR. Under the postulate that all possible error polynomials are equally likely, the probability of masking for an r-stage LFSR is $1/2^r$. Adding one stage to the LFSR reduces the probability of masking to $1/2^{r+1}$, a factor of 2 reduction from a modest increase in hardware overhead.

Repeating the Test. One obvious technique is to repeat the test sequence twice, using a different feedback polynomial on the signature analyzer each time. (Alternatively, two signature analyzers with different polynomials can be used on a single test pass, trading off hardware costs for reduced test time.) Another alternative is to provide an adjustable length LFSR, perhaps length r and length $r + 1$, with a facility for changing the length for the second repeat of the test. Third, it may be possible to reverse the test sequence, perhaps by using the reciprocal of the test generator, so as to apply the same tests in reverse order for the second test pass. A masking error polynomial $E(x)$ on the first test pass is replaced by its reciprocal on the second pass, with a significant potential for reduction in masking probability.

Taking More Signatures. Except for the error cancellation discussed earlier in connection with MISRs, masking occurs when an error already "trapped" in the signature analyzer is "corrected" by a later test error. It is possible to improve the error coverage by interrupting the test sequence periodically and checking the current signature, of course at the cost of some little test time and added good signature storage.

For example, suppose a data stream of length L bits is being compressed in an analyzer of degree r. It was shown earlier that the probability of masking, assuming equally likely error polynomials, was $1/2^r$. Now let the data stream be divided into two portions of lengths L_1 and L_2 and let the analyzer signature be checked at the end of each portion. The fault-free signatures are S_1 and S_2, respectively. The number of L_1 sequences that will produce S_1 is

$$2^{L_1 - r}$$

For each of these sequences, the number of L_2 sequences that will produce S_2 is

$$2^{L_2 - r}$$

Hence the number of L-bit sequences that will produce both S_1 and S_2 is

$$2^{L_1 - r} \times 2^{L_2 - r} = 2^{L - 2r}$$

Then if all error polynomials are equally likely, the probability of masking for both S_1 and S_2 is

$$P(M) = \frac{2^{L - 2r} - 1}{2^L - 1} \simeq \frac{1}{2^{2r}} = \frac{1}{4^r}$$

Monitoring the Quotient Bit. The LFSR shown in Figure 5.1 continuously divides the input bit stream by a divisor $G(x)$, the quotient being shifted off the right-hand end of the LFSR and the remainder being left in the LFSR stages after the input stream has completed shifting. If on every shift cycle the quotient bit were collected in some auxiliary storage, the quotient and the remainder could obviously be used to reconstruct the input bit stream. For a single-input signature analyzer there is no advantage in this process over a normal test process that continuously checks each circuit output bit for correctness. Compressing the quotient bits from a single-input m-stage signature analyzer in a separate n-stage analyzer has no advantage over merely increasing the length of the initial analyzer from m to $m + n$ stages. Checking the quotient bit on a MISR, on the other hand, does have some advantages [see Sridhar et al. (1982)]. Alternatively, if multiple MISRs are used, the quotient bits from each MISR can be compressed in an auxiliary MISR [Hassan et al. (1983)].

Variable-Shift MISR. An error captured in the MISR during a particular test cycle can be cancelled by an error appearing at a later test cycle on a different MISR input as was shown in the earlier discussion of error cancellation. These correlated errors can occur because of the repeated use of a faulty circuit that fans out to several circuit outputs. A reduction in masking probabilities for such errors is suggested in J. L. Carter (1982), where the MISR is modified to permit either one or two shifting clock cycles between each test response from the circuit. The choice of a single or double shift is made randomly by a signal on an added control line.

Further studies in the area of improving the masking characteristics of signature analysis are given in Agarwal (1983), Bhavsar and Krishnamurthy (1984), W. C. Carter (1982b), and Hassan and McCluskey (1984).

5.13   SEQUENTIAL CIRCUITS

Up to this point signature analysis has been considered for combinational circuits. There is no serious problem in using signature analysis for sequential circuits if the circuit can be initialized to a known (or at least repeatable) state so that the "good" signature is consistent across different test encounters with the same circuit part number [see Bhavsar and Heckelman (1981)]. Even this requirement may not be absolutely necessary if, with power-on, a sequential circuit has only a few possible starting states. The final circuit signature can then be compared to a short list of good signatures, one for each of the possible starting states. This obviously affects the masking probability of the signature compression, but this slight increase may be a small price to pay for the simplicity of initialization and, in any event, can be overcome by increasing the length of the signature register.

Diagnosis using a moveable signature probe in a backtrace procedure is also feasible with sequential circuitry if care is taken to avoid getting trapped by a fault in a global feedback loop that causes all probe nodes in the loop to exhibit incorrect signatures. An obvious cure is to break the loops before beginning the backtrace.

5.14   GENERATION OF THE GOOD MACHINE SIGNATURE

Signature testing on a newly built part requires knowing the correct or good machine signature for the part. The signature can be obtained in several ways. From a test engineering approach the simplest is to take a part which has already passed functional testing in the system, run the part against the actual test patterns, and save its signature as the reference for new production.

However, if testing is required before system installation, the simplest approach (in concept at least) is to simulate both the circuit and the signature analyzer against the actual test patterns. While this is only good machine simulation, the sheer number of test patterns coupled with a large circuit may make it prohibitively expensive in terms of computer time.

An alternative to complete simulation is to simulate up to some breakpoint, say at the first 1000 tests, and to obtain the expected signature at that point. Now the first 1000 tests are applied to actual hardware and the hardware signature is checked against the simulated breakpoint signature. A mismatch between the two requires that the hardware be reworked until it passes the checkpoint test. When the hardware passes the checkpoint test, there is some greater-than-zero probability that it is, in truth, fault-free. Now the remainder of the test patterns are applied to obtain the final hardware signature which is saved as the putative good machine signature.

If three copies of the hardware are available, bootstrapping can be used to obtain the signature. The first n tests are applied to each of the copies and their signatures are checked against one another. If all three signatures match, then the next n tests are applied. If two out of three signatures match, the third copy is assumed to be faulty and is reworked until its signature matches that of the other two. If all three signatures are different, the number of tests n is reduced until there is a match on at least two out of the three. (The end of the line for a bootstrap approach is when one test is applied and all three copies have different signatures. At that juncture the test engineer either simulates the first test to get a known good signature or asks production for a fourth copy of the part.)

5.15 SUMMARY

Signature analysis using selective feedback of various stages of a shift register fed by the data stream being created by a circuit is a powerful technique for coping with a large volume of built-in test response data. The major concern with signature analysis, or indeed with any form of data compression, is the possibility that an error pattern created by faults will be mapped inadvertently into the correct signature. A theoretical analysis of this failure mechanism depends upon knowledge, usually unavailable, of the error patterns caused by every possible circuit fault. Experience with signature analysis has shown, however, that it catches most, if not all, real errors, and despite the lack of a satisfying theoretical basis it is the favored data compression technique for built-in test.

6

Special Purpose Shift-Register Circuits

In Chapter 3 the properties of linear feedback shift registers used as sequence generators were examined in detail. However there are many special cases of linear and nonlinear feedback shift-register implementations that have been used in special situations. A sampling of those examples that have promise for use as stimulus generators for built-in test are described in the following.

6.1 LINEAR FEEDBACK SHIFT REGISTERS

The general properties of LFSRs have been discussed previously, but will be briefly reviewed here. A standard implementation was used for purposes of illustration. This is repeated below as Figure 6.1(a). In Figure 6.1(b), the LFSR implements the reciprocal characteristic polynomial. It was noted earlier that the sequences generated by the two circuits are the exact reversal of one another. Note that the sequence generated by the LFSR in Figure 6.1(a) is $\cdots 00001101010010001011111101100111 \cdots$ and the sequence of Figure 6.1(b) (reciprocal polynomial) is $\cdots 00001110011011111101000100101011 \cdots$, the exact reversal, as exemplified by the reversal about the unique marker '11111'. In an LFSR having a characteristic polynomial with many terms, it is sometimes advantageous to use a different form to reduce the delay that is introduced by a conventional cascade of two-input EXCLUSIVE ORs that are needed in the form shown in Figure 6.1. In this case, the LFSR form of Figure 6.2 can be used. This is the form of LFSR shown in Figure 5.1 to perform division of an incoming bit stream except that the EXCLUSIVE OR at the input of stage 1 is missing. The divider circuit in Figure 6.2 exhibits the exact output sequence as the circuit in Figure 6.1(b). In other words, the feedback is

145

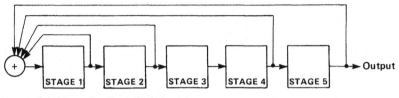

(a) **Linear Feedback Shift Register**

Characteristic Polynomial: $x^5 + x^4 + x^2 + x + 1$

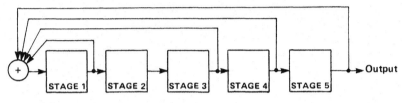

(b) **LFSR with Reciprocal Characteristic Polynomial**

Characteristic Polynomial: $x^5 + x^4 + x^3 + x + 1$

Figure 6.1. Linear feedback shift registers with reciprocal characteristic polynomials.

inserted at the points that would have been tapped for the reciprocal implementation in the standard LFSR form. In the case of LFSRs with only two
taps (those implementing trinomial characteristic polynomials) the circuit is
such that a circuit of the type shown in Figure 6.1(b) is topologically
equivalent to that in Figure 6.2 and can be put into that form by a planar
rearrangement. Circuits with more than two taps cannot be so reconfigured in
a planar manner.

To facilitate the discussion to follow, a convenient octal notation to
describe specific binary polynomials will be introduced. While it is easy to
write

$$f(x) = x^5 + x^2 + 1$$

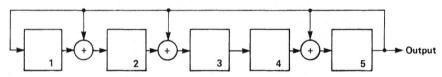

Characteristic Polynomial: $x^5 + x^4 + x^3 + x + 1$

Figure 6.2. Alternate implementation of LFSR.

it gets clumsy to write

$$f(x) = x^{16} + x^{12} + x^9 + x^7 + 1$$

or, worse yet,

$$f(x) = x^{34} + x^{32} + x^{30} + x^{27} + x^{24} + x^{22} + x^{21} + x^{19} + x^{17} + x^{15}$$
$$+ x^{12} + x^{11} + x^{10} + x^8 + x^6 + x^5 + x^4 + x^3 + x^2 + x + 1$$

If the binary coefficients $\{c_i\}$ of the previous 16th-degree polynomial are listed in the same order as in the polynomial (high order first) we have

$$\{c_i\} = 10001001010000001$$

If these coefficients are now grouped in threes (from the right or low-order position), each group can be replaced by an octal digit:

$$\{c_i\} = 010|001|001|010|000|001$$

$$= 2 \quad 1 \quad 1 \quad 2 \quad 0 \quad 1$$

Thus a standard octal notation for the example polynomial is 211201_8. In like manner, the 5th-degree polynomial is denoted 45_8 and the 34th-degree polynomial is 251132516577_8.

6.2 DE BRUIJN COUNTER

A maximal-length LFSR sequence has length $m = 2^n - 1$, where n is the number of stages in the LFSR. As discussed in Chapter 3, the all-zero n-tuple is missing. It is often desirable to convert the m-sequence to a de Bruijn sequence (see Section 3.4) by adding the all-zero n-tuple. If the complements of the $n - 1$ stages $1 \cdots n - 1$ are ANDed together and EXCLUSIVE ORed with the feedback used to implement the m-sequence, the all-zero n-tuple is inserted between the $00 \cdots 01$ state and the $100 \cdots 00$ state, thus realizing a de Bruijn sequence. Recognizing that $\overline{A} \cdot \overline{B} = \overline{(A + B)}$, Figure 6.3(a) is a direct implementation of a de Bruijn sequence. If it is desired to keep the fan-in to the EXCLUSIVE OR the same as for the m-sequence implementation, the complements of the $n - 1$ stage values can be fed to an AND and that signal ORed with any one of the feedback lines. This composite signal is then fed into the EXCLUSIVE OR. Figure 6.3(b) is such a circuit. [The connection from the complemented output of the first (leftmost) stage of the shift register shown in Figure 6.3 is actually redundant, as can be seen by simple logic minimization procedures.]

Often it is required to have the all-zero n-tuple as part of a sequence exactly 2^n bits long. In those cases the implementation of Figure 6.3 is suitable.

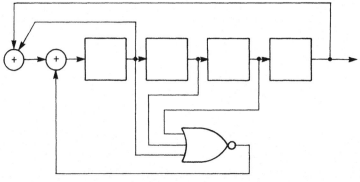

(a) **NOR Implementation**

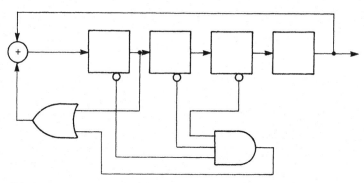

(b) **Constant XOR Implementation**

Figure 6.3. de Bruijn sequence generator. (a) NOR implementation. (b) Constant XOR fan-in implementation.

Obviously, the all-zero n-tuple is also present in an m-sequence of length $2^{n+1} - 1$ obtained by increasing the length of an LFSR from n stages to $n + 1$ stages. In many applications a satisfactory sequence can be obtained with a suitable starting seed in such an augmented LFSR. This often results in a much simpler circuit than a de Bruijn generator.

Another requirement is sometimes encountered where a sequence exactly 2^n bits long is required, and it must be exactly reversible. This suggests switching between feedback connections that implement reciprocal polynomials. Such a circuit is shown in Figure 6.4.

In this circuit, the feedback is selected by the FWD/REV signal. When FWD/REV is 1, the feedback is selected from stage 3, implementing the polynomial $x^4 + x^3 + 1$. When the FWD/REV signal is 0, the feedback is obtained from stage 1, thus implementing the reciprocal polynomial $x^4 + x + 1$. In this way, the sequence can be exactly reversed at any point.

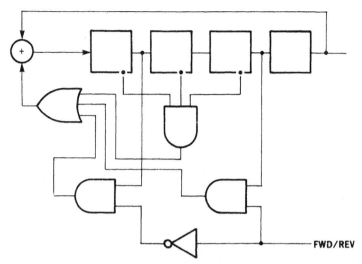

Figure 6.4. Reversible de Bruijn sequence generator.

6.3 CONCATENATABLE LFSRS

With the development of bit-sliced processors and other variable word size systems, it is often necessary to use different length LFSRs in different parts of a system. Bhavsar (1985) recognized a property of certain primitive polynomials that allows a building block approach by concatenation of short LFSRs to form longer LFSRs. Given two polynomials A and B of degrees r and s, respectively, if the polynomial formed by either of the concatenations

$$C = x^s(A + 1) + B$$
$$C = x^r(B + 1) + A$$

is primitive, the operation is called a *primitive concatenation*. There are no 3rd- or 4th-degree primitive polynomials that exhibit this property, but the 5th-degree polynomial 45_8 does. When 45_8 is concatenated with itself three times, a primitive concatenation is formed as follows:

$$
\begin{aligned}
C_1 &= x^5(x^5 + x^2 + 1 + 1) + x^5 + x^2 + 1 \\
&= x^{10} + x^7 + x^5 + x^2 + 1 \\
C_2 &= x^5(x^{10} + x^7 + x^5 + x^2 + 1 + 1) + x^5 + x^2 + 1 \\
&= x^{15} + x^{12} + x^{10} + x^7 + x^5 + x^2 + 1
\end{aligned}
$$

or 112245_8, which is primitive. The 8th-degree primitive polynomial 545_8 (and its reciprocal 515_8) has the property that two such polynomials form a primitive concatenation and implement the 16th-degree primitive polynomial

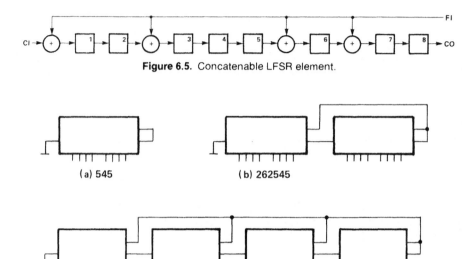

Figure 6.5. Concatenable LFSR element.

(a) 545   (b) 262545

(c) 54531262545

Figure 6.6. Primitive concatenations.

262545_8 (306515_8). Four such polynomials also concatenate to form a primitive 32nd-degree polynomial 54531262545_8 (51523246515_8). Seven of these polynomials form a primitive 56th-degree concatenation, but unfortunately, a 64th-degree primitive polynomial cannot be formed from 545_8 in this manner. However, the polynomial 545_8 is a useful building block from which 8-, 16-, and 32-degree primitive polynomials can be formed. Consider the circuit in Figure 6.5. If the terminals CO (concatenate output) and FI (feedback input) are joined, an alternative implementation (one with the EXCLUSIVE ORs in-line) of the LFSR with characteristic polynomial 545_8 is realized. The terminal CI (concatenate input) is tied to a 0 value as shown in Figure 6.6(a). To implement the 16th-degree LFSR associated with 262545_8, two such circuits are connected as shown in Figure 6.6(b). Likewise, the 32nd-degree concatenation of four circuits is shown in Figure 6.6(c). Suitable initialization of such circuits can be in parallel if the implementation allows or the feedback connection (CO–FI) can be broken (gated) and a suitable initial state can be scanned in through the CI input that is normally held at 0.

6.4 *k*-CASCADE SEQUENCE GENERATORS

At times it is desirable to bias the number of 1's in a pseudorandom bit sequence to some value other than the nearly uniform distribution of an *m*-sequence (further elaboration on the use of biased sequences is given in Section 7.6). Recall that the number of 1's in an *m*-sequence is $(m + 1)/2$ and

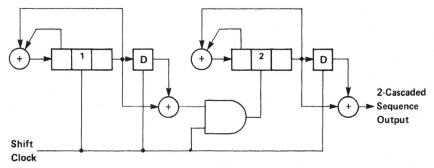

Figure 6.7. 2-cascaded sequence generator.

the frequency of 1's is $(m + 1)/2m$. For large m, this frequency approaches .5. One way to generate sequences with probability of a 1 approaching .25 is to cascade two sequence generators as described in the following. In fact, k such generators may be cascaded to achieve a frequency of 1's that approaches 2^{-k}.

The class of cascaded sequence generators described in this section has the property that an output transition of the ith LFSR is used to gate the clocking of the $(i + 1)$th LFSR. Such an arrangement is shown in Figure 6.7 for a 2-cascade. This circuit is sometimes called a "squitter" generator because one of its original uses was as a squitter in a TACAN beacon.[1] An arbitrary number of generators can be cascaded in this manner, thus the name k-cascade generator.

Note that in Figure 6.7 generator 2 is shifted only when a 1 is provided by the EXCLUSIVE OR connected to the output of generator 1. The output of each generator is fed through and around a delay, denoted D, to an EXCLUSIVE OR, so that only when there is a change in the last (output) stage of a generator will the output of the EXCLUSIVE OR be a 1. The delay is implemented as a register stage in Figure 6.7. Only three stages are shown in the LFSRs, but any degree LFSR can be cascaded.

The period of a k-cascade sequence generated by the cascade coupling of k LFSRs of arbitrary degree, each independently generating a maximum length sequence, is now derived.

Theorem 6.1. *Given a k-cascade sequence generated by the cascade coupling of k LFSRs (G_i), where n_i is the degree of G_i and $p_i = 2^{n_i} - 1$ is the period of G_i. The period of such a k-cascade sequence is $p_1 p_2 \cdots p_k$ and the number of output transitions is*

$$2^{-k} \prod_{i=1}^{k} (p_i + 1)$$

[1] Tactical Airborne Navigation system (TACAN) is a short range UHF navigation system that provides accurate slant range and bearing information. Part of its operation is governed by a squitter that randomly fires the transponder transmitter in the absence of interrogation.

Proof. The proof will be by induction on k. First consider a 2-cascade. After every p_1 shifts, G_1 returns to its initial state. In so doing, its output undergoes $(p_1 + 1)/2$ transitions (Property VI, Section 3.5). Likewise, after every p_2 gated shifts, G_2 returns to its initial state. When both generators have returned to their initial states simultaneously, the 2-cascade has completed one cycle (or period). Thus $p = p_1 p_2$. Let the number of full cycles G_1 has undergone be q, and r the number by G_2. Then

$$\frac{q(p_1 + 1)}{2} = rp_2 \tag{6.1}$$

The problem is to find the minimum integers q and r that satisfy the preceding equation. Since all the G_i's generate m-sequences, $(p_i + 1)/2$ is always prime relative to p_j, that is $(p_i + 1)/2 = 2^{n_i - 1}$ and $p_j = 2^{n_j} - 1$; therefore, the solution to (6.1) is $q = p_2$ and $r = (p_1 + 1)/2$. Thus the period of a 2-cascade sequence is $p_1 p_2$, and the output of G_2 has made

$$\frac{r(p_2 + 1)}{2} = \frac{(p_1 + 1)(p_2 + 1)}{4} = 2^{-2} \prod_{i=1}^{2} (p_i + 1)$$

transitions.

Now assume that the period of a $(k - 1)$-cascade sequence is

$$P = \prod_{i=1}^{k-1} p_i$$

and the number of output transitions is

$$T = 2^{-(k-1)} \prod_{i=1}^{k-1} (p_i + 1)$$

After every P shifts, the $k - 1$ cascade will return to its initial state. After every p_k gated shifts, G_k will return to its initial state. As before,

$$qT = rp_k \tag{6.2}$$

and $q = p_k$ and $r = T$ is the solution. Thus the period of a k-cascade sequence is

$$\prod_{i=1}^{k} p_i$$

The number of transitions that the output of the k-cascade undergoes in a period is given by $T(p_k + 1)/2$ or

$$2^{-k} \prod_{i=1}^{k} (p_i + 1) \qquad \text{Q.E.D.}$$

The number of 1's in the k-cascade sequence is given by the following corollary.

Corollary 6.1. *The number of 1's in a k-cascade sequence is*

$$2^{-k} \prod_{i=1}^{k} (p_i + 1)$$

Proof. By construction, there is a 1 output for every transition of a k-cascade generator output.   Q.E.D.

Thus k-cascaded sequences of varying period can be produced. The resultant frequency of 1's can be made arbitrarily close to 2^{-k} by increasing the length of each LFSR in the cascade.

The reader who desires more detail on the properties of k-cascade sequences is referred to Tretter (1974) and Kjeldsen and Andresen (1980).

6.5 WEIGHTED PATTERN GENERATOR

Another generator of weighted or *biased* pseudorandom sequences can be constructed by nonlinear operations combining outputs from various stages of a single LFSR. Recall that on the average, a stage of an LFSR has a 1 output half of the time. Stated somewhat differently, the probability that the output signal is a 1 is .5, or more precisely, $2^{n-1}/(2^n - 1)$. The probability that two statistically independent signals are both at 1 is the product of the probabilities of each being a 1. Thus, when two such signals are fed into an AND gate, the signal probability (probability that the line has value 1) at the output of an AND gate is the product of the signal probabilities at its inputs $P_{AND} = P_A P_B$. Thus if the outputs of two LFSR stages are fed into an AND gate, the output will have a signal probability approaching .25. The circuit shown in Figure 6.8 illustrates this concept. A more complex circuit operating on the same principle is detailed in Figure 6.9. The weight, or signal probability, of the output can be controlled by the word in the control register. The AND gate whose output is labeled 2 has inputs from two stages of the LFSR. Consequently, when position 3 of the control register contains a 1, AND gate 2 will feed a 1 to the OR gate $\frac{1}{4}$ of the time (signal probability of .25). When position 3 of the

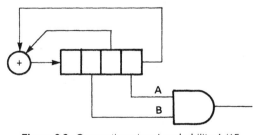

Figure 6.8. Generating signal probability 4/15.

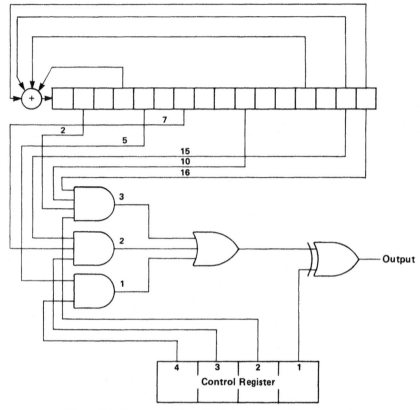

Figure 6.9. Programmable weighted random pattern generator.

TABLE 6.1 Programming PRPG

Control Register State	Approximate Signal Probability
1000	.5
0100	.25
0010	.125
0101	.75
0011	.875
1100	.625
1010	.5625
0110	.3437
⋮	

control register contains a 0, AND gate 2 is disabled and contributes nothing to the signal probability of the output. The outputs of the three AND gates, which contribute .5, .25, or .125, respectively, are selected by the value of the control register. The EXCLUSIVE OR driven by position 1 of the control register serves as an inverter. Table 6.1 shows the control register contents that are required to program the output signal probability to various values.

This relatively simple generator is an economical source of weighted pseudorandom sequences.

6.6 TWO-DIMENSIONAL SEQUENCE GENERATORS

Many built-in test schemes utilize one or the other of the scan-path features described in Chapter 2. The common complaint of long test application times due to loading long scan strings is addressed by a parallel approach. The scan strings are broken into several shorter strings to speed the test. In these structures, it is necessary to feed several scan paths with test stimuli in parallel. If the test stimuli are to be pseudorandom sequences, the characteristics of parallel sequence generators must be looked at. The problem addressed in this section is how to feed several scan paths in parallel so that the resulting array of pseudorandom patterns is an objective stimulus for the digital network that is contained within the scan path.

An *objective stimulus* is one that has the property that no conditions imposed by the structure or the sequence generator will preclude the inputs to the logic under test from observing all possible binary patterns, if enough patterns are generated. This could be called a "philosophy of possible exhaustion," in that as a k-input subnetwork is driven by a generator of this type, eventually all 2^k binary k-tuples will be presented to the network inputs. While in practice one may choose not to wait until this event is assured, this condition does not exclude the possibility at the start.

Another condition that is desirable is that no special conditions be imposed on the logic or the assignment of shift-register latches (SRLs) to specific logic inputs (no special encoding of the inputs). In particular, any circuit can be fed by any combination of the scan latches so a two-dimensional array of pseudorandom bits is required. With this condition in mind, several such generators will be discussed in the following sections.

6.6.1 LFSRs as Parallel Sequence Generators

The first implementation that comes to mind is simply to take the output from each stage of an LFSR and feed it into one of the scan paths. In this scheme, the LFSR shifting clocks and the scan-path shifting clocks are the same (Figure 6.10). This direct feed implementation results in an array of pseudorandom bits that has a *structural dependency*. If adjacent scan paths are fed from contiguous LFSR stages, the same values appear in the adjacent scan

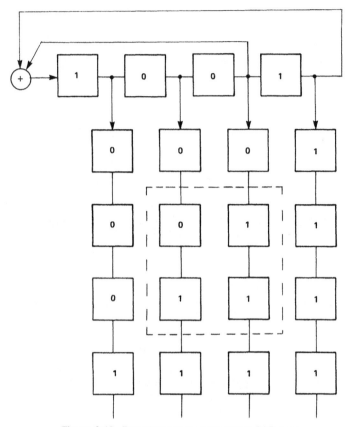

Figure 6.10. Pseudorandom array from LFSR feed.

paths offset by one shift. In particular, if a 2 × 2 window is placed anywhere on the array formed by the scan paths as shown in Figure 6.10, only half of the possible 2 × 2 binary arrays will appear as the LFSR is shifted through its cycle. If a 3 × 3 window is considered, only one-fourth of the possible binary arrays appear. Larger windows have proportionally smaller coverages. This is a serious structural dependency for built-in test applications. Consider the example shown in Figure 6.11. It is desired to test the AND gate input 1, stuck-at-1. If it is driven by the scan-path connections shown, the tests [011] and [101] never appear. Thus the circuit connected in this manner sees only half of its exhaustive test set. While this simple example could be remedied by a reassignment of scan latches for the inputs, such a solution is not allowed under the ground rules set out at the beginning of this section. Recall from that discussion, the sequence generator must be such that reassignment is not necessary to insure adequate coverage of the pseudorandom pattern set. Thus something more than simple LFSR parallel feed is required as a parallel sequence generator.

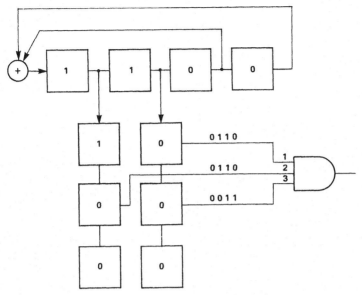

Figure 6.11. Effect of structural dependency.

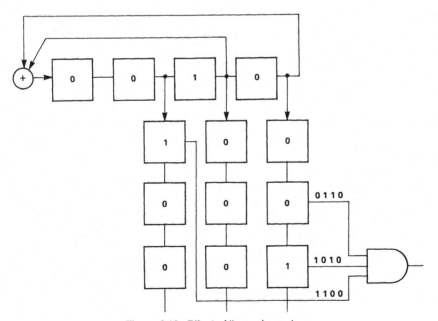

Figure 6.12. Effect of linear dependency.

In addition to the class of structural dependencies just shown, the m-sequence itself has *linear dependencies* which have their origin in the fundamental recursion that generates the sequence. Consider the example in Figure 6.12. The recursion that generates the sequence is

$$a_n = a_{n-3} + a_{n-4}$$

with characteristic polynomial

$$f(x) = x^4 + x^3 + 1$$

When the network is connected as shown, the test [111] does not occur since the third input is always the sum (modulo 2) of the first two. Here again, reassignment of the inputs to the circuit under test would avoid the problem, but a more general solution is desired.

When evaluating parallel sequence generators, the treatment of both structural dependencies and linear dependencies must be considered.

6.6.2 Segmented LFSR Generators

An alternate implementation of an LFSR is described in Hurd (1974). An example of one such generator is shown in Figure 6.13. Note that the shift-register stages have been arranged in small groups and that the feedback is structured in a particular way. This *factoring* of the latches into smaller chains and the limited feedback between chains can be of considerable utility in implementation when it is necessary or desirable to distribute the sequence generator over several different areas of a VLSI structure. The primary interest in these generators, however, arises from the fact that the phase shift between sequences generated by the small chains is generally quite large. This property will be explored in the following text.

Consider the shift-register configuration of Figure 6.13. The output bits of the first group of latches follow the equation

$$a^{(0)}_{k+q_0} = a^{(0)}_k + a^{(2)}_{k+d_2} \qquad (6.3)$$

where $a^{(0)}_k$ denotes the output bit of register (0) at time k, q_0 is the length of group 0 ($q_0 = 3$), and d_2 is the tap point of group 2 ($d_2 = 3$). Using the variable x as a delay operator to represent the time shift, $a^{(0)}_{k+q_0}$ becomes $x^{q_0} a^{(0)}_k$. Remembering that addition is modulo 2, Equation (6.3) can be written as

$$a^{(0)}_k = x^{q_0} a^{(0)}_k + x^{d_2} a^{(2)}_k \qquad (6.4)$$

Likewise, for the remaining groups:

$$a^{(1)}_k = x^{q_1} a^{(1)}_k + x^{d_0} a^{(0)}_k$$
$$a^{(2)}_k = x^{q_2} a^{(2)}_k + x^{d_1} a^{(1)}_k$$

If a set of nonzero initial conditions is assumed, this system of equations can

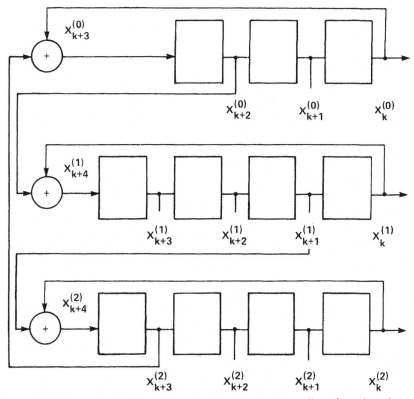

Figure 6.13. Segmented LFSR generator implementing $p(x) = x^{11} + x^8 + x^6 + x^3 + 1$.

be solved simultaneously, and the determinant yields the characteristic polynomial for the system with x as the variable:

$$f(x) = (x^{q_0} + 1)(x^{q_1} + 1)(x^{q_2} + 1) + x^{d_0 + d_1 + d_2} \qquad (6.5)$$

It can be shown that the formula for $f(x)$ generalizes to any number of *factors* N,

$$f(x) = \prod_{i=0}^{N-1} (x^{q_i} + 1) + x^{\sum d_i} \qquad (6.6)$$

Note that the terms of the characteristic polynomial are determined in large part by the factoring, and the tap points define one term which can add to the others or annihilate one of them. If the resulting polynomial has an odd number of terms, it is possibly irreducible (and possibly primitive).

As an example, consider the implementation of a 16th-degree polynomial with three factors. That is, three outputs are wanted that are large phase shifts apart. The factors (q_i's) are chosen as 5, 5, and 6. This gives a starting

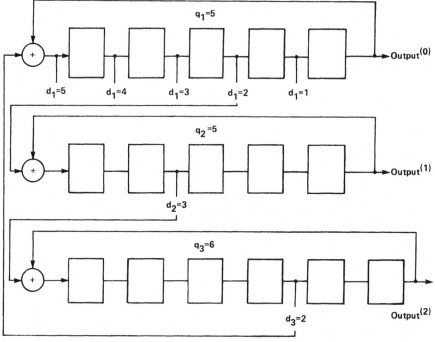

Figure 6.14. Segmented LFSR generator realizing 202301_8.

polynomial which corresponds to the product terms in Equation (6.6),

$$h(x) = x^{16} + x^{10} + x^6 + 1$$

Notice that by adding a term x^7, $h(x)$ becomes primitive ($202101_8 \rightarrow 202301_8$). This requires that the sum of the tap points (d_i's) must equal 7. One way is to use 2, 3, 2 as in Figure 6.14. The characteristics of the output sequences of this configuration will be analyzed.

In the preceding, each segment had one input from another segment. If each segment has inputs from two other segments, somewhat more freedom in determining the characteristic polynomial exists. The interested reader is referred to Bardell and Spencer (1984) for a detailed treatment of such generators.

Now consider the phase shift between the various sections of the factored shift register. If one shifts the generator of Figure 6.14 (202301_8) through its cycle and observes when a unique marker (e.g., 16 consecutive 1's) appears at each output, the following relationship is seen:

> Output 0 lags output 3 by 55047 shifts
> Output 1 lags output 0 by 55047 shifts
> Output 2 lags output 1 by 20976 shifts.

TABLE 6.2   Dependency Polynomials of 202301_8 of Degree Less Than or Equal to 16

$$g_1 = 1 + x + x^{34125}$$
$$g_2 = 1 + x^2 + x^{2715}$$
$$g_3 = 1 + x^3 + x^{43257}$$
$$g_4 = 1 + x^4 + x^{5430}$$
$$g_5 = 1 + x^5 + x^{55049}$$
$$g_6 = 1 + x^6 + x^{20979}$$
$$g_7 = 1 + x^7 + x^{52669}$$
$$g_8 = 1 + x^8 + x^{10860}$$
$$g_9 = 1 + x^9 + x^{51360}$$
$$g_{10} = 1 + x^{10} + x^{44563}$$
$$g_{11} = 1 + x^{11} + x^{8233}$$
$$g_{12} = 1 + x^{12} + x^{41958}$$
$$g_{13} = 1 + x^{13} + x^{60154}$$
$$g_{14} = 1 + x^{14} + x^{39803}$$
$$g_{15} = 1 + x^{15} + x^{58848}$$
$$g_{16} = 1 + x^{16} + x^{21720}$$

$$g(x) = 0 \bmod 202301_8$$

Upon constructing the expression for the input to the third shift register, one finds

$$x^0 = x^6 + (x \text{ raised to the phase shift coming from group 1})$$
$$= x^6 + x^{20976 + 3}$$

This phase relationship is predictable since

$$x^{20979} + x^6 + 1 = 0 \bmod f(x) \tag{6.7}$$

where $f(x)$ is the characteristic polynomial of the sequence (202301_8). The relation displayed by Equation (6.7) is a fundamental consequence of the recursion used to form the sequence. It is one of the dependency trinomials (see Section 3.6.3) of the sequence. All of the dependency polynomials of 202301_8 that might result from factoring the LFSR into segmented LFSR generators are shown in Table 6.2. While there are 32767 such dependency polynomials associated with the sequence in question, recall that Equation (6.4) is of the form $1 + x^{q_0} = x^B$. Since q_i is less than n for a segmented generator, only dependency polynomials of the form

$$g(x) = 1 + x^A + x^B$$

where $A < 16$ can be of interest in the analysis of these segmented LFSR generators with 16th-degree characteristic polynomials. Looking at the middle register in Figure 6.14, one observes that the phase relation at its input follows the equation

$$x^0 = x^5 + x^{55047 + 2}$$

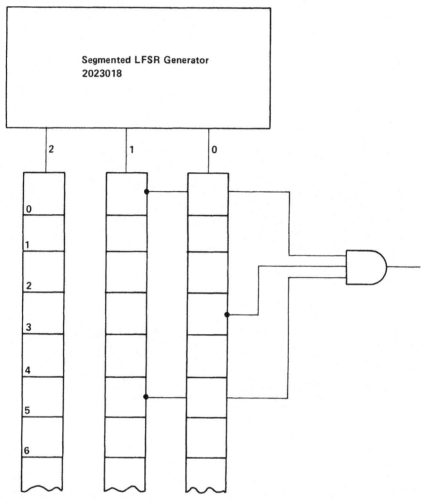

Figure 6.15. Linear dependencies in segmented LFSR output.

Notice also that the polynomial

$$g_5(x) = 1 + x^5 + x^{55049} = 0$$

is listed in Table 6.2.

When used as a parallel sequence generator to load scan paths, the segmented LFSR generator is free of structural dependencies as discussed in Section 6.6.1, but linear dependencies are still a problem, since the sequence is indeed folded about the linear dependencies. To see this effect, consider Figure 6.15. A linear dependency exists between the zeroth and fifth latches of the scan path fed by generator output 1 and the third latch of the path fed by output 0. Therefore the pattern [111] can never be obtained at the inputs of the gate shown. This may cause trouble when used as a source of stimuli for built-in test.

6.6.3 Fast Parallel Generator

A shift-register circuit designed to output an m-sequence in a parallel fashion is proposed in Hsiao (1969). An nth degree m-sequence can be generated n times as fast as the conventional serial generator if n is relatively prime to $2^n - 1$, the length of the sequence. This parallel sequence generator can be implemented directly from the nth power of the companion matrix of the characteristic polynomial. For the polynomial

$$g(x) = c_n x^n + c_{n-1} x^{n-1} + \cdots + c_1 x + c_0, \qquad c_n = 1$$

the companion matrix $\mathbf{C}_g$ [see Birkhoff and MacLane (1953)] is defined as

$$\mathbf{C}_g = \begin{bmatrix} 0 & 1 & 0 & \cdots & 0 & 0 \\ 0 & 0 & 1 & \cdots & 0 & 0 \\ 0 & 0 & 0 & \cdots & 0 & 0 \\ & & & \ddots & & \\ 0 & 0 & 0 & \cdots & 1 & 0 \\ 0 & 0 & 0 & \cdots & 0 & 1 \\ c_0 & c_1 & c_2 & \cdots & c_{n-2} & c_{n-1} \end{bmatrix}$$

Note that $\mathbf{C}_g$ is an $n \times n$ matrix with 1's on the first superdiagonal, the coefficients of $g(x)$, c_0 to c_{n-1}, as the bottom row, and zeros everywhere else. In the analysis of fast parallel generators, the transpose of $\mathbf{C}_g$ is needed. The transpose of a matrix is formed by interchanging rows and columns. Thus the transpose of $\mathbf{C}_g$, $\mathbf{C}_g^t$,

$$\mathbf{C}_g^t = \begin{bmatrix} 0 & 0 & 0 & \cdots & 0 & c_0 \\ 1 & 0 & 0 & \cdots & 0 & c_1 \\ 0 & 1 & 0 & \cdots & 0 & c_2 \\ & & & \ddots & & \\ 0 & 0 & 0 & \cdots & 0 & c_{n-2} \\ 0 & 0 & 0 & \cdots & 1 & c_{n-1} \end{bmatrix}$$

As an example, a fast parallel generator with characteristic polynomial

$$f(x) = x^5 + x^2 + 1$$

is implemented. The corresponding companion matrix is

$$\mathbf{C}_g^t = \begin{bmatrix} 0 & 0 & 0 & 0 & 1 \\ 1 & 0 & 0 & 0 & 0 \\ 0 & 1 & 0 & 0 & 1 \\ 0 & 0 & 1 & 0 & 0 \\ 0 & 0 & 0 & 1 & 0 \end{bmatrix}$$

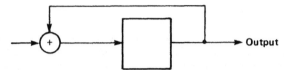

Figure 6.16. Basic elements of fast parallel generator.

and the fifth power of $\mathbf{C}_g^t$ is

$$[\mathbf{C}_g^t]^5 = \begin{bmatrix} 1 & 0 & 0 & 1 & 0 \\ 0 & 1 & 0 & 0 & 1 \\ 1 & 0 & 1 & 1 & 0 \\ 0 & 1 & 0 & 1 & 1 \\ 0 & 0 & 1 & 0 & 1 \end{bmatrix}$$

The elements of a fast parallel generator are single latches with feedback as shown in Figure 6.16. The rows of $[\mathbf{C}_g^t]^5$ indicate which present states are

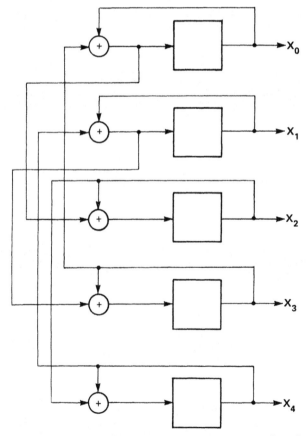

Figure 6.17. Fast parallel generator implementing $p(x) = x^5 + x^2 + 1$.

combined to get the next state. For example, in $[C_g^t]^5$, the first row specifies that the first output x_0 is obtained from stages 0 and 3. That is, the present outputs of stages 0 and 3 are summed to form the next input of stage 0. In like manner, the input to x_1 is obtained from the sum of the present states of stages 1 and 4, and the next input to x_2 is obtained by summing the present outputs of stages 0, 2, and 3. Since stages 0 and 3 were summed as the input to stage 1, that sum is fed into the EXCLUSIVE OR at the input to stage 2. The input to stage 3 is formed by summing the present outputs of stages 1, 3, and 4. The final input x_4 is formed by summing the outputs of stages 2 and 4. Figure 6.17 shows the resulting construction.

The output is a parallel group of r bits (five in the example) that are the first r bits of the m-sequence. On the next shift, the next r bits of the m-sequence appear at the outputs. Thus the m-sequence is advanced r bits each shift cycle. Table 6.3 shows the resulting pattern.

Looking at the sequence emitted by a single output it can be seen to be a decimation by 5 (every fifth symbol chosen) of the output of a stage of the corresponding LFSR. Since 5 is prime to 31 ($2^5 - 1$), it is a proper decimation. By Property IX (Section 3.5) the decimated sequence is an m-sequence. Thus it would appear that a suitable parallel stimulus generator could be formed with a fast parallel generator. However, some structural dependency problems exist. Consider the sequences emitted by adjacent outputs. Note that for the

TABLE 6.3   Output of Fast Parallel Generator

$$p(x) = x^5 + x^2 + 1$$
m-sequence: 00001|01011|10110| $\cdots$

	LFSR	Fast Parallel		LFSR	Fast Parallel
1	10000	00001	19	01111	00100
2	01000	01011	20	00111	00101
3	10100	10110	21	10011	01110
4	01010	00111	22	11001	11000
5	10101	11001	23	01100	11111
6	11010	10100	24	10110	00110
7	11101	10000	25	01011	10010
8	01110	10101	26	00101	00010
9	10111	11011	27	10010	10111
10	11011	00011	28	01001	01100
11	01101	11100	29	00100	01111
12	00110	11010	30	00010	10011
13	00011	01000	31	00001	01001
14	10001	01010	32	10000	00001
15	11000	11101	33	01000	.
16	11100	10001	.	10100	.
17	11110	11110	.	01010	.
18	11111	01101	.	10101	.

example shown in Table 6.3, adjacent outputs are only phase shifted by six shifts. Thus problems of structural dependencies could develop when a fast parallel generator is used as a parallel pattern generator for built-in test.

6.6.4 Two-Dimensional Window Generator

A generator with a two-dimensional *window property* will now be described. Many constructions of two-dimensional pseudorandom arrays and their applications have appeared in the literature [see MacWilliams and Sloane (1976), Nomura et al. (1971), and Reed and Stewart (1962)]. In this section, it is proposed that the array be implemented as the states of a group of i-bit shift registers. Consider an array of k shift registers, each of i stages, as shown in Figure 6.18. Call the $i \times k$ array A_m. The m-sequence is to be fed from an n-stage LFSR, where $n = rh$, into the shift-register array so that the following property exists:

Two-Dimensional Window Property. *If an $r \times h$ subspace (window) is defined on $A_m(i \times k)$, each of the $2^{rh} - 1$ nonzero binary $r \times h$ arrays will appear in the window exactly once in $2^{rh} - 1$ shifts of the LFSR.*

This window property is different from that defined in MacWilliams and Sloane (1976). but it is defined in a way that is useful for built-in test

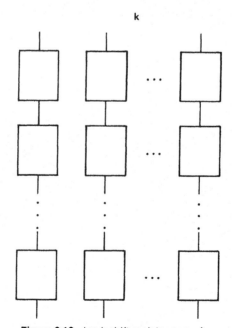

Figure 6.18. $i \times k$ shift register array A_m.

structures like STUMPS (Chapter 8) which have parallel scan paths. As will be shown, parallel bit streams can be shifted into the scan paths in such a way as to form this two-dimensional window.

To construct a pseudorandom array with this property, one chooses a tap point in the sequence, say a_j, to feed into the top of column 0 (shift register 1). A tap shifted s shifts, a_{j+s}, is fed into the second column, and so forth. At a given instant in time, the sequence appearing at the top of the columns,

$$\{a\} = a_j, a_{j+s}, a_{j+2s}, \ldots, a_{j+(h-1)s}$$

can be expressed as a polynomial

$$f_m(x) = a_j x^0 + a_{j+s} x^s + \cdots + a_{j+(h-1)s} x^{(h-1)s}$$

As the pattern is shifted down one stage and a new row appears at the top, the expression becomes

$$x f_m(x) = a_j x + a_{j+s} x^{s+1} + a_{j+2s} x^{2s+1} + \cdots + a_{j+(h-1)s} x^{(h-1)s+1}$$

After $r - 1$ shifts, to fill the h columns r deep, the result is

$$x^{r-1} f_m(x) = a_j x^{r-1} + a_{j+s} x^{s+r-1}$$
$$+ a_{j+2s} x^{2s+r-1} + \cdots + a_{j+(h-1)s} x^{(h-1)s+r-1}$$

at the bottom of the $r \times h$ window and

$$f'_m(x) = a_{j+r-1} + a_{j+s+r-1} x^s + a_{j+2s+r-1} x^{2s} + \cdots + a_{j+(h-1)s+r-1} x^{(h-1)s}$$

at the top. Table 6.4 shows the resulting array. The entire $r \times h$ window can be described by

$$f(x) = a_{j+r-1} + a_{j+r-1+s} x^s + \cdots + a_{j+r-1+(h-1)s} x^{(h-1)s}$$
$$+ a_{j+r-2} x + a_{j+r-2+s} x^{s+1} + \cdots + a_{j+r-2+(h-1)s} x^{(h-1)s+1}$$
$$+ \cdots + a_j x^{r-1} + a_{j+s} x^{s+r-1} + \cdots + a_{j+(h-1)s} x^{(h-1)s+r-1}.$$

TABLE 6.4 Construction of Two-Dimensional Pseudorandom Array

m-sequence: $a = a_0 a_1$	$\cdots$ $a_j a_{j+1}$	$\cdots$ a_{j+s}	$\cdots$ a_{j+2s}	$\cdots$ a_{2^m-2} $\cdots$ a_0 $\cdots$
a_{j+r-1}	$a_{j+s+r-1}$	$a_{j+2s+r-1}$	$\cdots$	$a_{j+(h-1)s+r-1}$
$\vdots$	$\vdots$	$\vdots$		$\vdots$
a_{j+2}	a_{j+s+2}	a_{j+2s+2}	$\cdots$	$a_{j+(h-1)s+2}$
a_{j+1}	a_{j+s+1}	a_{j+2s+1}	$\cdots$	$a_{j+(h-1)s+1}$
a_j	a_{j+s}	a_{j+2s}	$\cdots$	$a_{j+(h-1)s}$

Thus the $r \times h$ window of interest can be described by the polynomial $f(x)$ which is of degree $t = (h - 1)s + r - 1$. Let $s = r$, then t becomes

$$t = rh - 1$$

since $rh = n$, $t < n$. Under these conditions, note that the $r \times h$ members of $\{a_i\}$ appear once and only once in the window. In effect the sequence has been *folded* into the array. The first n bits were *pushed* into column 0, the second n bits into the next, and so on.

Since the LFSR with primitive characteristic polynomial $p(x)$ of degree n will generate all $2^n - 1$ sequences, all $2^n - 1$ distinct $r \times h$ binary arrays will appear in the window.

The window is a subspace of the m-sequence generated by $p(x)$, and is described by $f(x)$, a polynomial of degree $t < n$. By definition the set polynomial (Section 3.6.1) $F(x) \neq 0 \bmod p(x)$. This is a necessary and sufficient condition for complete coverage of the subspace described by $f(x)$ (Section 3.6.2).

Figure 6.19 shows a simple example of a 2×2 window array driven by a four-stage LFSR. As shown, it is not particularly interesting, but the reader can verify that indeed all 15 nonzero 2×2 binary arrays do appear in the window as the LFSR is cycled through its states. However, if instead of a 2×2 array of shift-register stages, a larger, say, 4×6 array is used, and it is still desired to have a 2×2 window at each position in the array, the construction becomes more interesting. Figure 6.20 shows such a construction using a 10-stage idler register to hold 10 shifted outputs of the LFSR.

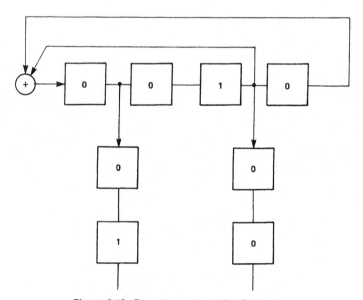

Figure 6.19. Two-dimensional (2×2) window.

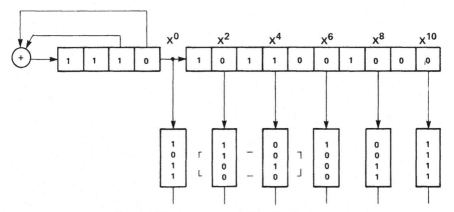

Figure 6.20. A 4 × 6 array with a 2 × 2 window.

It should be noted that an arbitrarily chosen window, such as that shown, contains four consecutive members of the m-sequence, so the window property holds. The columns of the array are driven by taps 0, 2, 4, 6, 8, and 10 of the idler register. Since the LFSR has only four stages, it is convenient to use the shift-and-add property of m-sequences to obtain the required inputs directly from the LFSR. The recursion relation is $a_n + a_{n-3} + a_{n-4} = 0$. As shown in Section 3.2, this recursion corresponds to the characteristic polynomial $p(x) = x^4 + x^3 + 1$. Writing the characteristic equation $p(x) = 0$ yields $x^4 = x^3 + 1$. Thus to generate the output labeled x^6 in Figure 6.20,

$$x^6 = x^2(x^4) = x^2(x^3 + 1)$$
$$= x^5 + x^2 = x(x^3 + 1) + x^2$$
$$= x^4 + x^2 + x$$
$$= x^3 + x^2 + x + 1$$

This implementation is shown in Figure 6.21. Likewise,

$$x^8 = x^4 + x^7 = 1 + x^6$$

It has been shown that

$$x^6 = x^3 + x^2 + x + 1$$

thus,

$$x^8 = x + x^2 + x^3$$

Likewise

$$x^{10} = x^2 x^8 = x^3 + x^4 + x^5$$
$$= x^3 + x(x^3 + x^4) = x^3 + x(x^3 + x^3 + 1)$$
$$= x + x^3$$

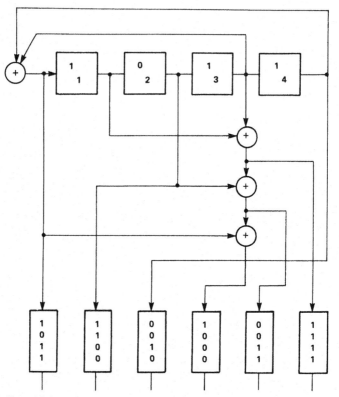

Figure 6.21. Compact construction of 4 × 6 array with 2 × 2 window.

A systematic construction technique based on the state transition matrix will be developed in the next subsection. This construction can be extended to any number of columns (of any depth). After $2^n - 1$ (or 15 in our example), the columns merely repeat. A 2 × 2 window placed anywhere on this plane will observe all $2^4 - 1$ nonzero binary patterns once and only once in $2^4 - 1$ consecutive shifts of the LFSR. Thus the construction of a pseudorandom array with the two-dimensional window property has been shown.

One apparent deficiency in the construction just shown is the absence of the all-zero n-tuple from the window. This property of m-sequences is avoided by some workers by using a nonlinear feedback network to achieve the all-zero state in the shift register. These de Bruijn sequences (Section 6.2) do not obey the shift-and-add property that is required to construct the window property just outlined. A simple alternative is to make the LFSR in the generator one stage longer. In the example shown in Figure 6.21, if the LFSR is increased to 5 stages, the all-zero 2 × 2 array would occur once every 31 shifts, while the other patterns would occur once each 15 shifts.

A corollary property of the array construction is as follows.

Corollary. *If the array A_m is $w \times w$ in extent, where $w = 2^n - 1$, an $r \times h$ window slid vertically over A_m in a cyclic manner (consider A_m a torus) will expose each of the $2^{rh} - 1$ nonzero $r \times h$ binary arrays exactly once. The property holds for horizontal shifts also, but not for combinations.*

Another way to implement a two-dimensional $r \times h$ window would be to have h distinct LFSRs with the same characteristic polynomial of degree at least rh. One could drive each of the h columns of the window with one of these generators. The window property would be obtained by seeding LFSR_i with a state that is r shifts advanced over that seeded in LFSR_{i-1}. As the h LFSRs cycle through their sequences, the h shift-register channels (each of length r) will display the window property. Thus if an array of h shift registers each r stages long is wanted to exhibit the two-dimensional window property, there is a choice of implementations, one LFSR of length rh or greater with its EXCLUSIVE OR network, or h of them, each rh stages long feeding the scan paths directly. The cost of implementation would dictate the single LFSR in most cases.

This section has been concerned with constructing a generator that will exhaust a k-subspace within an $r \times h$ window where $k < rh$. It was shown that an LFSR with rh stages is required. If the window is required to approach the size of the entire scan-path structure or the $i \times k$ array A_m, a large number of stages are required in the LFSR. In the limit, one stage in the LFSR would be required for each stage in the scan path. For a network with 500 scan latches, 500 stages would be required in the LFSR.

6.6.5 Linear Array Generator

Given a digital network with a parallel scan-path structure (such as STUMPS, Chapter 8) consisting of h channels each r SRLs deep, it was shown that it takes an LFSR of $r \times h$ stages to implement a two-dimensional window over the structure. In general this is too expensive, so an alternative approach is presented. By constructing a network that spaces the inputs to each of the h channels by r shifts (by utilizing the shift-and-add property) one can essentially fold the m-sequence into the h channels of the STUMPS structure. This implementation consists of an LFSR of degree n, where

$$n \ll r \times h \ll 2^n - 1$$

If one considers the structure as a matrix with r rows and h columns ($r \times h$), one sees that each column contains r elements of the sequence, running from bottom to top. Thus for the ith column, at the bottom in position (r, i), bit j of the m-sequence appears; at the top of that column, in position $(1, i)$, bit $j + r - 1$ appears; at the bottom of the $(i + 1)$th column, bit $j + r$ appears in position $(r, i + 1)$; bit $j + 2r - 1$ appears in $(1, i + 1)$ at the top of column $i + 1$. This folding continues until the array is full.

This construction proceeds in an orderly fashion. One first chooses an output that will be used for channel 1. Say this is the output corresponding to x^i. The next channel must be fed by $x^{(i+r)}$, the next by $x^{(i+2r)}$, and so forth. The problem reduces to finding

$$x^k \bmod f(x)$$

where $f(x)$ is the characteristic polynomial of the sequence. This can be done explicitly as before or an alternate approach using the state space transition matrix S can be utilized.

Consider an LFSR with characteristic polynomial of degree n

$$f(x) = c_n x^n + c_{n-1} x^{n-1} + \cdots + c_2 x^2 + c_1 x + c_0, \, c_n = 1$$

Let the state of an n-stage LFSR at time k be written as the vector

$$\mathbf{v}_k = [v_{1k}, v_{2k}, \ldots, v_{nk}] \tag{6.8}$$

At time $k + 1$, the state vector is

$$\mathbf{v}_{k+1} = [F_k, v_{1k}, v_{2k}, \ldots, v_{n-1k}] \tag{6.9}$$

where

$$F_k = \sum_{1=1}^{n} c_i v_{ik} \pmod 2 \tag{6.10}$$

is the feedback function that is fed into the first stage of the shift register. In other words, the values in the register stages at time $k + 1$ are

$$v_{1k+1} = F_k$$
$$v_{2k+1} = v_{1k}$$
$$v_{3k+1} = v_{2k}$$
$$\vdots$$
$$v_{nk+1} = v_{n-1k}$$

This system of equations leads to the transition matrix S such that

$$\mathbf{v}_{k+1} = \mathbf{v}_k S \tag{6.11}$$

where S is the transition matrix. S is an $n \times n$ matrix where the first column is formed by the coefficients c_i in ascending order. The remainder of the matrix consists of 1's on the first superdiagonal and 0's everywhere else as shown in Equation (6.12)

$$
\mathbf{S} =
\begin{bmatrix}
c_1 & 1 & 0 & \cdots & 0 & 0 \\
c_2 & 0 & 1 & \cdots & 0 & 0 \\
c_3 & 0 & 0 & \cdots & 0 & 0 \\
\vdots & \vdots & \vdots & \ddots & \vdots & \vdots \\
c_{n-2} & 0 & 0 & \cdots & 1 & 0 \\
c_{n-1} & 0 & 0 & \cdots & 0 & 1 \\
c_n & 0 & 0 & \cdots & 0 & 0
\end{bmatrix}
\tag{6.12}
$$

Equation (6.11) leads to

$$
\mathbf{v}_{k+1} = \mathbf{v}_1 \mathbf{S}^k
$$

where $\mathbf{v}_1$ is the initial state of the register. Thus the $(k+1)$th state of the register can be obtained by applying $\mathbf{S}$ to $\mathbf{v}_1$ k times. A selection vector $\mathbf{b}$ is defined as $\mathbf{b}_j = [0 \cdots 1 \cdots 0]$ where there is a single 1 in the jth $(1 \le j \le n)$ column of the vector. The presence of the 1 in the jth column of $\mathbf{b}$ means that the output sequence $\{u_k\}$ is taken from the jth register stage.

The kth bit u_k of the m-sequence is given by

$$
u_k^{(j)} = \mathbf{v}_k \mathbf{b}_j'
$$
$$
= \mathbf{v}_1 \mathbf{S}^{k-1} \mathbf{b}_j'
$$

The shifted sequence may be described by

$$
u_{k+q}^{(j)} = \mathbf{v}_1 \mathbf{S}^{k+q-1} \mathbf{b}_j'
$$

On the other hand, the sequence can be written

$$
u_{k+q}^{(j)} = \sum_{i=1}^{n} d_i u_k^{(i)} \bmod 2
$$

by the shift-and-add property of m-sequences. That is, the shifted sequence is some linear combination of the n unshifted sequences available at the n stages of the shift register. Combining these last two expressions,

$$
\mathbf{v}_1 \mathbf{S}^{k+q-1} \mathbf{b}_j' = \sum_{i=1}^{n} d_i u_k^{(i)} \bmod 2
$$

Recalling that $u_k^{(i)} = v_k b_i'$, the expression becomes

$$v_1 S^{k+q-1} b_j' = v_k \sum_{i=1}^{n} d_i b_i' \bmod 2$$

$$v_1 S^{k-1} S^q b_j' = v_1 S^{k-1} \sum_{i=1}^{n} d_i b_i'$$

or finally,

$$S^q b_j' = \sum_{i=1}^{n} d_i b_i' \tag{6.13}$$

which relates the transition matrix of the shifted sequences S^q to the weighting coefficients d_i. Let the linear combination of sequences available from the n stages of the register which combine to form a sequence shifted by q from the jth stage output, namely

$$\sum_{i=1}^{n} d_i b_i$$

be denoted by $\mathbf{p}_q^{(j)}$. Then Equation (6.13) can be written

$$S^q b_j' = \mathbf{p}_q'^{(j)} \tag{6.14}$$

Return now to the problem of spacing the outputs of an LFSR. Assume that the LFSR contains n stages [so that $f(x)$ is of degree n]. Also, channel 1 will be fed by output i ($i < n$). To calculate the tap points for channel 2, use Equation (6.14) with $q = r$ and $\mathbf{b}_j = \mathbf{b}_i$ ($1 \le i \le n$). The resulting vector $\mathbf{p}_r^{(i)}$ will contain a 1 at each position that must be added (modulo 2) to generate the sequence to feed channel 2 so that it is spaced r shifts down the m-sequence from channel 1. It is noted that

$$S^{2r} b_j' = S^r (S^r b_j') = S^r \mathbf{p}_r'^{(i)}$$

Thus, the procedure is repeated for each channel, starting with one channel's vector and applying S r times to get the next channel's tap vector.

Turning now to the previous example, construct a linear window generator for the sequence generated by

$$f(x) = x^4 + x^3 + 1$$

over a six column by four row array. The appropriate transition matrix is

$$S = \begin{bmatrix} 0 & 1 & 0 & 0 \\ 0 & 0 & 1 & 0 \\ 1 & 0 & 0 & 1 \\ 1 & 0 & 0 & 0 \end{bmatrix}$$

Since the columns have four rows, we need to shift each tap by 4. We let the initial state be defined by $\mathbf{b}_1 = [1000]$ and proceed as previously outlined. Now $\mathbf{p}_4^{(1)}$, the state shifted four times from $\mathbf{b}_1$, is given by [where the superscript (1) is dropped for simplicity]

$$\mathbf{p}_4' = \mathbf{S}^4 \mathbf{b}_1'$$
$$\mathbf{p}_4 = [1011]$$

where

$$\mathbf{S}^4 = \begin{bmatrix} 1 & 1 & 0 & 0 \\ 0 & 1 & 1 & 0 \\ 1 & 0 & 1 & 1 \\ 1 & 0 & 0 & 1 \end{bmatrix}$$

Likewise a shift of 4 from $\mathbf{p}_4$ is given by

$$\mathbf{p}_8' = \mathbf{S}^4 \mathbf{p}_4'$$
$$\mathbf{p}_8 = [1110]$$

continuing,

$$\mathbf{p}_{12}' = \mathbf{S}^4 \mathbf{p}_8'$$
$$\mathbf{p}_{12} = [0001]$$
$$\mathbf{p}_{16}' = \mathbf{S}^4 \mathbf{p}_{12}'$$
$$\mathbf{p}_{16} = [0011]$$
$$\mathbf{p}_{20}' = \mathbf{S}^4 \mathbf{p}_{16}'$$
$$\mathbf{p}_{20} = [0101]$$

Using the preceding state vectors to specify the taps to feed the array columns, the compact generator shown in Figure 6.22 is obtained.

Having laid the m-sequence out on the STUMPS structure, it can now be seen that if the channels are thought of as a long contiguous chain of registers, channel 1 top to channel 2 bottom, and so forth, the m-sequence is intact. Thus all of the properties of an m-sequence apply. The STUMPS structure forms, in effect, a linear window $r \times h$ positions long through which the entire m-sequence is observed in turn. Since all elements of the k-subspace are by definition confined to a portion of the m-sequence at most rh long, the maximum degree that the polynomial representation of the k-subspace may have is rh. The k-subspace is said to have a span of rh. The occurrence of linear dependencies in m-sequences has been discussed in Chapter 3. As an example, given a 16th-degree m-sequence and a $k = 12$ subspace, it can be shown that there is a probability of .06 that a randomly placed k-subspace will

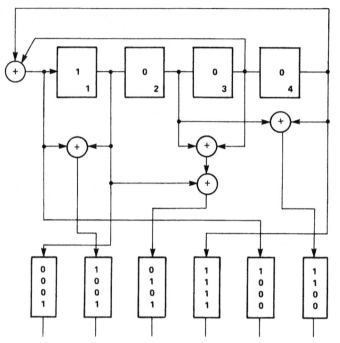

Figure 6.22. Compact construction of linear array.

contain a linear dependency. Strikingly lower probabilities occur for modest increases in the degree of the m-sequence. For $k = 12$ and $n = 23$, the probability of a linear dependency occurring is .00048. Thus the linear array generator allows the linear dependency analysis to proceed as for a known span of an unmodified m-sequence.

6.7 CONCLUSIONS

The construction of pseudorandom array generators that have use in built-in test schemes has been shown. The problems of local correlations, both structural and those inherent in the sequence, have been pointed out. A two-dimensional window generator has been shown which insures that all binary patterns occur within the window as the generator cycles. A method of folding the m-sequence on the columns of shift registers so that the array could be treated as an idler register was shown for use where it is undesirable to have a generator of the required length to provide a window.

7

Random Pattern Built-In Test

Random pattern testing has been a common practice in industry for a long time. It was recognized quite early in the development of test generation strategies that there is a lot to be gained if a short random test precedes the long and laborious deterministic test. The rationale behind this was based on experience gained while working with deterministic tests. It was found that it is quite easy to detect the first, say, 70% of the possible single stuck-at faults, and much more difficult to close up the gap to 100%. Thus, in order to catch the first 70% of the single fault population, it was advantageous to utilize a random test, which was less costly to generate. The deterministic test, which would follow this random test, would pick up the majority of the rest of the faults.

A *random test* is defined as a random selection of input vectors. The input vectors may or may not be equally likely. A random test is said to be *uniform* if all input vectors are equally probable, and said to be *biased* otherwise. It is important to note that no test is truly random. The reason is that no ideal random number generator exists. However, it is quite easy to generate a *pseudorandom test*, which is an approximation to the truly random one. As shown in Chapters 3 and 6, a pseudorandom test can be generated in a variety of ways, and it has properties which make it resemble a random test. The major difference between a random test and a pseudorandom test is that the latter is fully repeatable, while the former is not. For test applications this is an advantage, since usually one wants to repeat the test in the exact same sequence it was first applied to ensure repeatable signatures. In this chapter we will use the term "random test" with the understanding that, in practice, it will be approximated by a pseudorandom test. It is also important to mention that the random test is generally far from being exhaustive. It will usually sample only a small portion of the entire input space.

In this chapter, unless otherwise specified, it is assumed that the random test is uniform (unbiased).

7.1   RANDOM PATTERN TEST

Figure 7.1 depicts a random test without data compression. In Figure 7.1 the source is some random pattern generator, which feeds in parallel both the circuit under test and a reference unit. The reference unit is either a "golden unit" (a known good unit) or a good machine simulator computing the correct response to the input stimulus. A comparator compares the actual response from the circuit under test and the expected response from the reference unit to determine whether or not an error has been observed. The random test may stop either on the first observed error or after a sufficient number of random tests yield a correct response. This test length depends on the level of quality desired by the test, and will be addressed in Section 7.5.

Data compression has been used in a variety of fields in order to reduce the amount of data that has to be stored and analyzed. For test applications, it has been found that a number of data compression techniques may serve as a useful tool. Some of the existing techniques were described in Chapters 4 and 5: signature analysis, one's count, syndrome testing, and transition count. Generally speaking, a data compressor would map a long stream of data into a relatively short word, called a *signature*. An ideal data compressor would have the property that a faulty input data stream will yield a faulty signature. However, no such data compressor exists. The action of mapping an erroneous data stream into a correct signature is called *error masking*. The best one would hope to get from a data compressor is a small probability of error masking. Thus, the quality of a data compressor may be measured by its error

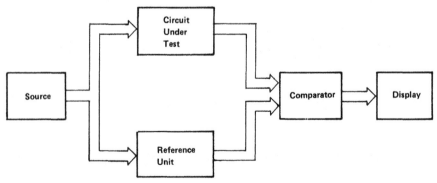

Figure 7.1. Random test with no data compression.

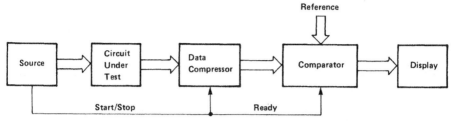

Figure 7.2. Random test with data compression.

masking probability, given some standard error distribution on its input data stream.

Figure 7.2 shows a schematic diagram of a random test using data compression. The source in Figure 7.2 supplies random patterns to the circuit under test. The source also has a start/stop control line which feeds the data compressor and the comparator. The start signal is issued with the generation of the first random pattern. This start signal tells the data compressor to start compressing output responses. The compression will go on until a stop signal is issued. The stop signal is issued by the source after the application of the last random pattern. A stop signal to the data compressor also constitutes a ready signal for the comparator. This ready signal informs the comparator that the signature stored in the data compressor is ready for comparison. For the scheme to work, a reference signature must be supplied. This reference signature is the expected signature from the fault-free circuit, and is usually computed beforehand by performing a good machine simulation. At the comparison time, the measured signature is compared to the reference signature. The circuit passes the test if the signatures are identical.

7.2 ERROR LATENCY FOR COMBINATIONAL AND SEQUENTIAL CIRCUITS

This section describes various statistical attributes of random testing in both combinational and sequential circuits.

7.2.1 Error Latency for Combinational Circuits

Let the circuit have n inputs. We use the following definitions.

Definition 7.1. The *error latency* EL_i of a fault f_i is the number of random input vectors applied to a circuit until the fault is detected at one of its primary outputs [Shedletsky and McCluskey (1975)].

The error latency is a random variable assuming values from the set of positive integers. The error latency is an attribute of the fault, and its value depends on the circuit and the fault in question.

Definition 7.2. The collection of all input vectors that detect the fault f_i is called the *affected subset* AF_i of the fault f_i. The set of the remaining input vectors, namely the collection of vectors which do not detect the fault, is called the *unaffected subset* UAF_i.

The affected subset of a fault is nothing but the complete test set for that fault. An application of any member of the affected subset AF_i to the circuit when f_i is present will result in at least one erroneous value at one of the outputs of the circuit. Note that AF_i and UAF_i are complements of one another with respect to the universe of all possible n-bit input vectors.

Definition 7.3. The *detection probability per pattern* (or simply the *detection probability*) q_i, of fault f_i, is the probability that a randomly selected input vector will detect the fault.

From this definition the detection probability is given by

$$q_i = \frac{|AF_i|}{2^n} \tag{7.1}$$

where $|AF_i|$ is the cardinality of the set AF_i.

Theorem 7.1. *The error latency of a fault has a geometric distribution.*

Proof. Under the assumption that the applications of input vectors are independent events, each such application is a Bernoulli trial with probability of success q_i. The probability that the error latency will assume the value j is the probability that the first $j - 1$ Bernoulli trials will result in failures, and the jth one in success. Thus,

$$\Pr\{EL_i = j\} = (1 - q_i)^{j-1} q_i \qquad \text{Q.E.D.} \tag{7.2}$$

The cumulative distribution function (cdf) of the error latency $F_{EL_i}(t)$ is the probability that the fault is detected at or before input vector t. The cdf of the error latency is sometimes called the *cumulative detection probability* (cdp), and we will use the terms interchangeably. The cdp can be obtained from (7.2):

$$F_{EL_i}(t) = \Pr\{EL_i \le t\} = \sum_{j=1}^{t} (1 - q_i)^{j-1} q_i$$

or

$$F_{EL_i}(t) = 1 - (1 - q_i)^t, \qquad t \ge 1 \tag{7.3}$$

The mean and variance of the error latency can be obtained from (7.2):

$$M_i = E(EL_i) = \sum_{j=1}^{\infty} j(1 - q_i)^{j-1} q_i = \frac{1}{q_i} \tag{7.4}$$

$$\text{var}(EL_i) = E(EL_i^2) - E^2(EL_i) = \frac{1 - q_i}{q_i^2} \tag{7.5}$$

where $E(x)$ denotes the expected value of the random variable x and $E(x^2)$ its second moment.

Since the smallest value q_i may assume in irredundant circuits is 2^{-n} (in a case where only one input vector detects the fault f_i), the mean and variance of the error latency are bounded from above by

$$M_i \le 2^n \tag{7.6}$$

and

$$\text{var}(EL_i) \le 4^n - 2^n \tag{7.7}$$

The quality of a random test is a measure of its capability to detect faults. The next definition is related to the test quality.

Definition 7.4. The *escape probability* of a fault f_i is the probability that the fault will go undetected after the application of t random input vectors. Similarly, the escape probability of a fault set $\{f_1, f_2, \ldots, f_m\}$ is the probability that at least one member of the fault set will be left undetected after the application of t random input vectors.

Notice that the escape probability and the cdp add up to 1. The cdp is a measure of the test quality. The higher the cdp the less likely it is that faults will escape detection. Saying it differently, the smaller the escape probability, the higher the test quality.

The escape probability of fault f_i is given by

$$\Pr\{\text{escape}\} = 1 - F_{EL_i}(t) = (1 - q_i)^t \tag{7.8}$$

The random test length required to detect a fault f_i with escape probability no larger than a given threshold e_i can be obtained from (7.8) as

$$T_i = \left\lceil \frac{\ln e_i}{\ln(1 - q_i)} \right\rceil \tag{7.9}$$

7.2.2 Error Latency for Sequential Circuits

This section considers synchronous sequential circuits that can be represented by a state diagram with transition assigned outputs or by its equivalent state table. The circuits are assumed to be *strongly connected*, namely that a path, not necessarily of length 1, exists between any two states. It is assumed that the circuits have an asynchronous preset or clear input and that the fault does not prevent the circuit from being initialized correctly to some initial state. It is also assumed that the fault does not increase the number of states of the machine. Before describing the error latency model for sequential circuits we briefly review the mathematical tools needed for it.

Basic Concepts and Definitions. The probability that a sequential circuit will be in a given state at time t depends, in general, on the state of the circuit at time $t - 1$ and on the input vector probabilities at time t. In an unbiased random test the inputs are equally likely and, therefore, the probability that the circuit will be in a given state at time t depends on its state at time $t - 1$. The operation of the circuit can be described by a stationary Markov chain.

Definition 7.5. The *state probability* $s_j(t)$ of state S_j at time t is the probability that the circuit is in state S_j after the application of input vector t. The *state probability vector* $s(t)$ is a row vector whose jth element is $s_j(t)$. Let m be the number of states of the sequential circuit. Clearly,

$$\sum_{i=1}^{m} s_i(t) = 1, \qquad t \geq 0 \tag{7.10}$$

Definition 7.6. The *t-step transition probability* $p_{ij}(t)$ is the conditional probability of being in state S_j after t steps, given that the current state is S_i. The *transition probability matrix* $\mathbf{P}(t)$ is an $m \times m$ matrix whose entries are the t-step transition probabilities.

The one-step transition probability $p_{ij}(1)$ can be obtained from the state table by summing the probabilities of all the input vectors that transfer the machine from state S_i to state S_j. For simplicity we will use p_{ij} to represent $p_{ij}(1)$ and $\mathbf{P}$ to represent $\mathbf{P}(1)$.

Each row of $\mathbf{P}(t)$ sums to 1:

$$\sum_{j=1}^{m} p_{ij}(t) = 1, \qquad i = 1, 2, \ldots, m, \, t \geq 1 \tag{7.11}$$

The transition probability matrix $\mathbf{P}(t)$ can be obtained from $\mathbf{P}$:

$$\mathbf{P}(t) = \mathbf{P}^t \tag{7.12}$$

The state probabilities can be obtained from

$$s(t) = s(t - 1)\mathbf{P} = s(0)\mathbf{P}(t) \tag{7.13}$$

where $s(0)$ is the initial state probability vector.

Definition 7.7. The *period* of a state S_j, written $d(j)$, is the greatest common divisor of all integers t for which $p_{jj}(t) > 0$.

Note that if the machine starts in state S_j, a return to this state is possible only in steps which are multiples of $d(j)$. All states of a strongly connected sequential circuit (which is the case in this discussion) have the same period. If all states of the machine have period d, the circuit is called *d-periodic*. A sequential circuit is called *aperiodic* if $d = 1$.

The limit

$$\pi = \lim_{t \to \infty} \mathbf{s}(t)$$

exists for aperiodic sequential circuits and is called the *stationary state distribution*. The stationary state distribution is independent of the initial state and indicates the relative proportion of time the machine is going to spend in its various states as the number of input vectors becomes very large.

The stationary state probability distribution

$$\pi = \left(\pi_1, \pi_2, \ldots, \pi_m \right)$$

can be obtained by solving the equation

$$\pi = \pi \mathbf{P} \tag{7.14}$$

jointly with the equation

$$\sum_{i=1}^{m} \pi_i = 1 \tag{7.15}$$

Notice that Equation (7.14) describes a system of m equations with m unknowns. One of these equations is linearly dependent on the other $m - 1$. Equation (7.15) is the mth equation needed to solve this system.

Example 7.1. Find the stationary state probabilities for the circuit of Figure 7.3.

SOLUTION. Table 7.1 displays the state table for this circuit. In order to compute the one-step transition probability p_{ij}, it is necessary to add the input vector probabilities that cause a transition from state S_i to state S_j. There are two input vectors that cause a transition from state S_1 to state S_4. Each input vector has a probability of $1/4$ of being applied. Thus, $p_{14} = 1/2$. Similarly, from state S_3 only a transition to state S_4 is possible. Therefore, $p_{34} = 1$. The one-step transition probability matrix that refers to the state table shown in

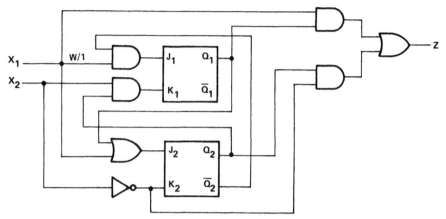

Figure 7.3. The sequential circuit of Example 7.1.

TABLE 7.1 The State Table M of the Sequential Circuit of Example 7.1

Present State Q_1Q_2	Next State, Z			
	$x_1 = 0, x_2 = 0$	$x_1 = 0, x_2 = 1$	$x_1 = 1, x_2 = 0$	$x_1 = 1, x_2 = 1$
$S_1 \leftarrow 00$	$S_1, 0$	$S_1, 0$	$S_4, 0$	$S_4, 0$
$S_2 \leftarrow 01$	$S_1, 1$	$S_2, 0$	$S_1, 1$	$S_2, 0$
$S_3 \leftarrow 10$	$S_4, 0$	$S_4, 0$	$S_4, 1$	$S_4, 1$
$S_4 \leftarrow 11$	$S_3, 1$	$S_4, 0$	$S_3, 1$	$S_2, 1$

Table 7.1 is therefore

$$
\mathbf{P} = \begin{bmatrix} 1/2 & 0 & 0 & 1/2 \\ 1/2 & 1/2 & 0 & 0 \\ 0 & 0 & 0 & 1 \\ 0 & 1/4 & 1/2 & 1/4 \end{bmatrix}
$$

The solution to Equations (7.14) and (7.15) is

$$\pi_1 = \pi_2 = \pi_3 = 1/5, \qquad \pi_4 = 2/5. \quad \square$$

The Error Latency Model. A detectable fault in a sequential circuit induces a state table which is not equivalent to the fault-free one. One way of keeping track of the difference between the fault-free and the faulty machine is to combine them into a *product machine*. The states of the product machine correspond to pairs of states, such that one state belongs to the fault-free machine and the other to the faulty machine. The product machine produces an output of 1 when the first discrepancy between the fault-free and faulty machine is observed. Procedure 7.1 gives the details of creating the product state table. This product state table is called the *error latency model* (ELM) [Shedletsky and McCluskey (1976)].

Procedure 7.1

Step 1: Obtain the state table M of the fault-free machine.

Step 2: Obtain the state table of the faulty machine M_f.

Step 3: Form the ELM. The state S_{ij} in the ELM refers to state S_i in M and state S_j in M_f. The states S_{ii} are the *initial states* of the ELM, since we assume that all circuits can be correctly initialized to a given state (this implies that some of the faults in the preset/clear circuitry are not handled by this model).

To construct the ELM start with a row for each initial state. For each row determine the successor states S_{ij} by consulting the state tables M and M_f. A transition is assigned an output of 1 if and only if the corresponding outputs in M and M_f differ. Any next state entry with a corresponding output of 1 is denoted S_{ab}. Add new rows to the table that reflect the states that are reachable from initial states. Fill out the next state transitions for these states and their corresponding outputs. Continue to add new rows until every state reachable from an initial state (not necessarily in one step) has been assigned a row. The next state entries for the state S_{ab} are all S_{ab} with a corresponding output of 0. The ELM is said to be in *standard form* if the reset initial state is listed as the first row and the state S_{ab} as the last row.

The state S_{ab} in the ELM is an *absorbing state*, namely, once the product machine reaches this state it can never leave. This absorbing state corresponds to a detection state and the error latency is, therefore, the time to absorption. The rest of the states are *transient states*, namely the probability that the product machine is in one of these states decreases with time and approaches zero for large t.

Let EL be the error latency of the fault f. The cdf of the error latency is given by

$$F_{EL}(t) = \Pr\{ EL \le t \} = \Pr\{\text{ELM is in state } S_{ab} \text{ at time } t\} = s_{ab}(t) \quad (7.16)$$

If the initial state probability vector $s(0)$ is known (by knowing which state is the reset state), then the state probability vector $s(t)$ can be obtained from (7.13) in which P is the one-step transition probability matrix of the ELM.

Example 7.2. Find the ELM and the one-step transition probability matrix of the circuit of Figure 7.3 with the fault $f = w/1$.

TABLE 7.2 The State Table M_f of the Sequential Circuit of Example 7.2

Present State Q_1Q_2	Next State, Z			
	$x_1 = 0, x_2 = 0$	$x_1 = 0, x_2 = 1$	$x_1 = 1, x_2 = 0$	$x_1 = 1, x_2 = 1$
$S_1 \leftarrow 00$	$S_3, 0$	$S_3, 0$	$S_4, 0$	$S_4, 0$
$S_2 \leftarrow 01$	$S_1, 1$	$S_2, 0$	$S_1, 1$	$S_2, 0$
$S_3 \leftarrow 10$	$S_4, 0$	$S_4, 0$	$S_4, 1$	$S_4, 1$
$S_4 \leftarrow 11$	$S_3, 1$	$S_2, 0$	$S_3, 1$	$S_2, 1$

TABLE 7.3 The ELM of the Sequential Circuit of Example 7.2 with fault $f = w/1$

Present State	Next State, Z			
	$x_1 = 0, x_2 = 0$	$x_1 = 0, x_2 = 1$	$x_1 = 1, x_2 = 0$	$x_1 = 1, x_2 = 1$
S_{11}	$S_{13}, 0$	$S_{13}, 0$	$S_{44}, 0$	$S_{44}, 0$
S_{22}	$S_{11}, 0$	$S_{22}, 0$	$S_{11}, 0$	$S_{22}, 0$
S_{33}	$S_{44}, 0$	$S_{44}, 0$	$S_{44}, 0$	$S_{44}, 0$
S_{44}	$S_{33}, 0$	$S_{42}, 0$	$S_{33}, 0$	$S_{22}, 0$
S_{13}	$S_{14}, 0$	$S_{14}, 0$	$S_{ab}, 1$	$S_{ab}, 1$
S_{42}	$S_{31}, 0$	$S_{42}, 0$	$S_{31}, 0$	$S_{ab}, 1$
S_{14}	$S_{ab}, 1$	$S_{12}, 0$	$S_{ab}, 1$	$S_{ab}, 1$
S_{31}	$S_{43}, 0$	$S_{43}, 0$	$S_{ab}, 1$	$S_{ab}, 1$
S_{12}	$S_{ab}, 1$	$S_{12}, 0$	$S_{ab}, 1$	$S_{42}, 0$
S_{43}	$S_{ab}, 1$	$S_{44}, 0$	$S_{34}, 0$	$S_{24}, 0$
S_{34}	$S_{ab}, 1$	$S_{42}, 0$	$S_{43}, 0$	$S_{42}, 0$
S_{24}	$S_{13}, 0$	$S_{22}, 0$	$S_{13}, 0$	$S_{ab}, 1$
S_{ab}	$S_{ab}, 0$	$S_{ab}, 0$	$S_{ab}, 0$	$S_{ab}, 0$

SOLUTION. We use the same state assignment as in Table 7.1. Table 7.2 shows the state table M_f of the faulty machine. Table 7.3 shows the ELM. Table 7.4 shows the one-step transition probability matrix P. □

Example 7.3. Compute the cdp of the fault $f = w/1$ in Figure 7.3 for test lengths $t = 2, 10, 20, 30, 40, 50$. Assume that S_1 is the reset initial state (namely the circuit is reset to this state before applying the random test).

SOLUTION. The initial state probability vector is $s(0) = (1, 0, \ldots, 0, 0)$. It can be seen from Equation (7.13) that $s_{ab}(t)$ is the entry in row 1 column m (in this example $m = 13$) of the matrix $\mathbf{P}(t)$. Since $\mathbf{P}(t) = \mathbf{P}^t$, it is necessary to raise the matrix $\mathbf{P}$ to the powers 2, 10, 20, 30, 40, and 50. The results obtained after performing these operations are

$$s_{ab}(2) = .25, \qquad s_{ab}(10) = .8178, \qquad s_{ab}(20) = .9631, \qquad s_{ab}(30) = .9925,$$
$$s_{ab}(40) = .9984, \quad \text{and} \quad s_{ab}(50) = .9996$$

Notice that the escape probability achieved after applying 50 input vectors is below .001. □

TABLE 7.4   The One-Step Transition Probability Matrix of the ELM of Table 7.3

	S_{11}	S_{22}	S_{33}	S_{44}	S_{13}	S_{42}	S_{14}	S_{31}	S_{12}	S_{43}	S_{34}	S_{24}	S_{ab}
S_{11}	0	0	0	1/2	1/2	0	0	0	0	0	0	0	0
S_{22}	1/2	1/2	0	0	0	0	0	0	0	0	0	0	0
S_{33}	0	0	0	1	0	0	0	0	0	0	0	0	0
S_{44}	0	1/4	1/2	0	0	1/4	0	0	0	0	0	0	0
S_{13}	0	0	0	0	0	0	1/2	0	0	0	0	0	1/2
S_{42}	0	0	0	0	0	1/4	0	1/2	0	0	0	0	1/4
$\mathbf{P} = S_{14}$	0	0	0	0	0	0	0	0	1/4	0	0	0	3/4
S_{31}	0	0	0	0	0	0	0	0	0	1/2	0	0	1/2
S_{12}	0	0	0	0	0	1/4	0	0	1/4	0	0	0	1/2
S_{43}	0	0	0	1/4	0	0	0	0	0	0	1/4	1/4	1/4
S_{34}	0	0	0	0	0	1/2	0	0	0	1/4	0	0	1/4
S_{24}	0	1/4	0	0	1/2	0	0	0	0	0	0	0	1/4
S_{ab}	0	0	0	0	0	0	0	0	0	0	0	0	1

The computation of the cdp of a fault in a sequential circuit is quite complicated. This was evidenced in Examples 7.2 and 7.3. The following theorem describes a way to compute a lower bound on the cdp, which is easier to compute.

Theorem 7.2. *Let G be the state diagram representing the* ELM. *Let h_j be the shortest path in G from state S_{jj} to the absorbing state S_{ab}. Let q_j be the probability of going from S_{jj} to S_{ab} along this shortest path. Define*

$$h = \max_j (h_j) \quad \text{and} \quad q = \min_j (q_j).$$

Then the cdp is bounded from below by

$$\Pr\{ EL_i \le kh \} \ge 1 - (1 - q)^k \tag{7.17}$$

Proof. The probability of detecting the fault in h steps is at least q. The escape probability in h steps is at most $1 - q$, and in kh steps is at most $(1 - q)^k$. Since the cdp is one minus the escape probability, it is at least $1 - (1 - q)^k$.   Q.E.D.

An interesting question is how long it takes from the time the fault strikes until an error is observed at one of the circuit outputs. Assuming that the circuit was performing its normal operation when the fault occurred, we can assume that the state probabilities follow the stationary probability distribution. Thus the cdf of the error latency can be computed from (7.13) using

$$\mathbf{s}(0) = \pi$$

By computing $s_{ab}(t)$ for various values of t, it is possible to find the error latency that will ensure the detection of the fault to a certain confidence level (measured by the cdf of the error latency or the escape probability).

7.3   SIGNAL PROBABILITY COMPUTATION

This section deals with a concept that will lead to an algorithm for computing detection probabilities of faults in combinational circuits. The detection probability profile of the fault set determines how long it will take for a random test to detect all faults with a certain degree of confidence. If this test length is unacceptable, the detection probability profile may indicate which faults are "random pattern resistant" and how to go about modifying the logic to make them easier to detect.

Definition 7.8. The *signal probability p* of a line *w* is the probability that the line will assume the value 1 as a result of a randomly applied input vector.

Section 7.3.1 describes the Parker–McCluskey algorithm [Parker and McCluskey (1975)] for computing signal probabilities. Section 7.3.2 describes the cutting algorithm that places bounds on the signal probability of a line. Both algorithms take advantage of the fact that a signal probability computation is quite simple in tree networks (circuits with no reconvergent fan-out). The following theorem establishes the basis for both algorithms.

Theorem 7.3. *For circuits with no reconvergent fan-out, the input/output signal probability relations for various gates are given in Figure 7.4.*

Proof. In tree networks all signals are independent. Consider a *k*-input AND gate with input signal probabilities p_i, $i = 1, 2, \ldots, k$. The probability of an intersection of independent events is the product of the probabilities of the isolated events. The output of the AND gate is 1 when all its inputs are 1. Thus, the signal probability at the output of this gate is

$$\prod_{i=1}^{k} p_i$$

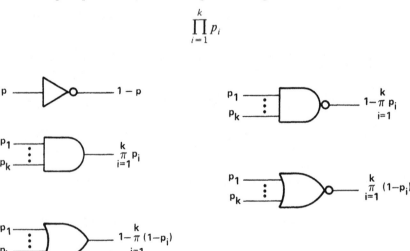

Figure 7.4. Input/output signal probability relations for tree networks.

Similarly, the signal probability at the output of an OR gate is

$$1 - \prod_{i=1}^{k} (1 - p_i)$$

If the signal probability at the input of an INVERTER is p, its output signal probability is $1 - p$. The remaining input/output relations follow in the same way.   Q.E.D.

7.3.1   The Parker – McCluskey Algorithm

Procedure 7.2

Step 1: Compute the signal probability expression E_i for each line in the circuit as a function of the input signal probabilities, $p_1, p_2, \ldots, p_n$, using the input/output relations of Figure 7.4.

Step 2: Suppress all high-order exponents (≥ 2) of the signal probability expressions.

Step 3: Substitute the values for the input signal probabilities in the expressions E_i and compute their value. These are the line signal probabilities.

Note that the high-order exponents generated in Step 1 can be suppressed as soon as they occur. Taking such an action will reduce the complexity of the algorithm.

Example 7.4. Compute the signal probabilities for all lines of Figure 7.5 assuming

1. unbiased random testing,
2. biased random testing with input signal probabilities $p_1 = 3/5$, $p_2 = 1/5$, $p_3 = 4/5$, and $p_4 = 2/5$.

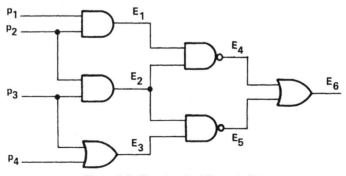

Figure 7.5. The circuit of Example 7.4.

SOLUTION. The signal probability expressions are

$$E_1 = p_1 p_2$$
$$E_2 = p_2 p_3$$
$$E_3 = 1 - (1 - p_3)(1 - p_4) = p_3 + p_4 - p_3 p_4$$
$$E_4 = 1 - E_1 E_2 = 1 - p_1 p_2^2 p_3$$

After suppressing the exponent in E_4 we get

$$E_4' = 1 - p_1 p_2 p_3$$
$$E_5 = 1 - E_2 E_3 = 1 - p_2 p_3 (p_3 + p_4 - p_3 p_4)$$

After suppressing the exponents in E_5 we get

$$E_5' = 1 - p_2 p_3$$
$$E_6 = 1 - (1 - E_4')(1 - E_5') = 1 - p_1 p_2^2 p_3^2$$

After suppressing the exponents in E_6 we get

$$E_6' = 1 - p_1 p_2 p_3$$

1. For the unbiased case all input signal probabilities are $1/2$. Thus the line signal probabilities are $E_1 = E_2 = 1/4$, $E_3 = 3/4$, $E_4 = 7/8$, $E_5 = 3/4$, and $E_6 = 7/8$.
2. For the biased case we get $E_1 = 3/25$, $E_2 = 4/25$, $E_3 = 22/25$, $E_4 = 113/125$, $E_5 = 21/25$, and $E_6 = 113/125$. $\square$

Theorem 7.4. *Procedure 7.2 computes the correct signal probabilities.*

Proof. [The proof follows Koren (1979).] The theorem is proved for an AND gate with inputs A and B and output C. The proofs for other kinds of gates are similar.

Dependency between input signals A and B is caused only by reconverging fan-out lines in the subnetworks feeding A and B. Let the Boolean function realized by A and B be

$$A = F(x_1, x_2, \ldots, x_k, x_{k+1}, \ldots, x_u)$$
$$B = G(x_1, x_2, \ldots, x_k, x_{u+1}, \ldots, x_v)$$

where the sets $\{x_1, x_2, \ldots, x_k\}$, $\{x_{k+1}, x_{k+2}, \ldots, x_u\}$, and $\{x_{u+1}, x_{u+2}, \ldots, x_v\}$ are disjoint.

Both functions F and G can be described in canonical sum of products form. Summing all products containing the same minterm of $(x_1, x_2, \ldots, x_k)$ yields

$$A = \bigcup_{i=0}^{2^k-1} F_i(x_{k+1}, \ldots, x_u) m_i(x_1, \ldots, x_k) \qquad (7.18)$$

$$B = \bigcup_{j=0}^{2^k-1} G_j(x_{u+1}, \ldots, x_v) m_j(x_1, \ldots, x_k) \qquad (7.19)$$

where $m_i(x_1, \ldots, x_k)$, $i = 0, 1, \ldots, 2^k - 1$, is the ith minterm.

No two products in (7.18) can be simultaneously 1; therefore, the signal probability of A is

$$p_A = \Pr\{A = 1\} = \sum_{i=0}^{2^k-1} \Pr\{F_i(x_{k+1}, \ldots, x_u) m_i(x_1, \ldots, x_k) = 1\}$$

Since the terms F_i and m_i are independent we have

$$p_A = \sum_{i=0}^{2^k-1} p_{F_i} p_{m_i} \qquad (7.20)$$

where

$$p_{F_i} = \Pr\{F_i(x_{k+1}, \ldots, x_u) = 1\}$$
$$p_{m_i} = \Pr\{m_i(x_1, \ldots, x_k) = 1\}$$

In a similar way we obtain

$$p_B = \sum_{j=0}^{2^k-1} p_{G_j} p_{m_j} \qquad (7.21)$$

where p_{G_j} is defined similarly to p_{F_i}.

In order to obtain a logical expression for the output C it is necessary to perform an AND operation on (7.18) and (7.19). Since the AND operation of $m_i(x_1, \ldots, x_k)$ and $m_j(x_1, \ldots, x_k)$ equals 0 if $i \neq j$, we get

$$C = AB = \bigcup_{i=0}^{2^k-1} F_i(x_{k+1}, \ldots, x_u) G_i(x_{u+1}, \ldots, x_v) m_i(x_1, \ldots, x_k)$$

Using the same reasoning as for p_A, the signal probability of C is given by

$$p_C = \sum_{i=0}^{2^k-1} p_{F_i} p_{G_i} p_{m_i} \tag{7.22}$$

We will now use Procedure 7.2 to derive an expression for the signal probability of C and show that it is equal to (7.22). Using the input/output relation for an AND gate and Equations (7.20) and (7.21) we get

$$p_A p_B = \left(\sum_{i=0}^{2^k-1} p_{F_i} p_{m_i} \right) \left(\sum_{j=0}^{2^k-1} p_{G_j} p_{m_j} \right) \tag{7.23}$$

Next we prove that if exponents are suppressed, the product $p_{m_i} p_{m_j}$ [denoted $(p_{m_i} p_{m_j})^*$] satisfies

$$(p_{m_i} p_{m_j})^* = \begin{cases} p_{m_i}, & \text{if } i = j \\ 0, & \text{otherwise} \end{cases} \tag{7.24}$$

In the logical expression for the minterm m_i, each of the input variables $x_1, \ldots, x_k$ appears either in its complemented or its uncomplemented form, hence

$$m_i(x_1, \ldots, x_k) = \bigcap_{r=1}^{k} x_r' \quad \text{where } x_r' \in \{x_r, \bar{x}_r\}$$

Since the input variables are independent we obtain

$$p_{m_i} = \prod_{r=1}^{k} \Pr\{x_r' = 1\}$$

where

$$\Pr\{x_r' = 1\} = \begin{cases} p_{x_r}, & \text{if } x_r' = x_r \\ 1 - p_{x_r}, & \text{if } x_r' = \bar{x}_r \end{cases}$$

If $i = j$, the product $p_{m_i} p_{m_j}$ equals

$$p_{m_i}^2 = \prod_{r=1}^{k} [\Pr\{x_r' = 1\}]^2$$

where

$$[\Pr\{x_r' = 1\}]^2 = \begin{cases} p_{x_r}^2, & \text{if } x_r' = x_r \\ 1 - 2p_{x_r} + p_{x_r}^2, & \text{if } x_r' = \bar{x}_r \end{cases}$$

Using exponent suppression we obtain

$$\left\{\left[\Pr\{x_r' = 1\}\right]^2\right\}^* = \begin{cases} p_{x_r}, & \text{if } x_r' = x_r \\ 1 - p_{x_r}, & \text{if } x_r' = \bar{x}_r \end{cases}$$

Thus,

$$\left\{\left[\Pr\{x_r' = 1\}\right]^2\right\}^* = \Pr\{x_r' = 1\}$$

If $i \neq j$ there exists at least one variable $x_{i0} \in \{x_1, x_2, \ldots, x_k\}$ which appears in its uncomplemented form in $m_i(m_j)$ and in its complemented form in $m_j(m_i)$. Hence

$$p_{m_i} p_{m_j} = W p_{x_{i0}}(1 - p_{x_{i0}})$$

where W is a product of probabilities not including $p_{x_{i0}}$. Using exponent suppression, we obtain

$$(p_{m_i} p_{m_j})^* = 0, \qquad i \neq j$$

thus

$$(p_A p_B)^* = \sum_{i=0}^{2^k - 1} p_{F_i} p_{G_i} p_{m_i} = p_C \qquad\qquad \text{Q.E.D.}$$

The Parker–McCluskey algorithm for computing signal probabilities is of exponential complexity in the worst case. To see this consider a cascade of two-input OR gates implementing the n-input OR function. Procedure 7.2 suggests that the output signal probability expression for this cascade will include $2^n - 1$ terms. This indicates a necessity for exponential storage capacity. This exponential storage capacity translates into exponential execution time.

7.3.2 The Cutting Algorithm

The objective of the cutting algorithm [Savir et al. (1984)] is to avoid the complexity of computing exact signal probability figures by using lower bounds. The way it is done is by cutting reconvergent fan-out branches and turning the circuit into a tree. The computational complexity of signal probabilities in tree networks is linear as implied in Theorem 7.3. In the process of cutting fan-out branches it is necessary to assign signal probability bounds to the cut points in order to ensure that all the signal probability bounds computed in the resulting tree enclose the true values. This section describes two versions of this algorithm. The *full-range cutting algorithm* assigns the signal probability range $[0, 1]$ to all cut branches. The *partial-range cutting algorithm* assigns a tighter range to the cut branches. This range depends on the topology of the circuit.

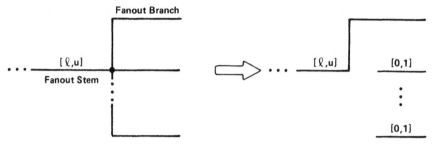

Figure 7.6. The full-range cutting process performed on an *n*-way reconvergent fan-out point (reconverging gates not shown).

The Full-Range Cutting Algorithm. The full-range cutting algorithm assigns a probability range of [0, 1] to all the cut points, and propagate the bounds to all the other lines of the circuit by using tree formulas. Notice that in order to cut an *n*-way reconvergent fan-out it is only necessary to cut $n - 1$ of its branches to turn it into a tree. Therefore, the line not cut receives the signal probability of its immediate ancestor. Figure 7.6 illustrates this point.

Definition 7.9. The *cone of influence* $c(w)$ of a line w is the logic that feeds, directly or indirectly, this line.

Definition 7.10. A *tree line* is a line w for which $c(w)$ is a tree.

The signal probabilities of all the tree lines can be computed without the cutting algorithm. These values will naturally be exact. The cutting algorithm computes signal probability bounds for all the nontree lines as described by the following procedure.

Procedure 7.3

Step 1: Assign signal probability of $1/2$ to all input lines.

Step 2: Compute the signal probabilities for all the tree lines.

Step 3: Turn the circuit into a tree by cutting reconvergent fan-out branches according to Figure 7.6.

Step 4: Propagate the signal probability bounds to all the nontree lines by using the tree formulas of Figure 7.7.

Figure 7.7 displays the formulas of propagating signal probability bounds through an INVERTER, AND, and OR gates. The formulas for NAND and NOR can be easily derived from Figure 7.7 by noting that NAND is AND-NOT and that NOR is OR-NOT.

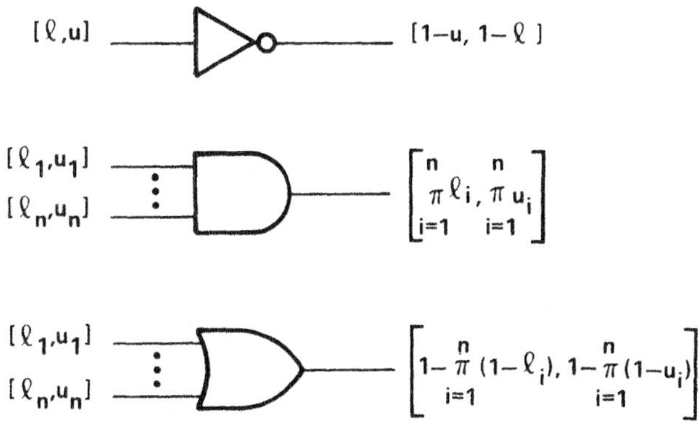

$[\ell,u]$ — (NOT gate) — $[1-u, 1-\ell\,]$

$[\ell_1,u_1]$ / $[\ell_n,u_n]$ — (AND gate) — $\left[\displaystyle\prod_{i=1}^{n}\ell_i,\ \prod_{i=1}^{n}u_i\right]$

$[\ell_1,u_1]$ / $[\ell_n,u_n]$ — (OR gate) — $\left[1-\displaystyle\prod_{i=1}^{n}(1-\ell_i),\ 1-\prod_{i=1}^{n}(1-u_i)\right]$

Figure 7.7. Tree formulas for propagating signal probability bounds (ℓ = lower bound, u = upper bound).

Example 7.5. Consider the circuit of Figure 7.8. Figure 7.8(a) illustrates the signal probabilities for all tree lines. Figure 7.8(b) illustrates the computation of signal probabilities for the nontree lines. The complete set of signal probabilities is shown in Figure 7.8(c). The only nontree lines in Figure 7.8 are lines $\{w_1, w_2\}$.

Note that the exact signal probabilities for the nontree lines can be computed using the Parker–McCluskey algorithm. The Parker–McCluskey algorithm verifies that the bounds computed here enclose the true values. The exact signal probability for line w_1 happens to be $19/32$ which is in the range $[1/4, 3/4]$, and the exact signal probability for line w_2 is $41/64$ which lies in the range $[17/32, 27/32]$. □

Theorem 7.5. *The full-range cutting algorithm computes true bounds.*

Proof. There is a one-to-one correspondence between set operations and logic gates. AND corresponds to intersection; OR to union; NOT to complementation, and so on. Let the universe be the set of all possible input vectors. Assume that we assign to each input line the subset of input vectors for which the line assumes the value 1 (called the "on-set") and that we perform the set operations that correspond to the logic gates that the circuit utilizes.

By performing those operations, the resulting set attached to each line in the circuit will be the subset of all input vectors that force that line to have the value 1. Thus, the normalized cardinality of each such set is nothing but the signal probability of that line (normalized with respect to the cardinality of the universe).

A cutting of a fan-out branch corresponds to a replacement of the on-set of that line by two sets: the null set and the set universe. Assume that Procedure 7.3 is followed to completion with the associated sets. Then all the set

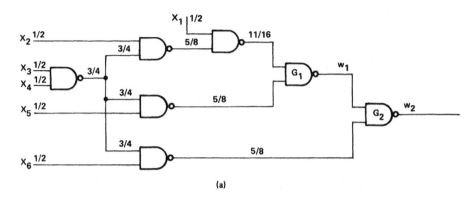

(a)

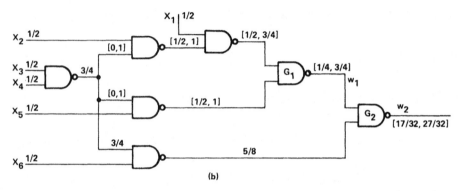

(b)

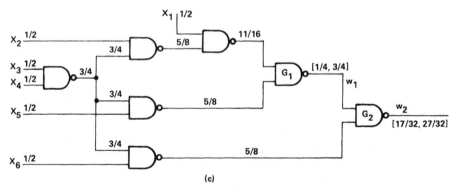

(c)

Figure 7.8. (a) Signal probabilities for tree lines. (b) Signal probabilities for nontree lines. (c) The final set of signal probabilities.

operations done on the modified circuit (which is a tree) will compute subsets or supersets of the on-sets of all the lines in the circuit. In other words, if the on-set of line w_i is $OS(w_i)$, then the cutting algorithm will compute the sets $OS'(w_i)$ and $OS''(w_i)$ such that $OS'(w_i) \subseteq OS(w_i) \subseteq OS''(w_i)$. The reason lies in the fact that all gates represent monotone functions and that the on-set relation above holds separately for each gate.   Q.E.D.

		Type of Reconverging Gate, H			
		AND/NOR	OR/NAND		
Parity of Cut Branch between w and h	Even	[g,1]	[0,g]	Equal	Inversion Parities of Reconverging Paths
	Odd	[0,g]	[g,1]		
	Even	[0,g]	[g,1]	Unequal	
	Odd	[g,1]	[0,g]		

Figure 7.9. The partial ranges assigned to the cut fan-out branch as a function of type of reconverging gate, the path inversion parity, and the signal probability of the stem.

It is important to note that divergent (nonreconvergent) fan-out branches do not have to be cut. These lines will get the signal probability of their ancestor during the computation process.

The Partial-Range Cutting Algorithm. The objective of the partial-range cutting algorithm is to take advantage of the inversion parity of the reconvergent fan-out branches to compute tighter bounds. It is shown that in some cases it is possible to assign one of the ranges [0, g] or [g, 1] to the cut fan-out branch, where g is the signal probability of the stem of the fan-out. If g is in itself a range [l, u], then one of the ranges [0, u] or [l, 1] will be assigned.

Theorem 7.6. *Let w be a fan-out stem with signal probability g, from which one pair of reconverging paths emanates. Let this pair of reconverging paths be the only ones in the network. Let H be the gate through which the pair of paths reconverges. Let h be the output of this gate. Then, if a fan-out branch is cut and assigned a range according to the appropriate case of Figure 7.9, all signal probabilities computed on the resultant tree will enclose their true values.*

Before we prove Theorem 7.6 we show an example.

Example 7.6. Consider the circuit of Figure 7.10. The two paths emanating from w and reconverging at h have unequal inversion parities and the type of reconverging gate is a NOR. Suppose we decide to cut the upper path, which has an odd inversion parity between w and h. According to Figure 7.9, the assigned range should be [3/4, 1]. The signal probability ranges computed for the resultant tree are shown in Figure 7.10. □

Proof of Theorem 7.6. For the purpose of this proof, let g be the Boolean function realized by the fan-out stem and $p(g)$ its signal probability. Similarly, let $p(x)$ be the signal probability of line x, the cut fan-out branch. We will prove one of the entries in Figure 7.9. The proofs of the rest are similar.

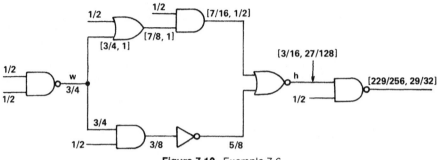

Figure 7.10. Example 7.6.

We prove it for the case where both paths have even inversion parities and where the reconverging gate is an AND. It is necessary to show that an assigned range of $[p(g), 1]$ for the cut branch will yield true bounds on all the other lines in the circuit. Because of the assumptions imposed in the theorem, it is sufficient to prove that the signal probability range computed for h encloses the true value.

The Boolean function of h expressed in terms of g can be written as

$$h = (A_1 g + B_1)(A_2 g + B_2)$$

where g is independent of A_1, A_2, B_1, and B_2.

If the first fan-out branch is cut and assumed to be a primary input x, then the function realized by h expressed in terms of x and g is

$$h' = (A_1 x + B_1)(A_2 g + B_2)$$

The question now is what should $p(x)$ be so that

$$p(h) = p(h')$$

The computation of $p(h)$ and $p(h')$ yields

$$p(h) = p[(A_1 A_2 \bar{B}_1 + A_1 \bar{B}_1 B_2 + A_2 B_1 \bar{B}_2)g] + p(B_1 B_2)$$
$$p(h') = p[A_1 \bar{B}_1(A_2 g + B_2)] p(x) + p(A_2 B_1 \bar{B}_2 g) + p(B_1 B_2)$$

Equating $p(h)$ and $p(h')$ yields

$$p(x) = \frac{p[A_1 \bar{B}_1(A_2 + B_2)g]}{p[A_1 \bar{B}_1(A_2 g + B_2)]}$$

Note that $p(g) = 0 \rightarrow p(x) = 0$ and $p(g) = 1 \rightarrow p(x) = 1$. Moreover,

$$p(x) - p(g) = \frac{p\left(A_1 A_2 \bar{B}_1 \bar{B}_2 \bar{g}\right) p(g)}{p\left[A_1 \bar{B}_1 (A_2 g + B_2)\right]} \geq 0$$

which completes the proof.   Q.E.D.

The partial-range cutting algorithm is described in the following procedure.

Procedure 7.4

Step 1: Assign a signal probability of $1/2$ (or another value, if the random test is biased) to all input lines and compute the signal probabilities of all the tree lines.

Step 2: Moving from inputs toward the outputs, cut reconvergent fan-out branches to turn the circuit into a tree. When a fan-out branch is cut, assign to it a partial range if all reconverging pairs of paths in which this line participates, along with their corresponding reconverging gates, yield the same restricted range according to Figure 7.9; otherwise assign a full range to the cut branch.

Step 3: Propagate the bounds on the resultant tree.

Example 7.7. Consider the circuit of Figure 7.8, redrawn in Figure 7.11. The tree line values were computed in Example 7.5. Figure 7.11(a) shows the computation of the signal probability bounds for the nontree lines $\{h_1, h_2\}$ based on partial ranges. Suppose we decide to cut line w_2. Line w_2 participates in two pairs of reconverging paths: the pair $\{G_1 G_4 G_5, G_2 G_5\}$ with G_5 the reconverging gate and the pair $\{G_1 G_4 G_5 G_6, G_3 G_6\}$ with G_6 the reconverging gate. In the first pair, w_2 lies in an odd path where the paths have unequal inversion parities. In the second pair, w_2 lies on an even path where the paths have equal inversion parities. According to Figure 7.9, in both these two cases, the range to be assigned to line w_2 should be $[0, g]$, where g is the signal probability of the stem w_1. Thus, in Figure 7.11(a), a range of $[0, 3/4]$ is assigned to line w_2.

After cutting line w_2, there is only one pair of reconverging paths left, namely, $\{G_2 G_5 G_6, G_3 G_6\}$. Suppose we decide to cut line w_4 next. This is the case where the paths have unequal inversion parities, and where w_4 lies on the even path. According to Figure 7.9, the range to be assigned to it is $[g, 1] = [3/4, 1]$. Figure 7.11(a) shows the propagation of the bounds to the nontree lines h_1 and h_2. Figure 7.11(b) shows the signal probability bounds for the entire circuit. $\square$

Note that Step 2 in Procedure 7.4 calls for cutting fan-out branches as you move from inputs to outputs. Thus, when a line w is considered for cutting, there should be no stem of reconvergent fan-out in $c(w)$. In other words, all

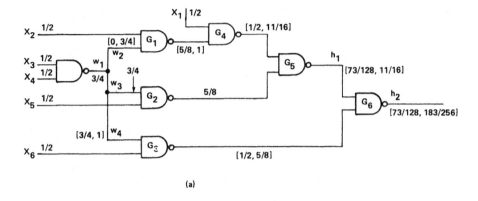

(a)

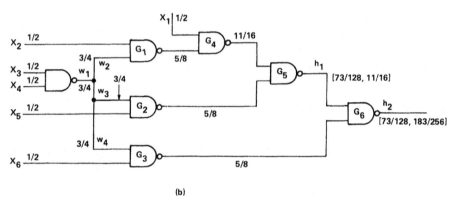

(b)

Figure 7.11. Example 7.7. (a) Nontree bounds computed with partial ranges. (b) The final set of signal probabilities.

fan-out lines prior to line w should be resolved by the time line w is considered.

Note also that there is room for optimization in Procedure 7.4 as far as the choice of the cut branches go. To see this, let a fan-out stem have a signal probability $g > 1/2$ and let the branches following it be such that if one is cut, a bound $[0, g]$ should be assigned; if the other one is cut, a bound $[g, 1]$ should be assigned. Then, at least locally, the second choice will result in a tighter bound.

Another point to note is that in Example 7.7 if we decided to cut line w_4 before cutting line w_2, we would have had to assign the full range to it. This choice probably would have resulted in looser bounds for lines h_1 and h_2.

Theorem 7.7. *The partial-range cutting algorithm computes true bounds.*

A rigorous proof of this theorem is beyond the scope of this book, and can be found in Markowsky (1987). An argument why the theorem holds is given here.

As outlined in Procedure 7.4, fan-out branches are cut starting from the inputs. Thus, when a fan-out branch has to be cut, there is no fan-out stem left in its cone of influence. If according to all possible pairs of reconverging paths in which this fan-out branch participates the required partial-range is unique, then such an assignment will result in signal probability bounds that enclose the true value for each and every line in the circuit. If, however, according to Figure 7.9 one pair of paths requires one type of partial range, while another pair of paths requires the other one, Procedure 7.4 would assign to that branch the full range, which would also yield true bounds, according to Theorem 7.5. Therefore, since the network is turned into a tree sequentially, and since each time a branch is cut its attached range is valid, then the final end result is that the computed bounds will enclose the true signal probability for each and every line in the circuit.

Cutting Algorithm Heuristics. In order to get tighter bounds using the cutting algorithm it is sometimes beneficial to use heuristics. Some insight into the importance of the ordering of the cuts was given previously in this section; here we continue in that vein. The following heuristics show how to improve the signal probability bounds based on multiple runs of the cutting algorithm, where each run makes a random selection of the cut branches in a q-way reconverging fan-out.

Theorem 7.8. *If the cutting algorithm is run $k \geq 1$ times and if the signal probability of a line is computed to be in the range $[l_i, u_i]$ for the ith trial, then the signal probability of the line is bounded by $[l, u]$ where*

$$l = \max_i \{ l_i \}, \qquad i = 1, 2, \ldots, k \tag{7.25}$$

$$u = \min_i \{ u_i \}, \qquad i = 1, 2, \ldots, k \tag{7.26}$$

Proof. Suppose the cutting algorithm is applied k times, each time recording the computed bounds for each line in the circuit. Let w be a line in the circuit and let the computed bounds for this line be $[l_i, u_i]$, $i = 1, 2, \ldots, k$. Then, since each bound constitutes a true range, the combined bound for this line may be set to $[l, u]$, according to Equations (7.25) and (7.26).   Q.E.D.

Coin Flipping. Coin flipping is an a priori unbiased way of cutting fan-out branches. Suppose w is a fan-out stem of a q-way reconverging fan-out. In any assignment of ranges, one fan-out branch will be assigned the signal probability of the stem g, while all the rest will be cut and assigned the appropriate bounds. Obviously, there are q ways in which the line assigned to g can be selected. The coin flipping method would flip a fair q-way coin to decide which line would be selected. Note that if full ranges are considered, there are q different cutting assignments associated with the q-way fan-out. If, however, partial ranges are used, more cutting assignments are possible.

Suppose the coin flipping method is applied k times. Then based on the results of these k trials, a tight bound can be computed for each line in the circuit by using (7.25) and (7.26).

Group Exhaustion. In this heuristic we group the fan-out stems in small groups to make it possible to exhaust all cutting patterns within a group and compute tight bounds based on (7.25) and (7.26). It is worthwhile to group fan-out stems that appear "close" in terms of circuit connectivity.

Adaptive Cuts. Suppose that after computing the bounds based on some cutting patterns, some lines still have unacceptable ranges. In the next cutting trial it is possible to orient the cuts in such a way that true values (or tight bounds) would be propagated toward those lines, thus improving the bounds.

Macros. Many regular structures constitute repetition of a basic macro. A ripple carry adder repeats the basic structure of a full adder. The number of repetitions depends on the width of the word. An EXCLUSIVE OR gate is used regularly in many digital designs, and also can be considered a macro.

It is beneficial to have an input/output signal probability bound relation for common macros. Having these input/output relations will simplify the calculation of bounds, since an expansion to gate primitives will not be necessary during the computation process.

Consider a two-input EXCLUSIVE OR gate. Let the signal probability bounds on the inputs be $[l_1, u_1]$ and $[l_2, u_2]$. Let the output signal probability bound be $[l, u]$. By performing the full-range cutting algorithm on an AND–OR implementation of the EXCLUSIVE OR the following relations are derived:

$$l = \max\{ l_1(1 - u_2), l_2(1 - u_1)\} \tag{7.27}$$

$$u = \min\{1 - l_1 l_2, u_1 + u_2 - u_1 u_2\} \tag{7.28}$$

7.4 DETECTION PROBABILITY CALCULATION

The task of computing detection probabilities can be translated into a signal probability computation problem [Savir et al. (1984)]. Figure 7.12 shows one way of doing this. The circuit N is the original circuit, and the circuit N_f is the circuit with the fault f injected into it. The corresponding outputs of N and N_f are EXCLUSIVE ORed. The outputs of the EXCLUSIVE ORs are fed into an OR gate whose output is denoted F. The detection probability of fault f is, therefore, the signal probability at the output F.

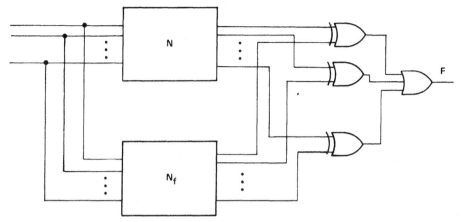

Figure 7.12. Computing the detection probability of a fault by signal probability techniques.

Example 7.8. Consider the circuit of Figure 7.13(a) with the fault $w/0$. Figure 7.13(b) shows the circuit that has to be analyzed by signal probability techniques in order to compute the detection probability of the fault.

Let p_i be the signal probability on input x_i. Using the Parker–McCluskey method on Figure 7.13(b) yields

$$E_1 = E_3 = p_1 p_2, \qquad E_2 = p_2 p_3$$
$$E_4 = E_1 + E_2 - E_1 E_2 = p_1 p_2 + p_2 p_3 - p_1 p_2^2 p_3$$

After suppressing the exponent in E_4 we get

$$E_4' = p_1 p_2 + p_2 p_3 - p_1 p_2 p_3$$
$$E_5 = E_2 + p_4 - E_2 p_4 = p_2 p_3 + p_4 - p_2 p_3 p_4$$

The output signal probability expression of a two-input EXCLUSIVE OR gate with input signal probabilities p_1 and p_2 is $p_1 + p_2 - 2 p_1 p_2$; therefore,

$$E_6 = E_3 + E_4' - 2 E_3 E_4'$$
$$= p_1 p_2 + p_1 p_2 + p_2 p_3 - p_1 p_2 p_3 - 2 p_1 p_2 (p_1 p_2 + p_2 p_3 - p_1 p_2 p_3)$$
$$= 2 p_1 p_2 + p_2 p_3 - p_1 p_2 p_3 - 2 p_1^2 p_2^2 - 2 p_1 p_2^2 p_3 + 2 p_1^2 p_2^2 p_3$$

After suppressing the exponents in E_6 we get

$$E_6' = p_2 p_3 - p_1 p_2 p_3$$
$$E_7 = E_5 + p_4 - 2 E_5 p_4 = p_2 p_3 + 2 p_4 - 3 p_2 p_3 p_4 - 2 p_4^2 + 2 p_2 p_3 p_4^2$$

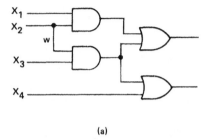

(a)

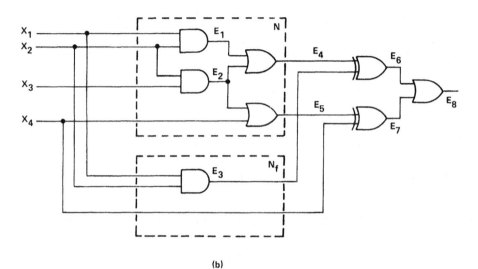

(b)

Figure 7.13. Example 7.8. (a) The original circuit with the fault $w/0$. (b) The circuit analyzed to compute the detection probability of $w/0$.

After suppressing the exponents in E_7 we get

$$E_7' = p_2 p_3 - p_2 p_3 p_4$$
$$E_8 = E_6' + E_7' - E_6' E_7'$$
$$= 2 p_2 p_3 - p_1 p_2 p_3 - p_2 p_3 p_4 - p_2^2 p_3^2$$
$$+ p_2^2 p_3^2 p_4 + p_1 p_2^2 p_3^2 - p_1 p_2^2 p_3^2 p_4$$

After suppressing the exponents in E_8 we get

$$E_8' = p_2 p_3 - p_1 p_2 p_3 p_4$$

thus

$$q = E_8' |_{p_1 = p_2 = p_3 = p_4 = 1/2} = 3/16 \quad \square$$

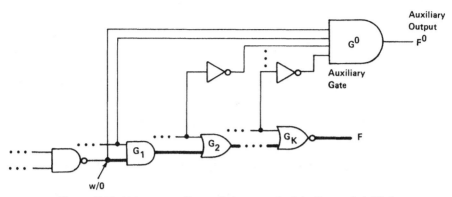

Figure 7.14. Using an auxiliary gate to compute detection probabilities.

If, rather than using the Parker–McCluskey method, it is desired to compute detection probabilities by using the cutting algorithm it is also possible to do it by computing the output signal probability of the network of Figure 7.12. Since the cutting algorithm only computes bounds on the signal probabilities this method will only compute bounds on the detection probabilities. However, for the purpose of computing a lower bound on the detection probability of a fault, a more efficient method is possible.

To describe the method, assume first that there exists only one propagation path from the site of the fault to a primary output as shown in Figure 7.14.

Assume that the fault is $w/0$. Detection of this fault requires having an input vector that will force both inputs of gate G_1 to 1; the input of G_2 that does not lie on the propagation path should be forced to a 0; the input of G_k that does not lie on the propagation path should be forced to a 0, and so forth. This set of conditions can be put together by introducing an auxiliary gate G^0, whose output $F^0 = 1$ if and only if the previous set of sensitizing conditions is satisfied. Thus, a line that has to be forced to a 0 (according to the sensitizing pattern) should be complemented when connected to G^0; and a line that has to be forced to a 1 should be left uninverted. With this arrangement the detection probability problem has been turned into a signal probability problem.

Suppose now that there are j paths from the site of the fault to a primary output. In this case there are, in general, $2^j - 1$ propagation patterns between the site of the fault and the primary output, counting all single and multiple paths. If an auxiliary gate G^r is introduced whose output $F^r = 1$ if and only if the fault propagates along the rth path (assume this path to be a single propagation path) to the primary output, then the signal probability at F^r is a lower bound on the detection probability of the fault, namely

$$q \geq p(F^r) \tag{7.29}$$

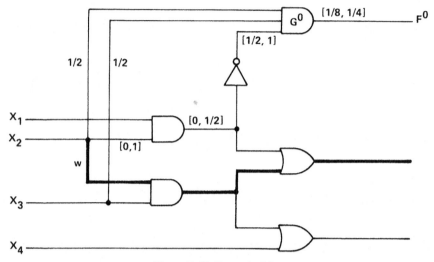

Figure 7.15. Example 7.9.

In order to compute a lower bound on the detection probability of a fault, it is possible, therefore, to use the cutting algorithm to compute the lower bound signal probability at the output of an auxiliary gate that displays the conditions needed to detect the fault along a single propagation path.

The reason for restricting the propagation path to a single path is to keep the auxiliary gate simple. The auxiliary gate for a single propagation path is an AND gate. Sensitization of a fault along multiple paths requires a few more gates to display the sensitization conditions. This method may be quite costly if the method is to be automated. The penalty paid by only resorting to single paths is the possibility of reporting a zero detection probability for a fault that can only be detected along multiple paths.

Example 7.9. Figure 7.15 shows the auxiliary gate needed to detect the fault $w/0$ in Figure 7.13(a). The propagation path selected for the fault is shown by a boldfaced line in the circuit of Figure 7.15.

In order to compute a lower bound on the detection probability of the fault $w/0$ it is necessary to compute the lower bound of the signal probability at F^0 in Figure 7.15. Figure 7.15 shows the calculation of the signal probability bound for F^0 by using the full-range cutting algorithm. Only the lower bound of the signal probability at F^0 is of interest, since the upper bound of this signal probability does not constitute, in general, an upper bound on the detection probability of the fault. Thus a lower bound on the detection probability of $w/0$ is $1/8$. □

Before describing the procedure of determining which faults are detectable by random patterns and which are not, the following terms are defined.

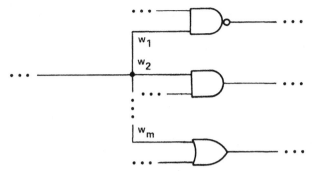

Figure 7.16. An illustration of the origin of fan-out branches, $\{w_1, w_2, \ldots, w_m\}$.

Definition 7.11. The *prime faults* are the faults on input lines and on the origin of fan-out branches.

In Figure 7.16 the origin of the fan-out branches are lines $\{w_1, w_2, \ldots, w_m\}$.

Definition 7.12. A fault w/i, $i \in \{0, 1\}$, is *random pattern testable* if its detection probability $q(w/i)$ is greater than a given threshold q_{th}.

The question of how to determine the threshold q_{th} will be addressed in Section 7.5.

Theorem 7.9. *A circuit is random pattern testable if all its prime faults are.*

Proof. Since all the prime faults are random pattern testable, then their detection probability is larger than the threshold q_{th}. However, since the prime faults dominate all the other single stuck-at faults in the circuit, their detection probabilities are at least q_{th}.   Q.E.D.

The following procedure uses the cutting algorithm to compute lower bounds on detection probabilities and identify the random pattern resistant faults.

Procedure 7.5

Step 1: Pick a prime fault which has not been considered yet. If all prime faults have been considered—stop.

Step 2: Choose a single propagation path, which has not yet been considered, from the site of the fault to a primary output. If all single paths have been considered, mark the fault as "hard fault" (or random pattern resistant) and go to Step 1.

Step 3: Introduce the auxiliary AND gate whose signal probability equals the detection probability along the selected path.

Step 4: Compute the lower bound of the signal probability on the output of the auxiliary gate. If the lower bound is acceptable (larger than the threshold), mark it off as an "easy fault" (or random pattern testable) and go to Step 1. Otherwise, go to Step 2.

Procedure 7.5 identifies the random pattern resistant faults among the class of prime faults. Since multiple propagation paths are not considered, faults that cannot be detected along single paths will also be marked as random pattern resistant. Later in this chapter we will address the question "What do we do about random pattern resistant faults?"

7.5 THE RANDOM PATTERN TEST LENGTH

Section 7.2 derived the formula for the test length required to detect a fault f_i with escape probability no larger than e_i. This section proceeds to compute the test length necessary to detect the entire fault set with escape probability no larger than a given threshold e_{th} [Savir and Bardell (1984)].

Definition 7.13. The *worst fault* is the fault having the lowest detection probability of any fault in the circuit.

The question naturally arises: Is it sufficient to detect the worst fault in order to ensure that the random pattern test will detect the entire fault set? In other words, will not all the faults be detected by the time the worst fault is detected? The answer to this question is no. It generally requires more tests to guarantee that the entire fault set is detected than to just detect the worst fault in the design.

In order to compute the escape probability of a general fault set, the problem of a fault set with only two members f_1 and f_2 is first examined. Figure 7.17 shows a Venn diagram of the test sets that detect the two faults in question.

As shown in Figure 7.17, q_1 is the probability that a randomly selected input vector will detect f_1 but not f_2; q_2 is the probability that the vector will detect f_2 but not f_1; and q_3 is the overlap probability, or the probability that the vector will detect both faults. If $q_3 = 0$, the corresponding test sets are disjoint. The detection probability of fault f_1 is $q_1 + q_3$, and that of fault f_2 is $q_2 + q_3$.

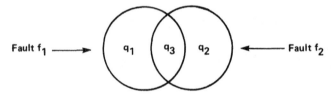

Figure 7.17. A Venn diagram showing the detection probabilities of two faults.

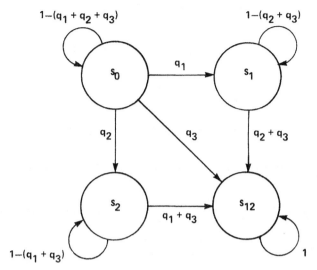

Figure 7.18. The Markov chain describing the process of detecting two faults.

Figure 7.18 shows the four states of the Markov chain describing the detection process for faults f_1 and f_2. In state S_0 neither fault has yet been detected. States S_1 and S_2 correspond to the detection of f_1 and f_2, respectively. In state S_{12} both faults have been detected. State S_{12} is an absorbing state since the objective is to detect both faults. The transition probabilities between states are labeled along the corresponding edges of the graph of Figure 7.18.

The initial state probabilities are

$$s_0(0) = 1, \qquad s_1(0) = s_2(0) = s_{12}(0) = 0 \tag{7.30}$$

where s_i, $i = 0, 1, 2$, are the state probabilities of S_i, $i = 0, 1, 2$, and s_{12} is the state probability of S_{12}.

The following set of recursive relations describes the relationship between the state probabilities:

$$s_0(t) = [1 - (q_1 + q_2 + q_3)]s_0(t - 1) \tag{7.31}$$

$$s_1(t) = [1 - (q_2 + q_3)]s_1(t - 1) + q_1 s_0(t - 1) \tag{7.32}$$

$$s_2(t) = [1 - (q_1 + q_3)]s_2(t - 1) + q_2 s_0(t - 1) \tag{7.33}$$

$$s_{12}(t) = s_{12}(t - 1) + q_3 s_0(t - 1) + (q_2 + q_3)s_1(t - 1)$$
$$+ (q_1 + q_3)s_2(t - 1) \tag{7.34}$$

The solution to this system of difference equations subject to the initial conditions in (7.30) is

$$s_0(t) = [1 - (q_1 + q_2 + q_3)]^t \tag{7.35}$$

$$s_1(t) = [1 - (q_2 + q_3)]^t - [1 - (q_1 + q_2 + q_3)]^t \tag{7.36}$$

$$s_2(t) = [1 - (q_1 + q_3)]^t - [1 - (q_1 + q_2 + q_3)]^t \tag{7.37}$$

$$s_{12}(t) = 1 - [1 - (q_1 + q_3)]^t - [1 - (q_2 + q_3)]^t + [1 - (q_1 + q_2 + q_3)]^t \tag{7.38}$$

The escape probability of this fault set can be computed from (7.38) as

$$\Pr\{\text{escape}\} = 1 - s_{12}(t)$$
$$= [1 - (q_1 + q_3)]^t + [1 - (q_2 + q_3)]^t - [1 - (q_1 + q_2 + q_3)]^t \tag{7.39}$$

Observe that (7.39) follows the inclusion/exclusion principle. The first term, $[1 - (q_1 + q_3)]^t$, is the escape probability of fault f_1 alone, ignoring the effect of fault f_2. The second term, $[1 - (q_2 + q_3)]^t$, is the escape probability of fault f_2, ignoring the effect of fault f_1. The third term, $[1 - (q_1 + q_2 + q_3)]^t$, is a correction term due to the fact that both the first and the second terms account for the possibility that both faults escape detection and compensates for this overcounting.

In order to compute the minimum random pattern test length that detects both faults with escape probability no larger than e_{th}, it is necessary to compute t that satisfies

$$\Pr\{\text{escape}\} \le e_{th}$$

Since there is no closed form solution for the random pattern test length, we proceed by searching for good approximations that will enable us to compute an upper bound for the length.

Table 7.5 considers two faults structured so that the detection probability of f_1 is .001, and the detection probability of f_2 varies between .001 and .002 in increments of .0001. The escape probability threshold e_{th} was assumed to be .001. In Table 7.5, T_{2D} is the test length for the disjoint case ($q_3 = 0$), and T_{2C} is the test length for the conjoint case with an overlap probability of .0005. T_{WF} is the test length defined by the worst fault. It is important to compare T_{WF} to the other test lengths to note how much more difficult it is to detect both faults than the fault which is the hardest of the two. T_{WF} can be

TABLE 7.5   The Minimum Test Length for the Disjoint Case (T_{2D}), For the Conjoint Case (T_{2C}),and the Minimum Test Length Contributed by the Worst Fault T_{WF}. The Worst Fault Has a Fixed Detection Probability of .001. The Escape Probability Threshold is .001. The Overlap Probability is .0005.

$q_1 + q_3$	$q_2 + q_3$	e_{th}	T_{WF}	T_{2D}	T_{2C}
.001	.001	.001	6905	7597	7586
.001	.0011	.001	6905	7298	7290
.001	.0012	.001	6905	7120	7115
.001	.0013	.001	6905	7019	7016
.001	.0014	.001	6905	6964	6963
.001	.0015	.001	6905	6935	6934
.001	.0016	.001	6905	6920	6920
.001	.0017	.001	6905	6913	6913
.001	.0018	.001	6905	6909	6909
.001	.0019	.001	6905	6907	6907
.001	.002	.001	6905	6906	6906

computed by the closed form formula [see Equation (7.9)]

$$T_{WF} = \left\lceil \frac{\ln e_{th}}{\ln(1 - q_{min})} \right\rceil \qquad (7.40)$$

where $q_{min} = \min(q_1 + q_3, q_2 + q_3)$.

A number of very important observations can be made based on Table 7.5. The first observation (which is intuitive) is that the disjoint test length is greater than or equal to the conjoint test length. Thus, an upper bound to the minimum test length can be computed by assuming that all faults have disjoint test sets. The second observation is that faults whose detection probabilities are not comparable to that of the worst fault (in this context comparable means within a factor of 2 of the worst detection probability), do not significantly affect the random pattern test length. Thus, a large network containing many thousands of potential faults may have very few faults whose detection probabilities are close to that of the worst fault, and therefore, the vast majority of these faults can be ignored in the test length computation.

Section 7.4 showed how to compute detection probabilities (or lower bounds of detection probabilities) of faults in the circuit. Having the detection probability profile of all faults in the circuit, one can compute an upper bound on the random pattern test length needed to detect all faults with an escape probability no larger than a prespecified threshold. According to the previous observations, only those faults which have a detection probability no larger than twice that of the worst fault can have any significant effect on the test length. Hence, it is possible to reduce the fault list by only considering the subset of faults whose detection probabilities are near that of the worst fault. As a further simplification, it is assumed that the test sets of this reduced fault list are all disjoint. These two simplifying actions will result in a test length which is an upper bound on the actual random pattern test length.

The computation can be further simplified by assuming that all faults in the reduced fault list have detection probabilities which are identical to that of the worst fault. In other words, if the class of faults that have detection probabilities close to that of the worst fault has been identified as $\{f_1, f_2, \ldots, f_k\}$ with detection probabilities $\{q_1, q_2, \ldots, q_k\}$, respectively, then for the purpose of computing the test length we will consider having k faults each with detection probability $q_{\min} = \min_i (q_i)$. This simplification will also result in a test length which exceeds its actual value. However, all these simplifications, while substantially reducing the complexity of the computation, also produce test lengths which are very close to the real length and are therefore very practical in an actual environment. The test length computation is based on the following theorem.

Theorem 7.10. *The escape probability of a class of k faults with disjoint test sets, each having a detection probability q is given by*

$$\Pr\{\text{escape}\} = \sum_{i=1}^{k} \binom{k}{i}(-1)^{i+1}(1 - iq)^t \qquad (7.41)$$

where t is the test length.

Proof. From (7.39) and the inclusion/exclusion principle, the expression $k(1 - q)^t$ deviates from the escape probability by the overconsideration of the escape probability of all pairs of faults. Conversely, the expression $k(1 - q)^t - \binom{k}{2}(1 - 2q)^t$ deviates from the escape probability by the underconsideration of the escape probability of all triplets. This argument continues until $i = k$, at which point the expression converges to the desired escape probability. Q.E.D.

The problem now is the computation of the test length necessary so that the escape probability determined by (7.41) will be at most a given threshold e_{th}. Before computing this test length we state two useful lemmas.

Lemma 7.1. *If t is large such that*

$$(1 - q)^t \le \delta \quad \text{and} \quad k\delta < 1$$

then the terms of the alternating series described by Equation (7.41) have decreasing absolute values.

Proof. The alternating series of (7.41) can be described as

$$e_t = \sum_{i=1}^{k} a_i, \qquad a_i = (-1)^{i+1}\binom{k}{i}(1 - iq)^t$$

The ratio of the absolute values of two consecutive terms in the series is computed to be

$$\left|\frac{a_{i+1}}{a_i}\right| = \frac{\binom{k}{i+1}[1-(i+1)q]^t}{\binom{k}{i}(1-iq)^t}$$

$$= \frac{k-i}{i+1}\left(1-q-\frac{iq^2}{1-iq}\right)^t$$

Thus,

$$\left|\frac{a_{i+1}}{a_i}\right| \le \frac{k-i}{i+1}(1-q)^t$$

Since $k-i < k$, $i+1 \ge 2$, $(1-q)^t < \delta$ and $k\delta < 1$ we have

$$\left|\frac{a_{i+1}}{a_i}\right| \le \frac{k\delta}{2} < 1 \qquad\qquad \text{Q.E.D.}$$

Lemma 7.2. *Given the same conditions as in Lemma 7.1, the alternating series described by Equation (7.41) is monotonically decreasing in t.*

Proof. The difference between two consecutive terms in (7.41) is

$$\Delta e_t = e_{t+1} - e_t = \sum_{i=1}^{k}(-1)^{i+1}\binom{k}{i}\left[(1-iq)^{t+1}-(1-iq)^t\right]$$

$$\Delta e_t = -q\sum_{i=1}^{k}(-1)^{i+1}i\binom{k}{i}(1-iq)^t$$

$$= -qk\sum_{i=1}^{k}(-1)^{i+1}\binom{k-1}{i-1}(1-iq)^t$$

Using similar arguments as in the proof of Lemma 7.1, we conclude that Δe_t is a sum of an alternating series in i, with decreasing terms in absolute value. However, since the first term of the series is negative, so is Δe_t.   Q.E.D.

Theorem 7.11. *The random pattern test length needed to achieve an escape probability not larger than e_{th} and which is due to k faults with disjoint test sets, where each fault has a detection probability q, is bounded by $[T_{kD}^{L}, T_{kD}^{U}]$ where*

$$T_{kD}^{U} = \left\lceil \frac{\ln(e_{th}/k)}{\ln(1-q)} \right\rceil \qquad\qquad (7.42)$$

$$T_{kD}^{L} = \left\lceil \frac{\ln(e_{th}/k) - \ln(1-e_{th}/2)}{\ln(1-q)} \right\rceil \qquad (7.43)$$

Proof. From Lemmas 7.1 and 7.2 we conclude that if we solve for t using the first term of (7.41) we get an upper bound for the desired test length. Furthermore, if we solve for t using the first two terms of (7.41) we get a lower bound for the test length. Thus T_{kD}^U is a solution of

$$k(1 - q)^t \le e_{th} \tag{7.44}$$

yielding (7.42). T_{kD}^L, on the other hand, is a solution to

$$k(1 - q)^t - \binom{k}{2}(1 - 2q)^t \le e_{th} \tag{7.45}$$

We can rewrite (7.45) as

$$k(1 - q)^t \left[1 - \frac{k - 1}{2}\left(1 - q - \frac{q^2}{1 - q}\right)^t \right] \le e_{th} \tag{7.46}$$

The inequality in (7.46) can be rewritten by dropping the $q^2/(1 - q)$ term yielding

$$k(1 - q)^t \left[1 - \frac{k - 1}{2}(1 - q)^t \right] \le e_{th} \tag{7.47}$$

However since

$$\frac{k - 1}{2}(1 - q)^t < \frac{k}{2}(1 - q)^t$$

inequality (7.47) can be written in the form

$$\frac{k^2}{2}(1 - q)^{2t} - k(1 - q)^t + e_{th} \ge 0 \tag{7.48}$$

The solution to (7.48) is

$$k(1 - q)^t \le 1 - (1 - 2e_{th})^{1/2} = \frac{2e_{th}}{1 + (1 - 2e_{th})^{1/2}} \tag{7.49}$$

Relation (7.49) can be further simplified to

$$k(1 - q)^t \le \frac{e_{th}}{1 - e_{th}/2}$$

whose final form is (7.43). Q.E.D.

TABLE 7.6   Bounds on the Test Length for Several Size Classes of Hard Faults. The Parameters are $q = 10^{-5}$ and $e_{th} = .001$.

k	T_{kD}^{L}	T_{kD}^{U}	T_{WF}
5	851,666	851,716	690,773
10	920,980	921,030	690,773
15	961,526	961,576	690,773
20	990,294	990,344	690,773
25	1,012,609	1,012,659	690,773
30	1,030,841	1,030,891	690,773
35	1,046,256	1,046,306	690,773
40	1,059,609	1,059,659	690,773
45	1,071,387	1,071,437	690,773
50	1,081,923	1,081,973	690,773

If in a large network k faults have been identified as having a detection probability no larger than twice the minimum detection probability in the profile, then Theorem 7.11 states that by applying T_{kD}^{U} random tests all faults will be detected with escape probability no larger than e_{th}.

Table 7.6 displays the important characteristics of Theorem 7.11. It shows the values of T_{kD}^{L} and T_{kD}^{U} as well as T_{WF}, the effect of a single fault with minimum detection probability on the test length, for several classes of hard faults with class sizes ranging from 5–50. The detection probability of all faults is $q = 10^{-5}$, and the escape probability threshold is $e_{th} = 10^{-3}$.

From Table 7.6 the first observation is that T_{kD}^{U} and T_{kD}^{L} track very closely. In fact, the difference between the two is independent of the class size k. Looking at (7.42) and (7.43), this may not seem so surprising. The "distance" between T_{kD}^{U} and T_{kD}^{L} can be computed from (7.42) and (7.43) as

$$\Delta T_{kD} = T_{kD}^{U} - T_{kD}^{L} \simeq \left\lceil \frac{\ln(1 - e_{th}/2)}{\ln(1 - q)} \right\rceil \tag{7.50}$$

For values of e_{th} and q which are much less than 1, (7.50) can be approximated by

$$\Delta T_{kD} \simeq \left\lceil \frac{e_{th}}{2q} \right\rceil \tag{7.51}$$

For most applications, ΔT_{kD} is going to be very small (say 1000 at the most). This suggests that T_{kD}^{U} can serve as a very good upper bound to the random pattern test length.

The second observation from Table 7.6 is that T_{kD}^{U} does not increase very rapidly with respect to increasing k. As shown, it takes about 700,000 random patterns to essentially ensure that one fault with detection probability 10^{-5} is detected, and less than 1,000,000 patterns to ensure that twenty such faults are detected. This suggests that the rate of increase is less than linear. In fact, (7.42) shows that the rate of increase is logarithmic in k.

For practical values of q and e_{th} (values much less than 1) the equation for T_{kD}^U can be approximated by

$$T_{kD}^U \simeq \left\lceil \frac{\ln k - \ln e_{th}}{q} \right\rceil \tag{7.52}$$

and for the case where $e_{th} = .001$, (7.52) can be rewritten as

$$T_{kD}^U \simeq \left\lceil \frac{6.9 + \ln k}{q} \right\rceil \tag{7.53}$$

From (7.53) (and also from Table 7.6), one can see that random pattern test length of $11/q_{min}$ can detect as many as 50 hard faults, each having a detection probability no larger than $2q_{min}$, with a confidence of at least 99.9%.

Equation (7.42) shows that the test length (in fact the upper bound on the test length) depends on the level of detection confidence required (measured by the escape probability); on the minimum detection probability in the profile q_{min}; and on the number of faults whose detection probabilities are in the range between q_{min} and $2q_{min}$. For a fixed escape probability threshold, say .001, the test length is still a function of two parameters: the minimum detection probability q_{min} and the number of faults whose detection probability is in the preceding range, that has been denoted by k.

Assuming a fixed escape probability threshold, the detection probability threshold q_{th} between "easy faults" and "hard faults" is a function of the allowable test length T and the number of faults whose detection probability lies between q_{th} and $2q_{th}$. Thus for $e_{th} = .001$, the detection probability threshold is given by

$$q_{th} = \frac{6.9 + \ln k}{T} \tag{7.54}$$

A conservative figure of q_{th} can be achieved by taking $k = 1$.

Example 7.10. Given a circuit with the following detection probability profile:

> 2 faults with detection probability 10^{-6}
>
> 25 faults with detection probability 10^{-5}
>
> All remaining faults have detection probabilities $> 10^{-4}$

The circuit has to be tested with no more than 10^6 random patterns, and the escape probability should be no larger than 10^{-3}. Assuming that the designer can alter the circuit to upgrade some detection probabilities, what should he do in order to meet his testing objectives?

SOLUTION. Without modifying the logic one needs

$$T = \frac{6.9 + \ln 2}{10^{-6}} = 7.6 \times 10^{6}$$

random patterns to detect all faults with escape probability no greater than 10^{-3}. Since this test length is larger than 10^6, it is necessary to upgrade some of the detection probabilities in the profile. Suppose the detection probabilities of the two hardest faults are enhanced to a value which is larger than 2×10^{-5}. In this case $q_{min} = 10^{-5}$ and $k = 25$. The test length needed to detect all faults has been decreased now to

$$T = \frac{6.9 + \ln 25}{10^{-5}} = 1.011 \times 10^{6}$$

Since this test length is still too high, one has to upgrade the detection probabilities of a few more faults. We next determine how many faults with detection probabilities 10^{-5} a test of length 10^6 can detect:

$$k \simeq \left\lfloor e_{th} e^{Tq_{min}} \right\rfloor = \left\lfloor 10^{-3} \times e^{10^6 10^{-5}} \right\rfloor = 22$$

Therefore, three more faults with detection probabilities 10^{-5} must be upgraded to at least 2×10^{-5}. $\square$

7.6 WEIGHTED RANDOM PATTERNS

The previous sections discussed ways to compute detection probabilities of faults in the circuit, as well as ways to determine the test length needed to detect all faults with a given confidence.

The question naturally arises: What do we do when the affordable test length is insufficient to detect all faults of interest? One possible solution to this problem is to try to test the logic with biased random patterns. As we will see in this section, quite a substantial improvement can be achieved if the test patterns are properly weighted.

Consider an AND gate with n inputs and one output. In order to detect all single stuck-at faults in this gate the following $n + 1$ tests have to be applied to it:

$$01 \cdots 11$$
$$10 \cdots 11$$
$$\cdots$$
$$\cdots$$
$$11 \cdots 10$$
$$11 \cdots 11$$

In the first n tests all the input lines except one have the value 1. In the last test all inputs are 1. This test set indicates that as far as faults in an AND gate are concerned, a random pattern test will be more efficient if it will generate more 1's than 0's. In order to substantiate this observation we will show how to compute the optimal signal probability assignment to the inputs of the AND gate. The optimality criterion chosen here is maximizing the cumulative detection probability, or equivalently minimizing the escape probability.

In order to simplify the discussion, refer to the first n vectors of the preceding test set as type 0 vectors and the last vector as a type 1 vector. Notice that every type 0 vector produces an output 0 and the type 1 vector produces an output 1 in normal operation. From symmetry considerations the optimal signal probability assignment is identical for all input lines. Let p be the signal probability assignment to an input line of the AND gate. The probability that a vector selected at random with signal probability p will be the type 1 vector is denoted v and is equal to

$$v = p^n$$

The probability of generating a type 0 vector is identical to all members of this class, and is given by

$$u = (1 - p)p^{n-1}$$

Figure 7.19 shows the Markov chain of detecting all faults in an AND gate. The states in this Markov chain are denoted S_{ij}, where the index i identifies how many type 1 vectors have been generated so far and the index j identifies how many type 0 vectors have been generated so far. Thus, there are $2n + 2$ states altogether including the initial state S_{00} and the success state S_{1n}. The system moves from state S_{00} to state S_{01} with the occurrence of a type 0

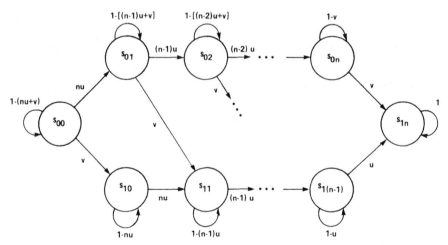

Figure 7.19. Markov chain of detecting all faults in an n-input AND gate.

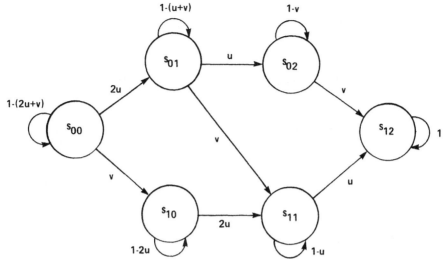

Figure 7.20. Markov chain of detecting all faults in a two-input AND gate.

vector. Since there are n such vectors, the probability of this transition is nu. The process moves from state S_{00} to state S_{10} if the random vector generated happens to be the type 1 vector. The probability of this transition is v. When in state S_{01}, the chain moves to state S_{02} with the occurrence of a second type 0 vector, and to state S_{11} with the generation of a type 1 vector. The probabilities of these transitions are $(n-1)u$ and v, respectively. The transition to state S_{02} occurs with probability $(n-1)u$ since there are only $n-1$ type 0 vectors that have not yet been generated, and the probability that the next vector will be one of them is $(n-1)u$. The success state is S_{1n}, which corresponds to a state where all the necessary $n+1$ vectors, to detect all faults, have been generated. This state is made an absorbing state in Figure 7.19 since we are interested in detecting all faults at least once. The cdp of all faults is, therefore, the probability of being in state S_{1n}. All the remaining transition probabilities between states are shown in Figure 7.19.

Figure 7.20 shows a special case of detecting all faults in a two-input AND gate. The transition probability matrix of the chain of Figure 7.20 is

$$
\mathbf{P} = \begin{array}{c} \\ S_{00} \\ S_{01} \\ S_{10} \\ S_{02} \\ S_{11} \\ S_{12} \end{array}
\begin{array}{c} S_{00} \\ \left[\begin{matrix} 1-(2u+v) \\ 0 \\ 0 \\ 0 \\ 0 \\ 0 \end{matrix} \right. \end{array}
\begin{array}{c} S_{01} \\ 2u \\ 1-(u+v) \\ 0 \\ 0 \\ 0 \\ 0 \end{array}
\begin{array}{c} S_{10} \\ v \\ 0 \\ 1-2u \\ 0 \\ 0 \\ 0 \end{array}
\begin{array}{c} S_{02} \\ 0 \\ u \\ 0 \\ 1-v \\ 0 \\ 0 \end{array}
\begin{array}{c} S_{11} \\ 0 \\ v \\ 2u \\ 0 \\ 1-u \\ 0 \end{array}
\begin{array}{c} S_{12} \\ 0 \\ 0 \\ 0 \\ v \\ u \\ 1 \end{array} \left. \begin{matrix} \\ \\ \\ \\ \\ \\ \end{matrix} \right]
$$

We are interested in $s_{12}(t)$, the probability of being in state S_{12} at time t. The boundary conditions for this state probability are

$$s_{12}(0) = s_{12}(1) = s_{12}(2) = 0, \qquad s_{12}(3) = 6u^2v,$$
$$s_{12}(4) = 12u^2v(2 - 2u - v)$$

The solution for $s_{12}(t)$ is

$$s_{12}(t) = 1 - (1 - v)^t + 2[1 - (u + v)]^t$$
$$- [1 - (2u + v)]^t - 2(1 - u)^t + (1 - 2u)^t \qquad (7.55)$$

The optimal signal probability assignment can be computed by finding the maximum of the function $s_{12}(t)$. Equating the first derivative of $s_{12}(t)$ with respect to p to 0 yields

$$p(1 - p^2)^{t-1} - (1 - p)^{t-1} + (1 - p)(1 - 2p + p^2)^{t-1}$$
$$+ (1 - 2p)(1 - p + p^2)^{t-1} - (1 - 2p)(1 - 2p + 2p^2)^{t-1} = 0 \quad (7.56)$$

The solution of (7.56) is a function of t, the intended test length. There is no closed form solution for the optimal assignment. However, an approximation to the optimal assignment can be computed by numerical methods like Newton–Raphson. Table 7.7(a) shows the optimal signal probability assignment for a two-input AND as a function of the test length t. Table 7.7(b) shows the optimal assignment for a three-input AND (these results were obtained by solving the corresponding eight-state Markov chain).

Several important observations can be obtained from the previous analysis and Table 7.7. The first observation is that the optimal assignment is defined only for test lengths of at least $n + 1$, where n is the number of inputs to the

TABLE 7.7   Optimal Assignment as a Function of the Test Length. (*a*) Optimal Assignment for a Two-Input AND. (*b*) Optimal Assignment for a Three-Input AND.

(a)		(b)	
P_{opt}	t	P_{opt}	t
.6667	3	.75	4
.6424	4	.7337	5
.6256	5	.7216	6
.6132	6	.6937	10
.584	10	.6717	20
.5552	20	.6667	100
.5189	100	.6667	∞
.5056	500		
.5	∞		

gate. The optimal assignment of signal probabilities to an AND gate for $t = n + 1$ is $p_{opt} = n/(n + 1)$. This optimal assignment decreases as the size of the test length increases. A second observation is that this optimal assignment of signal probabilities reaches an asymptotic value as the test increases. The optimal signal probability assignment asymptotically approaches $(n - 1)/n$ for large t. These results can be explained by looking at the type 0 and type 1 tests that are needed to detect all faults. The initial optimal assignment $p_{opt} = n/(n + 1)$ is also the ratio of the number of bits with a value 1 to the total number of bits in the entire test set. The asymptotic optimal assignment $p_{opt} = (n - 1)/n$ is the ratio between the bits with a value 1 within the type 0 tests and the total number of bits in the type 0 family. What this means is that if one plans to use a relatively short test, the type 1 test has an influence on the optimal assignment. If, on the other hand, one plans to use a relatively long test, the optimal assignment is mostly influenced by the type 0 tests. In other words, a long test is very likely to pick up the single type 1 test quite early, and therefore, for long tests only the density of 1's in the type 0 tests counts. The optimal assignment of signal probabilities for other types of gates is stated in the following two theorems.

Theorem 7.12. *The optimal signal probability assignment to the inputs of an AND or NAND gate with n inputs is $n/(n + 1)$ for $t = n + 1$ and $(n - 1)/n$ for $t = \infty$.*

Theorem 7.13. *The optimal signal probability assignment to the inputs of an OR or NOR gate with n inputs is $1/(n + 1)$ for $t = n + 1$ and $1/n$ for $t = \infty$.*

The proof of Theorems 7.12 and 7.13 is beyond the scope of this book. Special cases, like $n = 2$ or $n = 3$, can be treated by solving Markov chains similar to Figure 7.19.

We proceed now to describe a heuristic for generating nearly optimal weights in a general combinational circuit. These nearly optimal weights are generated by computing the set of weights needed to impose an optimal assignment at the input of each gate in the circuit, as described in Theorems 7.12 and 7.13. Since these "local optimum" requirements generally lead to contradictory signal probability assignments at the primary inputs, the nearly optimal assignment is determined as their average. The details of the algorithm are spelled out in the next procedure.

Procedure 7.6. For each gate in the circuit do:

Begin:

Step 1: Assign to the inputs of the gate the asymptotic optimal assignment for $t = \infty$ according to Theorems 7.12 and 7.13.

Step 2: Moving backward from the lines being assigned in Step 1, compute the signal probability assignment at the primary inputs by propagating signal probability values from outputs toward inputs according to the following formulas:

For a k-input AND gate with output signal probability p_o:

$$p_i = (p_o)^{1/k}$$

For a k-input OR gate with output signal probability p_o:

$$p_i = 1 - (1 - p_o)^{1/k}$$

For an INVERTER with output signal probability p_o:

$$p_i = 1 - p_o$$

For a k-input NAND gate with output signal probability p_o:

$$p_i = (1 - p_o)^{1/k}$$

For a k-input NOR gate with output signal probability p_o:

$$p_i = 1 - (p_o)^{1/k}$$

For a fan-out with branch signal probabilities p_i, $i = 1, 2, ...k$, the stem signal probability p_s is

$$p_s = \frac{1}{k} \sum_{i=1}^{k} p_i$$

Step 3: Record the signal probability assignments computed for the primary inputs.

End;

Step 4: To each primary input assign a weight which equals the average of all signal probability assignments recorded in Step 3. Use this weight as the signal probability of the primary input in a biased random test.

Example 7.11. Consider the circuit of Figure 7.21. Figure 7.21(a) shows the result of backtracing the local requirements of gate G_4 to the primary inputs. A value .5 is assigned to the inputs of gate G_4. This value is the asymptotic optimal signal probability assignment for a two-input NAND gate. These values are backtraced via gates G_3, G_2, and G_1 to the primary inputs using the

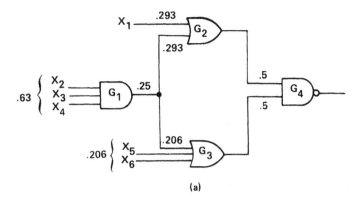

(a)

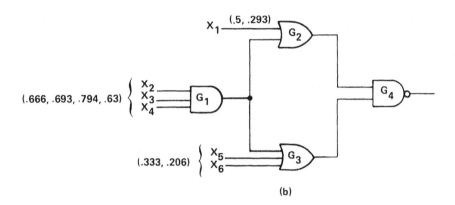

(b)

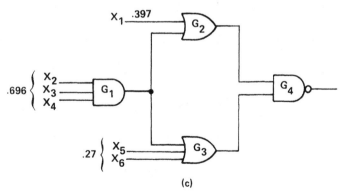

(c)

Figure 7.21. (a) Backtracing the local requirements for gate G_4. (b) The computed weights for the primary inputs. (c) The nearly optimal signal probability assignment.

formulas of Procedure 7.6, Step 2. This operation is repeated for the other three gates. As a result of this, six vectors of weights are computed for the primary inputs. The vectors associated with x_1, x_5, and x_6 have two components each, while the vectors associated with x_2, x_3, and x_4 have four components each. These vectors are shown in Figure 7.21(b). The average of the components of these vectors are the nearly optimal signal probability assignment for the primary inputs. This assignment is shown in Figure 7.21(c).

It is interesting to see the improvement in test length due to the weighting of the inputs. Consider the fault $x_2/0$. The affected subset of this fault is {0111xx, 111100}, where x denotes a "don't care." The detection probability of this fault in a uniform random test is .078, while in the biased test computed here it is .275. Using Equation (7.9) with an escape probability of .001 yields test lengths of 85 for the uniform case and 22 for the biased case.
□

Multiple Sets of Weights. In many circuits the nearly optimal signal probability assignment yields a test length which is too high to be acceptable. To see this consider the circuit of Figure 7.22.

Let both the AND gate and the OR gate of Figure 7.22 have n inputs. When Procedure 7.6 is applied to the circuit of Figure 7.22 we get a vector of two components for the input line x. Since the two components of this vector are $(n - 1)/n$ and $1/n$, the weight assignment to line x, according to Procedure 7.6, is the average of these two components $p(x) = 1/2$. Thus, although line w_1 is highly biased toward 1 and w_2 is highly biased toward 0, the nearly optimal assignment to the stem x, that feeds lines w_1 and w_2, is $1/2$. It might happen, therefore, that in a circuit with many fan-outs (whether reconvergent or divergent) quite a few input lines will be assigned a signal probability close to $1/2$.

Although this assignment might be very close to the optimal signal probability assignment, it will very likely tend to yield a relatively large random pattern test.

One way to solve this problem is to use an adaptive set of weights. We begin by using the global weights computed by Procedure 7.6, and run the test

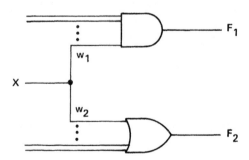

Figure 7.22. The problem with the nearly optimal weights.

using these weights for a predetermined number of vectors. We then simulate the vectors against the set of faults and identify the faults which were not detected by the test (this identification can also be done using the cutting algorithm). Having found the list of faults not detected by the test, we pick one of them and generate a test for it. The computed test vector for the fault indicates how to weigh the inputs in order to detect this fault. Suppose the computed test vector is $1001x0 \cdots x1$. Then an input with a corresponding value of 1 is assigned a signal probability close to 1 (say .9), and an input with a corresponding 0 value is assigned a signal probability close to 0 (say .1). Inputs with corresponding "don't care" values (denoted x in the previous vector) are applied with a signal probability of $1/2$. The test is then run with this new set of weights for a prespecified number of patterns. The faults not covered by the previous test are simulated against the new patterns, and the undetected faults are recorded. The procedure continues by computing the input signal probability assignments in an adaptive fashion. Experience indicates that the process, in most cases, converges very fast, namely, after a few sets of weights most, if not all, faults have been detected.

If fault simulation is chosen as the method to evaluate the adequacy of a particular set of weights, it is very important to have a fast simulator that will be capable of simulating a large number of patterns in a relatively short period of time.

It is also worthwhile to mention that it is always possible to store the tests needed to detect the leftover faults in a read-only memory and use these supplemental tests before or after the weighted random test. This will make the whole test procedure semirandom, because it uses partly random and partly deterministic tests.

Generation of Weighted Inputs. Equally likely patterns are easily generated with an LFSR. Each bit in the LFSR has a signal probability of approximately $1/2$. If two bits from an LFSR are connected to an AND gate, the signal probability at the output of this gate is approximately $1/4$. Similarly, an AND gate with three LFSR bits feeding it yields a signal probability of approximately $1/8$. On the other hand, an OR gate with two (three) LFSR bits feeding it yields a signal probability of approximately $3/4$ $(7/8)$. Thus, different weights can be generated by connecting LFSR stages to either AND or OR gates.

When building a weighted random pattern generator from an LFSR with AND/OR gates it is important to provide weighted outputs which are independent. If lines x_1 and x_2 in the circuit under test have to be assigned to signal probability $3/4$ you may need, in general, two OR gates connected to two different LFSR bits each (sometimes three bits may be sufficient when the two OR gates share one LFSR bit). The outputs of these OR gates may then be connected to lines x_1 and x_2.

For more detailed discussion of generation of weighted inputs refer to Chapter 6.

7.7 LOGIC MODIFICATION

A different approach to enhance the random pattern testability of logic is to modify the logic to upgrade the detection probabilities of the hard-to-detect faults.

Since logic modification to enhance random pattern testability may adversely affect the circuit performance, it is bound to face resistance from the logic design community. Thus it is of utmost importance to have a logic modification algorithm that has little or no effect on the circuit speed. The algorithm must be such that it will not require the designer's intervention. It should be implementable on a computer as a design-aid tool and should have the capability of suggesting logic modification actions that will enhance the detection probability of the hard-to-detect faults. The user of such a design-aid tool would then be able to weigh one possible logic modification proposal against another and make a decision as to which approach is better.

7.7.1 Logic Modification Operations

The following definitions are commonly used terms in test generation and serve in describing logic modification operations.

Definition 7.14. The *controllability* of a line is the probability that a randomly applied input vector will set the line to a given value. The *0-controllability* is the probability that a randomly applied input vector will set the line to 0; the *1-controllability* is the probability that a randomly applied input vector will set the line to a value 1.

Notice that the signal probability is the 1-controllability, and 1 minus the signal probability is the 0-controllability.

Definition 7.15. The observability of a line is the probability that a randomly applied input vector will sensitize one or more paths from that line to a primary output.

Controllability and observability are important factors in enhancing the random pattern testability of a design. Although, in general, designs which have high values of controllability and observability also tend to have high degrees of testability, this is not always the case.

Consider the circuit of Figure 7.23. In order to detect the fault $w/1$ it is necessary to apply 0 to line w and propagate the effect of the fault to the output F. The 0-controllability of line w is $1/2$. The observability of line w is $1/4$. One may be tempted, therefore, to think that the fault $w/1$ is testable. The truth is, however, that the fault $w/1$ is undetectable. As a matter of fact, the circuit of Figure 7.23 is redundant. The connection between input line x_2 and the AND gate can be eliminated, and the function F can be realized using

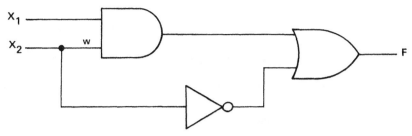

Figure 7.23. A circuit with high controllability and observability, and poor testability.

only one INVERTER and one OR gate, implementing the function $F = x_1 + \bar{x}_2$.

Although the concepts of controllability and observability are useful, one should always remember that the bottom line is the detection probability. A circuit is random pattern testable if all its faults have a detection probability above a given threshold.

Controllability Enhancement. Consider the circuit of Figure 7.24. Assume that the technology permits wired logic in the circuit (usually one type of wired logic is supported by a given technology, namely, either a wired OR or a wired AND). Assume that line w combining gates G_1 and G_2 has a very low 0-controllability. By performing a wired AND connection on line w, and either making this line a primary input or an output of an SRL (if LSSD is used), the 0-controllability can be pushed close to a value $1/2$. To see this assume that the signal probability of line w before the wired connection was $1 - \varepsilon$, $\varepsilon \ll 1$. The signal probability of line w after the connection is $(1 - \varepsilon)/2 \simeq 1/2$. Thus, this action brought both the 0- and 1-controllability close to $1/2$. By similar arguments we can show that if line w has a very poor 1-controllability a wired OR connection to that line will bring its signal probability close to $1/2$.

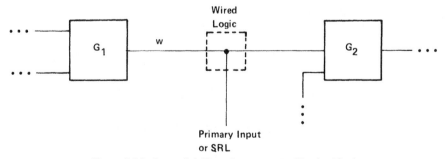

Figure 7.24. Controllability enhancement with wired logic.

If the wired logic connection needed to enhance the controllability of line w is not supported by the technology, it may be possible (although not always) to add an extra input to the gate G_1, so that by applying random patterns to the inputs the controllability of that line will be close to $1/2$.

Consider again Figure 7.24. Let the 1-controllability of line w be very low and assume that the technology does not support wired OR connections. If G_1 is a NAND gate, then by attaching an extra input to it and having it serve as a primary input, the controllability of line w (whether 0 or 1) becomes close to $1/2$.

Notice that if G_1 is an AND or a NOR gate it is impossible to enhance the 1-controllability of line w by attaching an extra input to it. The possible solution of placing an extra OR gate between G_1 and G_2 controlled by an extra primary input is usually not acceptable since the addition of logic levels drastically degrades the machine speed. Different solutions must therefore be used in this case, such as trying to affect the controllability of the line indirectly through other gates.

Also notice that the 0-controllability and the 1-controllability add up to 1. Thus, a logic modification action tending to affect one of these controllabilities will also affect the other. Experience shows that by keeping the signal probabilities of internal nodes close to $1/2$, the design is generally more testable (this assumes that the random test is uniform).

Observability Enhancement. It is possible to enhance the observability of a line by inserting test points in the logic. The simplest way of doing this is by adding an extra output pin for every test point. This solution is quite costly since pins are very often scarce. If LSSD is used in the circuit, one can make the test points observable at a shift-register latch (SRL). In this way the need for extra pins is avoided, and is traded off by using extra hardware and longer shift-register chains. If the number of test points needed to enhance the observability of a design is small, this solution is generally acceptable (provided you have a control over the circuit implementation). If, however, the number of test points is large, it is possible to connect the test points to an EXCLUSIVE OR tree whose output feeds an SRL. In this way one SRL is shared by many test points. The expense of this approach is, therefore, the addition of the EXCLUSIVE OR tree. The EXCLUSIVE OR tree has the property that any odd number of faults propagated through it will yield an output error. It will, however, mask the appearance of even numbers of faults. Thus, it may be advisable to have a number of EXCLUSIVE OR trees with independent signals feeding each one of them. This requires clustering independent test points in one group and connecting them to an EXCLUSIVE OR tree. Figure 7.25 illustrates the use of EXCLUSIVE OR trees to enhance observability.

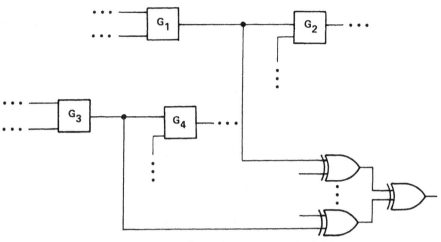

Figure 7.25. Enhancing observability by EXCLUSIVE OR tree.

7.7.2 Logic Modification Procedure

Before describing the procedure we make some comments.

The first is that observability enhancement, as described in this section, is a benign operation. No degradation of the detection probability of faults will occur as a result of enhancing observability. Thus, the observability enhancement operation may increase the observability of some lines in the circuit and, at the same time, it will not decrease the detection probability of the rest of the faults for which this operation was not intended. This is certainly not true when controllability enhancement is employed, as described in the following theorem.

Theorem 7.14. *An insertion of an extra input to enhance the controllability of a line may decrease the detection probability of a fault by a factor of* 2.

Proof. Suppose a control line x is added to the circuit to enhance the controllability of line w. Let f be a fault in $c(w)$ whose detection requires the sensitization of a path going through line w. Since one of the possible values that can be assigned to x will block the sensitized path for this fault, it is necessary to have the line x set to the opposite value for the effect of the fault to propagate to the output. Thus, a new condition arises that is required in order to detect the fault f and that did not exist before the insertion of the control input x. The probability that a random input vector will satisfy this new condition is $1/2$. The detection probability of the fault f has, therefore, degraded by a factor of 2.   Q.E.D.

Assume that k controllability enhancement operations have been performed on the circuit resulting in k additional inputs. According to Theorem 7.14 a fault may suffer, in the worst case, a degradation of a factor of 2^k in its detection probability. Thus, although the controllability enhancement operation is intended to upgrade the controllability of some faults, it may adversely affect other faults. The logic modification procedure will have to examine this effect to see whether any faults that were detectable by random patterns before the modification are still detectable after the modification. If some of those faults that were detectable are now resistant (namely, their detection probability fell below the threshold), the procedure will have to look at these faults also and upgrade their detection probability by controllability/observability operations.

A second comment regards the observability enhancement operation and is called *the problem of the split personality*. Assume that a line w required an observability enhancement operation and that a test point is connected to this line. The test point insertion split the line w into two lines w_1 and w_2. Assume that w_1 is the predecessor of line w_2, namely w_2 is closer to the output than w_1. Since faults on both w_1 and w_2 have to be considered, we observe that although the test point has enhanced the observability of w_1, the observability of line w_2 is the same as the observability of line w before the test point insertion. The end result is, therefore, that a line w with unacceptable observability has been transformed into two lines where only one of them achieved an acceptable observability, while the other one remained as bad as the original one.

The logic modification procedure is described next.

Procedure 7.7

Step 1: Compute all detection probabilities (exact values or lower bounds) and signal probabilities in the network.

Step 2: Determine the threshold detection probability below which the circuit has to be modified.

Step 3: For the faults that have to be modified and have nonzero detection probability (meaning that they are not redundant), determine which operation to perform. To do this use the following priority guidelines:

- If the controllability is acceptable and detection probability is not, try an observability enhancement operation.
- If both controllability and detection probability are unacceptable and are of the same order of magnitude, try a controllability enhancement operation.
- If both controllability and detection probability are unacceptable but the latter is much worse than the first, try both a controllability and an observability enhancement operation.

Step 4: Determine the faults that might be drawn to an unacceptable detection probability level due to the controllability operations done in Step 3. In order to determine these faults, raise the detection probability threshold by a factor of 2^k, where k is the number of controllability enhancement operations done in Step 3. The faults whose detection probability lies between the old and the new threshold are the candidate faults that have to be reexamined to see if they are still testable.

Step 5: Compute the signal probability and detection probability of the candidate faults derived in Step 4.

Step 6: Mark the faults identified in Step 5 as having a detection probability below the new threshold and repeat the process on this subset by going to Step 3.

As was described earlier, faults which require an observability enhancement operation may suffer from a split personality problem. Only one member of the split personality pair is fixed by going through a single path of Procedure 7.7. The other member of this pair should be handled in the next iteration by trying to perform another observability enhancement operation on a line closer to the output along which the signal propagates.

As noticed in Procedure 7.7, the faults which have zero detection probability are not handled. If the detection probability is exact, then these faults are redundant by definition. Redundant faults may cause some problems as may be evidenced in the following example.

Example 7.12. How would you modify the logic to make the circuit of Figure 7.26 random pattern testable?

SOLUTION. The faults $w_1/0$, $w_2/0$, and $w_3/0$ are undetectable in Figure 7.26(a). These three lines are quite easily controllable and observable at F. One way of solving the zero detection probability problem is to modify the logic as in Figure 7.26(b). Notice that an extra input was inserted to a gate which happens to be located in a different place than where the problem was. As a matter of fact, local operations on lines w_1, w_2, and w_3 would not have solved all zero detection probability problems that exist in this circuit (the faults $x_3/0$ and $x_3/1$ are also undetectable in the original circuit). Additional inputs to either gate G_2 or G_3 would not have solved the problem either. The conclusion to be drawn from this example is that global detection probability enhancement operations may be necessary in the presence of redundancies. □

7.8 RANDOM PATTERN TESTABILITY IN STRUCTURED DESIGNS

Most data flow structures employ a certain amount of regularity that can be utilized in designing an efficient random pattern test [Illman (1985)]. These

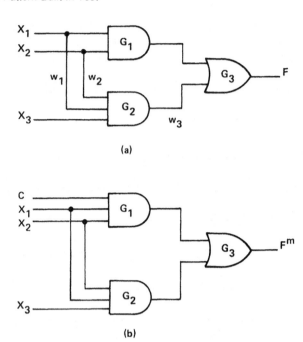

(a)

(b)

Figure 7.26. (a) The circuit to be modified in Example 7.12. (b) The modified circuit.

data flow structures, like an arithmetic logic unit (ALU) for example, are constructed by cascading a basic building block, like a byte-wide ALU. Investigation of these regular structures allows a rather easy manual computation of the lower bounds of fault detection probabilities.

Consider the 16-bit parallel adder shown in Figure 7.27. The adder is constructed out of four 4-bit blocks with ripple carry between the blocks. From simple arguments concerning the cone widths, it can be seen that the

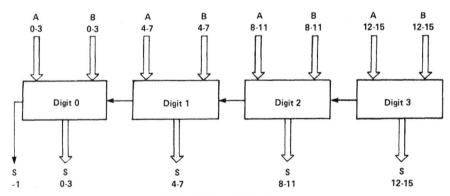

Figure 7.27. A 16-bit adder.

detection probability of a testable fault within block 3 must be at least 2^{-8}, and for a fault in block 2 it must be at least 2^{-16}. This seems to indicate that the detection probabilities decrease from the least significant to the most significant blocks. A careful study of the signal probability of the carry line gives a better assessment of these probabilities. If the two most significant bits in a digit block are both zero, then the carry-out of the block will also be zero. On the other hand if these bits are both 1, then the carry line will have the value 1. Since both these patterns have a probability of occurrence of .25, we can safely say that the 0-controllability and the 1-controllability of the carry line to any digit block are at least .25. Thus the 0- and 1-controllabilities of any line within a digit block must be at least $.25 \times 2^{-8}$, since there are eight primary inputs to a digit block in addition to the carry-in line.

Any detectable fault within a digit block is either detectable at its immediate sum outputs or its effect is propagated through the carry line to succeeding outputs. In order to assess the detection probability of a fault within a digit block we still have to compute the observability of an arbitrary line within a block. Notice that the observability of a sum output is 1. The observability of a carry-out is also 1. To see this observe that any transition on the carry line has the effect of adding or subtracting 1 from the sum formed by the succeeding bits. Thus, no matter whether a fault is detectable at the digit block outputs or at a succeeding one, its detection probability is at least 2^{-10}.

The advantage of this functional analysis is that it allows easy assessment of the minimum detection probability of faults within structured logic without having to perform detailed detection probability analysis. The disadvantage of it is that it does not provide information on how many faults have a detection probability close to this minimum value and, therefore, will only enable an approximate determination of the test length necessary to achieve a given fault coverage.

7.9   OTHER RANDOM PATTERN TESTABILITY EVALUATION METHODS

Chapter 2 described some testability measures that may give an indication of how easy it is to detect faults in the circuit. Although the methods reported in Chapter 2 are not primarily oriented toward random patterns, they can still be used to get a rough estimate of how easy or difficult it is to detect the faults by using random patterns. In this section we describe other methods that may assist in evaluating the random pattern testability of a design.

7.9.1   Fault Simulation

The detection probabilities of faults in the circuit can be estimated by fault simulating a large number of randomly generated vectors. Let f be a fault in the circuit whose detection probability is to be estimated. Let T be the number of random patterns that are used to estimate the detection probability of f. For each random pattern the logic is simulated in order to determine whether or not the fault is detected. Let N be the number of times the fault f has been detected $N \leq T$. The detection probability q of the fault f can be estimated to be

$$q = \frac{N}{T} \qquad (7.57)$$

This estimate, however, includes a random error that will approach zero as T approaches infinity. For large T the sample mean given in (7.57) is approximately normally distributed. The 95% confidence interval for this point estimate lies between $q - 2e$ and $q + 2e$, where

$$e = \sqrt{\frac{N}{T^2}\left(1 - \frac{N}{T}\right)} \qquad (7.58)$$

Fault simulation to determine random pattern testability is an expensive process in most cases. To see this let τ be the simulation time needed to propagate values from a gate input to the output of the same gate. Let G be the number of gates in the circuit and let T be the number of random patterns that have to be simulated. Let Ω be the number of faults whose testability has to be determined. Using this notation, the time it takes to simulate one random pattern on the entire circuit is $\alpha\tau G$, where the factor α depends on the simulation method and indicates what fraction of the logic is simulated, on the average, in order to determine whether or not a given vector detects a given fault. If for every new pattern the entire logic is simulated, then $\alpha = 1$. On the other hand, if the method of simulation is such that only the gates which experience a change in the values applied to their inputs (due to the propagation of errors from the fault site) are simulated, then $\alpha < 1$. The coefficient α is called the *simulation reduction factor*, and it may range between 0.01 and 0.9 depending on the number of levels in the circuit, and on whether or not the logic has disjoint partitions. The faults of interest are usually all the single stuck-at faults and, therefore, $\Omega = k_1 G$. If no fault reduction is being made, then k_1 is approximately twice the average number of the gate fan-in plus fan-out. With fault reduction this coefficient may range between 1 and 3. Notice that if the fault reduction includes a process of identifying fault equivalence classes, the coefficient k_1 can even be less than 1. This process of fully identifying the fault equivalence classes is, however, very expensive, and is hardly worthwhile in an automatic environment. The number of patterns T is usually proportional to the number of gates in the circuit, and in many cases is quite high because it takes many random patterns to reach an acceptable fault coverage. Thus $T = k_2 G$, where k_2 is this coefficient of proportionality. It is important to note that a reduction in simulation time may be achieved if more than one pattern is simulated against a given fault for each simulation pass. This is possible since a 32-bit machine can conceptually simulate 32 patterns in parallel. If this is done it will decrease the coefficient k_2 significantly [Waicukauski et al. (1986)]. Notice that it is much more advantageous to perform the fault simulation so that many patterns are simulated in parallel against a single fault in each pass, rather than performing it on many faults in parallel and a single pattern per pass. The reason is that when many patterns are simulated in parallel against a single fault, it is possible to take advantage of the fact that the fault affects a relatively small number of gates yielding a relatively small α. This localized effect of the fault can be utilized in the simulation process resulting in a lower CPU time. If, on the other hand, the simulator simulates many faults in parallel for each pattern, a great portion of the network has to be simulated each simulation pass since the faults are generally spread all over the network.

Combining all the preceding arguments yields a formula for the running time R_t:

$$R_t = \alpha k_1 k_2 G^3 \tau \qquad (7.59)$$

Example 7.13. Compute the running time of a deductive simulation on a circuit having the parameters $G = 10000$, $\alpha = 0.02$, $k_1 = 2$, $k_2 = 5$, and $\tau = 100$ ns.

SOLUTION. Substituting the parameter values into (7.59) yields

$$R_t = 0.02 \times 2 \times 5 \times 10000^3 \times 10^{-7} = 2 \times 10^4 \text{ s} \approx 5.55 \text{ hr} \qquad \square$$

7.9.2 Statistical Fault Analysis (STAFAN)

STAFAN [Jain and Agrawal (1985)] uses results on controllability of lines, as well as statistics on sensitization frequency of lines computed by simulating a given set of patterns against the logic, to deduce fault detection probabilities in the network. Simulation is only done on the good machine, and therefore the expensive stage of fault simulation is avoided.

In order to show the procedure we denote by $C_0(w)$ $[C_1(w)]$ the 0-controllability (1-controllability) of line w. We also define the observability of a line a little differently than in Definition 7.15.

Definition 7.16. The *0-observability* (*1-observability*) of line w, denoted $B_0(w)$ $[B_1(w)]$, is the probability of observing a 0 (1) on line w at a primary output.

Notice that 0-observability (1-observability) of line w is the conditional probability of sensitizing at least one path from line w to a primary output, given that the value of line w is 0 (1).

STAFAN also computes the one-level path sensitization probability from the simulation results. This probability is defined as follows.

Definition 7.17. The *one-level path sensitization probability* $U(w)$ of a line w is the probability that a given input vector will sensitize the value on the line to the output of the gate it is connected to. The one-level path sensitization of an output line or a fan-out stem is defined as 1.

The parameters $C_i(w)$ and $U(w)$ are estimated by keeping a count of how many times during the simulation a line w has the values 0 and 1, and how many times a sensitization pattern occurs on the inputs to a gate. An n-input AND gate, for example, has a sensitization pattern of at least $n - 1$ ones. For an EXCLUSIVE OR gate every vector is a sensitization pattern. The observability values $B_i(w)$ are then approximated using the simulation results concerning $C_i(w)$ and $U(w)$. Consider, for example, an n-input AND gate where one of its inputs is denoted w_1 and the output is w_2. Assume that the observabilities $B_0(w_2)$ and $B_1(w_2)$ have been computed, and that it is necessary to compute the observabilities $B_0(w_1)$ and $B_1(w_1)$. If no fan-out exists in the circuit then the probability of propagating a 1 at w_1 to line w_2 is the ratio $C_1(w_2)/C_1(w_1)$. Under these conditions the observability $B_1(w_1)$ can be computed from $B_1(w_2)$ by the formula

$$B_1(w_1) = \frac{B_1(w_2)C_1(w_2)}{C_1(w_1)} \tag{7.60}$$

Similarly, if no fan-out is present, the observability $B_0(w_1)$ can be computed by

$$B_0(w_1) = B_0(w_2)\frac{U(w_1) - C_1(w_2)}{C_0(w_1)} \tag{7.61}$$

Similarly, if w_1 is an input to an OR gate and w_2 its output, the observabilities on line w_1 are derived by STAFAN using the relations

$$B_0(w_1) = \frac{B_0(w_2)C_0(w_2)}{C_0(w_1)} \tag{7.62}$$

and

$$B_1(w_1) = B_1(w_2)\frac{U(w_1) - C_0(w_2)}{C_1(w_1)} \tag{7.63}$$

The observability relations in an INVERTER are also derived in a similar fashion. Let w_1 be the INVERTER input and w_2 its output. Then,

$$B_0(w_1) = B_1(w_2) \tag{7.64}$$
$$B_1(w_1) = B_0(w_2) \tag{7.65}$$

Similar relations can be derived for a NAND and a NOR gate.

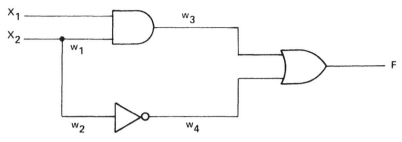

Figure 7.28. Example 7.14.

STAFAN uses Equations (7.60)–(7.63) even in the presence of fan-out and the results obtained can, therefore, only be considered approximate.

In order to complete the STAFAN algorithm it is necessary to show how to compute the observabilities of the fan-out stem given the observabilities of its branches. Let a fan-out stem w_1 split into two branches denoted w_2 and w_3. STAFAN computes the observabilities of the stem w_1 via a weighted sum with an arbitrary coefficient α, $0 \leq \alpha \leq 1$, according to

$$B_i(w_1) = (1 - \alpha) \max_j \left\{ B_i(w_j) \right\} + \alpha \left[B_i(w_2) + B_i(w_3) - B_i(w_2) B_i(w_3) \right]$$

$$i = 0, 1; \ j = 2, 3 \quad (7.66)$$

The detection probability of a fault is estimated by the product of the corresponding controllability and observability, namely, the detection probability for a stuck-at-0 fault is the product of its 1-controllability and 1-observability, while the detection probability of a stuck-at-1 fault is computed by the product of its 0-controllability and 0-observability.

Example 7.14. Refer to the circuit of Figure 7.28 and compute the detection probabilities of all the single stuck-at faults according to STAFAN, using a fan-out factor $\alpha = 0$.

SOLUTION. Table 7.8 shows the controllabilities, observabilities, and detection probabilities as computed by STAFAN. In constructing Table 7.8 it was assumed that STAFAN collected statistics on the controllability and sensitization for a long period of time, so that their estimated values are identical to the exact values. The observabilities $B_i(\cdot)$, $i = 0, 1$, were then computed according to the STAFAN formulas. The columns S-SA0 and S-SA1 refer to STAFAN's estimation of the stuck-at fault detection probabilities. The columns SA0 and SA1 refer to the exact fault detection probabilities. Notice that STAFAN's estimation is not one-sided. There are cases where it overestimates the detection probability (see $w_1/1$) and there are cases where it underestimates it (see $x_1/1$). □

TABLE 7.8   STAFAN Results Versus Exact Results for Example 7.14

Line	C_0	C_1	U	B_0	B_1	S-SA0	S-SA1	SA0	SA1
x_1	1/2	1/2	1/2	1/6	1/2	1/4	1/12	1/4	1/4
x_2	1/2	1/2	1	1	1/2	1/4	1/2	1/4	1/4
w_1	1/2	1/2	1/2	1/6	1/2	1/4	1/12	1/4	0
w_2	1/2	1/2	1	1	1/2	1/4	1/2	1/4	1/2
w_3	3/4	1/4	1/2	1/3	1	1/4	1/4	1/4	1/4
w_4	1/2	1/2	3/4	1/2	1	1/2	1/4	1/2	1/4
F	1/4	3/4	1	1	1	3/4	1/4	3/4	1/4

7.9.3   Probabilistic Estimation of Digital Circuit Testability (PREDICT)

The PREDICT algorithm [Seth et al. (1985)] is, in principle, identical to the STAFAN algorithm. The difference between the two lies in the way they compute the controllabilities and the probabilities of the one-level path sensitization of the different lines. Unlike STAFAN, which uses good machine simulation to collect these statistics, PREDICT uses analytical methods to derive them. The observability values are calculated as in STAFAN, with the exception that a fan-out factor of $\alpha = 0$ is always used for fan-out stems. Like STAFAN, the results obtained from PREDICT can only be considered approximate, and they are, in general, not one-sided.

PREDICT computes the signal probabilities (which are also the 1-controllabilities) using formulas which are similar to those used by the cutting algorithm (discussed in Section 7.3). The computation of the signal probabilities of the nontree lines is, however, exact. This exact computation of signal probabilities results in a much higher computational complexity.

In order to show the way PREDICT computes the signal probabilities we denote by $FS = \{w_1, w_2, \ldots, w_m\}$ the collection of all the reconvergent fan-out stems in the circuit, where $m > 0$ (the case where $m = 0$ corresponds to a tree network that has already been considered in Section 7.3). Consider an assignment of binary values to the elements of FS, represented by the m-tuple $A = (a_1, a_2, \ldots, a_m)$, $a_i \in \{0, 1\}$ for all i. Let g be the nontree line whose signal probability is to be computed. For each such assignment A, PREDICT computes the conditional signal probability $p(g|A)$. To do this, each independent input is assigned a signal probability of $1/2$ and each fan-out stem its corresponding a_i value. The signal probability $p(g|A)$ is then computed by using the tree formulas (see Figure 7.4). The signal probability of line g is then computed from $p(g|A)$ using the formula

$$p(g) = \sum_A p(g|A)\text{Pr}(A) \qquad (7.67)$$

Notice that when the fan-out stems are independent, the probability $\text{Pr}(A)$ is the product of $\text{Pr}(w_i = a_i)$ for all $i = 1, 2, \ldots, m$. Notice also that the

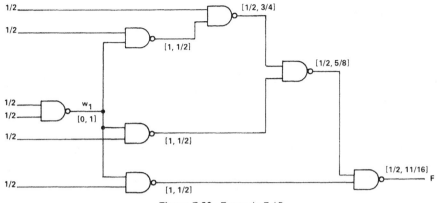

Figure 7.29. Example 7.15.

1-controllability is the signal probability and the 0-controllability can be computed as 1 minus the 1-controllability. We illustrate the PREDICT way of computing signal probabilities in the next example.

Example 7.15. Consider the circuit of Figure 7.8 redrawn in Figure 7.29. The only reconvergent fan-out stem in the circuit is w_1. Thus, there are only two possible assignments to it: $w_1 = 0$ and $w_1 = 1$. Figure 7.29 shows the propagation of signal probabilities in these two cases using the bracket notation. The assignment $[0, 1]$ to line w_1 means that we are trying to handle both assignments in one propagation pass through the network. It does not mean a bound, as in the cutting algorithm. The resultant pair of values at the output means that $p(F|w_1 = 0) = 1/2$ and that $p(F|w_1 = 1) = 11/64$. The probability of the two possible assignments to the fan-out stem are $p(w_1 = 0) = 1/4$ and $p(w_1 = 1) = 3/4$. Thus, the signal probability of the output F is

$$p(F) = \frac{1}{2} \times \frac{1}{4} + \frac{11}{16} \times \frac{3}{4} = \frac{41}{64}$$

which is the exact value. □

In the case where the number of elements in FS exceeds the number of inputs to the circuit, the PREDICT algorithm for computing the signal probabilities of the nontree lines becomes more expensive than exhaustive good machine simulation. In these cases it is better to perform the exhaustive true value simulation. Since the computational complexity of signal probabilities is generally exponential, PREDICT uses a heuristic that simplifies this computation at the expense of computing only approximate results. The heuristic is based on assuming independence between signals that meet some topological conditions. The decision as to which signals to consider independent is based on the distance property defined next.

Definition 7.18. The *distance* $d(w_1, w_2)$ *between lines* w_1 *and* w_2 is defined as the minimum number of gates that a signal at w_1 has to traverse before it reaches line w_2.

The heuristic uses an integer n as a threshold distance in order to determine which signals can be considered independent. The heuristic would trace back all paths up to distance n from the nontree line and assume that the signals at the lines so arrived as being independent. In this way, the effect of reconvergent fan-out within a distance n of the line of interest will be taken into account, and the effect of those which are further apart will be ignored. At one extreme, when the threshold is $n = 1$, the circuit is handled as if it was a tree structure. At the other extreme, when the threshold exceeds the number of levels in the circuit, all fan-outs are handled properly, and the calculation is therefore exact.

Example 7.16. Repeat the calculation of the signal probability of the output F in Figure 7.29, using the PREDICT heuristic with distance threshold $n = 3$.

SOLUTION. Figure 7.30 shows in boldface the lines that are included within a distance 3 from the output F. Notice that although the fan-out stem w_1 is included within a radius of 3 from the output F, only two of its three fan-out branches are. Figure 7.30 shows the computation of the conditional signal probabilities $p(F|w_1 = 0)$ and $p(F|w_1 = 1)$. Notice also that the signal on line w_2 in Figure 7.30 is considered independent of the signal on line w_1 according to the heuristic. The PREDICT heuristic will therefore conclude that the signal probability of the output F is

$$p(F) = \frac{11}{16} \times \frac{1}{4} + \frac{43}{64} \times \frac{3}{4} = \frac{173}{256} \quad \square$$

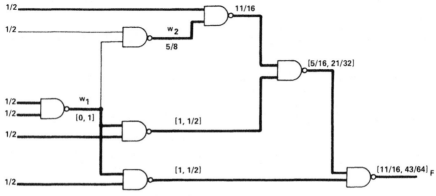

Figure 7.30. Example 7.16.

In order to show how PREDICT computes the one-level sensitization probability we denote by $U(g|A)$ the conditional sensitization probability of line g given that an assignment A has been made to the elements of FS. Then, the one-level sensitization probability of line g is given by

$$U(g) = \sum_{A} U(g|A)\Pr(A) \tag{7.68}$$

7.9.4   Controllability Observability Program (COP)

Like PREDICT, COP's objective is to assess the detection probabilities of faults by using analytical tools [Brglez et al. (1984)]. In order to keep the procedure as simple as possible it does not take into account the statistical dependence of signals at a reconvergent point. Therefore COP will not compute exact results for the detection probability. It will also not compute one-sided results as in the case of the cutting algorithm.

COP propagates 1-controllabilities using the tree formulas for signal probabilities (see Figure 7.4) and computes the 0-controllabilities as 1 minus the 1-controllabilities. To see how COP computes observabilities, let w_1 and w_2 be two input lines to a gate, and let w_3 be the output of the gate. The observability $B(w_1)$ of w_1 is computed from $B(w_3)$ according to the following formulas:

For two-input AND or NAND gates,

$$B(w_1) = C_1(w_2)B(w_3) \tag{7.69}$$

For two-input OR or NOR gates,

$$B(w_1) = C_0(w_2)B(w_3) \tag{7.70}$$

For an INVERTER with input w_1 and output w_2,

$$B(w_1) = B(w_2) \tag{7.71}$$

If w_1 is a fan-out stem whose branches are w_2 and w_3, the observability of the stem is computed by the formula

$$B(w_1) = 1 - \left[1 - B(w_2)\right]\left[1 - B(w_3)\right] \tag{7.72}$$

As in PREDICT, the observability of the primary outputs is defined to be 1. Formulas (7.69), (7.70), and (7.72) can be similarly extended to higher input gates.

TABLE 7.9   COP Results Versus Exact Results for Example 7.17

Line	C_0	C_1	B	C-SA0	C-SA1	SA0	SA1
x_1	1/2	1/2	1/8	1/16	1/16	1/4	1/4
x_2	1/2	1/2	13/16	13/32	13/32	1/4	1/4
w_1	1/2	1/2	1/4	1/8	1/8	1/4	0
w_2	1/2	1/2	3/4	3/8	3/8	1/4	1/2
w_3	3/4	1/4	1/2	1/8	3/8	1/4	1/4
w_4	1/2	1/2	3/4	3/8	3/8	1/2	1/4
F	3/8	5/8	1	5/8	3/8	3/4	1/4

The detection probabilities for stuck-at faults are then computed by the following formulas:

For stuck-at-0 faults,

$$q(w/0) = C_1(w)B(w) \tag{7.73}$$

For stuck-at-1 faults,

$$q(w/1) = C_0(w)B(w) \tag{7.74}$$

Example 7.17. Consider again the circuit of Figure 7.28 and compute the detection probabilities of all the single stuck-at faults according to COP.

SOLUTION. Table 7.9 shows the results obtained using COP. The columns C-SA0 and C-SA1 refer to the detection probabilities of the stuck-at-0 fault and the stuck-at-1 fault as predicted by COP. The columns SA0 and SA1 correspond to the correct results. □

7.10   RANDOM PATTERN TESTING OF RANDOM ACCESS MEMORIES

This section discusses the problem of testing for stuck-at storage cells in an embedded memory. Figure 7.31 shows the memory and its embedding logic. The data-in lines, the address lines, and the read/write control line are not driven directly by primary inputs. The data-out lines of the memory are assumed to be directly observable. The memory has n data-in lines, m address lines, and a single read/write control line. The m-input 2^m output address decoder is assumed to be internal to the memory. The memory outputs are assumed not to be latched, so that during memory write operations the outputs display a fixed value.

Since the hardest stuck fault to detect with random test patterns is one which affects a single-bit storage location [McAnney et al. (1984)], this section concentrates on detecting these faults. A test designed to detect the single

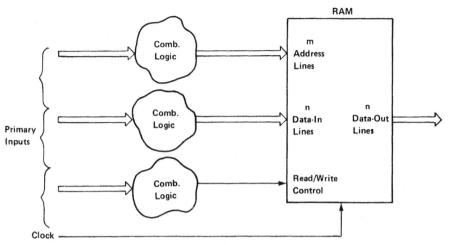

Figure 7.31. An embedded random access memory.

stuck cells with a given confidence will also detect easier faults with equal or greater confidence.

Since a decoder stuck fault is easier to detect than a single stuck-at cell, we will not consider such faults in our analysis. With this ground rule each input vector selects only one address.

We further assume that reading from the memory is nondestructive and that the logic preceding the memory does not contain any sequential circuits. In the following analysis we make the simplifying assumption that each of the data-in lines (at the memory boundary) has the same signal probability p_d and that the signals on these lines are independent. Similarly, it is assumed here that each of the address lines has a signal probability of p_a and that the signals on these lines are also independent. Since the memory has a single read/write control line, the read and the write operations are mutually exclusive. Let p_c be the signal probability of the read/write control line. Assume that the memory performs a write operation when this control line has the value 1 and performs a read operation when it is 0. Thus the probability of writing into the memory at any given clock is p_c and the probability of reading from the memory is $1 - p_c$.

Before applying the random test, the memory is initialized to a given state, perhaps by write operations or a power-on reset. The discussion to follow analyzes the effect of initializing the memory to all 0's or all 1's. Regardless of the initialization method, a stuck-at storage cell will remain at its stuck value during the initialization process.

In the binary representation of an arbitrary memory address, there are h 1's and $m - h$ 0's. The probability r of selecting this address is

$$r = p_a^h (1 - p_a)^{m-h}$$

The probability that a random vector writes to a particular address is

$$rp_c = p_c p_a^h (1 - p_a)^{m-h}$$

The probability of writing a 1 to an arbitrary bit in a given address location is

$$p_1 = p_d p_c r$$

Similarly, the probability of writing a 0 to an arbitrary bit in a given address location is

$$p_0 = (1 - p_d) p_c r$$

The probability of reading from a given address is

$$p_r = (1 - p_c) r$$

Since at any given time either a read, a write 0, or a write 1 operation is occurring,

$$r = p_1 + p_0 + p_r$$

Let an arbitrary word in memory have a single cell which is stuck-at-0. Assume that there are no other faults in the memory. We wish to calculate the cdp of a random pattern test of length t, denoted $Q_0(t)$.

Figure 7.32 describes the detection process of a stuck-at-0 cell in an arbitrary word. State S_0 represents the storage of a logical 0 in the cell. State S_1 is entered from state S_0 by writing 1 to the cell and represents a storage of logical 1. Detection of a stuck-at-0 fault occurs from state S_1 on a read operation, moving the system into state S_2. State S_2 is made an absorbing state since we are only interested in the first detection of this fault. The transition probabilities between states are labeled along their corresponding edges in Figure 7.32. The difference equations describing the transitions between states are

$$s_0(t) = (1 - p_1) s_0(t - 1) + p_0 s_1(t - 1) \qquad (7.75)$$
$$s_1(t) = p_1 s_0(t - 1) + (1 - p_0 - p_r) s_1(t - 1) \qquad (7.76)$$
$$s_2(t) = p_r s_1(t - 1) + s_2(t - 1) \qquad (7.77)$$

Let the memory cells be initialized to all 1's with probability i_0. Thus, the

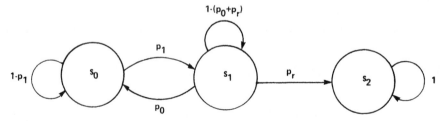

Figure 7.32. Markov chain of detecting a stuck-at-0 fault.

initial state probabilities are

$$s_0(0) = 1 - i_0, \qquad s_1(0) = i_0, \qquad s_2(0) = 0$$

The cdp of a test of length t is the state probability $s_2(t)$, the solution of which is

$$Q_0(t) = 1 - \frac{C_1(A - B)^t + C_2(A + B)^t}{2B} \qquad (7.78)$$

where

$$A = 1 - \frac{r}{2}, \qquad B = \frac{1}{2}\sqrt{r^2 - 4p_1 p_r}$$

$$C_1 = -1 + A + B + p_r i_0, \qquad C_2 = 1 - A + B - p_r i_0$$

By substituting the values of A, B, C_1, and C_2 in terms of p_d, p_c, and r we get

$$Q_0(t) = 1 - \frac{1 - 2(1 - p_c)i_0 + \alpha}{2\alpha}\left(1 - \frac{r}{2} + \alpha\frac{r}{2}\right)^t$$
$$+ \frac{1 - 2(1 - p_c)i_0 - \alpha}{2\alpha}\left(1 - \frac{r}{2} - \alpha\frac{r}{2}\right)^t \qquad (7.79)$$

where $\alpha = \sqrt{1 - 4p_d p_c(1 - p_c)}$.

The test length needed to detect a stuck-at-0 storage cell can be derived from (7.79). Assuming that the test length is large, we observe that for the case in which $\alpha \neq 0$ the dominant term in (7.79) is $(1 - r/2 + \alpha r/2)^t$. Let e_{th} be the desired escape probability. Then, the value of t that satisfies the equation

$$Q_0(t) = 1 - \frac{1 - 2(1 - p_c)i_0 + \alpha}{2\alpha}\left(1 - \frac{r}{2} + \alpha\frac{r}{2}\right)^t = 1 - e_{th}$$

is the test length T_0. Thus,

$$T_0 \approx \left| \frac{\ln\left[\dfrac{2\alpha e_{th}}{1 + \alpha - 2(1 - p_c)i_0}\right]}{\ln\left[1 - \dfrac{(1 - \alpha)r}{2}\right]} \right| \qquad (7.80)$$

Figure 7.33 describes the Markov chain for detecting a stuck-at-1 fault in a storage cell. State S_0 corresponds to storing a 0 in a cell. State S_1 corresponds to storing a 1 in a cell and state S_2 is the detection state. Let the memory cells be initialized to all 1's with probability i_0. Thus, the initial state probabilities are

$$s_0(0) = 1 - i_0, \qquad s_1(0) = i_0, \qquad s_2(0) = 0$$

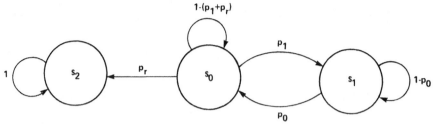

Figure 7.33. Markov chain of detecting a stuck-at-1 fault.

By carrying out an analysis similar to the one shown for the stuck-at-0 case, the cdp of a test of length t is computed to be

$$Q_1(t) = 1 - \frac{1 - 2(1 - p_c)(1 - i_0) + \beta}{2\beta}\left(1 - \frac{r}{2} + \beta\frac{r}{2}\right)^t$$

$$+ \frac{1 - 2(1 - p_c)(1 - i_0) - \beta}{2\beta}\left(1 - \frac{r}{2} - \beta\frac{r}{2}\right)^t \qquad (7.81)$$

where $\beta = \sqrt{1 - 4(1 - p_d)p_c(1 - p_c)}$.

Similarly, the test length necessary to detect a stuck-at-1 storage cell (provided β is not equal to 0) is approximately

$$T_1 \simeq \left|\frac{\ln\left[\dfrac{2\beta e_{\text{th}}}{1 + \beta - 2(1 - p_c)(1 - i_0)}\right]}{\ln\left[1 - \dfrac{(1 - \beta)r}{2}\right]}\right| \qquad (7.82)$$

Since there is no advance information as to whether a faulty RAM has a stuck-at-1 fault or a stuck-at-0 fault, the test length to detect a stuck-at fault should be the larger of the two. Thus the test length T that will detect any stuck-at storage cell with escape probability no larger than e_{th} is

$$T = \max\{T_0, T_1\} \qquad (7.83)$$

A random pattern test of length T can therefore constitute a pass/fail test for the embedded memory with confidence of at least $1 - e_{\text{th}}$.

There is a point regarding the test length formulas that has to be clarified. The point is that the address selection probability r is generally different for different addresses (except for the case where $p_a = 1/2$, in which case $r = 1/2^m$), and the question therefore arises as to which r to use. If the signal probability p_a on an address line is less than $1/2$, the address which is most difficult to access is $111 \cdots 1$, and the minimum r is therefore p_a^m. Conversely,

TABLE 7.10   Test Length Required to Detect Stuck-at-0, Stuck-at-1, and Either Stuck-at-0 or Stuck-at-1 Storage Cells in a 32-Word RAM Initialized To All 0's for Various Values of p_c and p_d. The Access Probability is $r = 1/32$ and Escape Probability Threshold is $e_{th} = .01$.

p_d	$p_c = .1$	$p_c = .3$	$p_c = .5$	$p_c = .7$	$p_c = .9$
(a) Test Length for Detecting a Stuck-at-0 Fault					
.1	16256	6898	5775	6898	16256
.3	5338	2215	1839	2215	5338
.5	3153	1273	1046	1273	3153
.7	2215	865	699	865	2215
.9	1693	633	497	633	1693
(b) Test Length for Detecting a Stuck-at-1 Fault					
.1	208	295	367	568	1651
.3	626	524	552	784	2161
.5	1194	849	853	1162	3078
.7	2344	1558	1531	2031	5213
.9	7841	5025	4877	6352	15881
(c) Test Length for Detecting any Stuck-at Fault					
.1	16256	6898	5775	6898	16256
.3	5338	2215	1839	2215	5338
.5	3153	1273	1046	1273	3153
.7	2344	1558	1531	2031	5213
.9	7841	5025	4877	6352	15881

if $p_a > 1/2$, the address which is most difficult to access is $000 \cdots 0$, and the minimum r is $(1 - p_a)^m$. If the address selection probabilities are not all equal, one has to compute the minimum r and use it in the formulas for T_0 and T_1. The test length computed in this manner will constitute an upper bound for the actual test length needed; in other words, the test length so computed will insure with confidence of at least $1 - e_{th}$ that any stuck-at storage cell will be detected.

Tables 7.10 and 7.11 display the test length required to detect stuck-at cells for various values of the RAM parameters.

7.11   FAULT PROPAGATION THROUGH RANDOM ACCESS MEMORIES

The memory model considered here is shown in Figure 7.31. The memory has n data-in lines and m address lines. It also has a single read/write control line so that read and write operations are mutually exclusive. The memory is written into when the value on the read/write control is 1 and is read from when the value on this line is 0. As in Section 7.10, it is assumed that the

TABLE 7.11   Test Length Required to Detect Stuck-at-0, Stuck-at-1, and Either Stuck-at-0 or Stuck-at-1 Storage Cells in a 32-Word RAM Initialized To All 1's for Various Values of p_c and p_d.The Access Probability is $r = 1/32$ and Escape Probability Threshold is $e_{th} = .01$.

p_d	$p_c = .1$	$p_c = .3$	$p_c = .5$	$p_c = .7$	$p_c = .9$
(a) Test Length for Detecting a Stuck-at-0 Fault					
.1	7841	5025	4877	6352	15881
.3	2344	1558	1531	2031	5213
.5	1194	849	853	1162	3078
.7	626	524	552	784	2161
.9	208	295	367	568	1651
(b) Test Length for Detecting a Stuck-at-1 Fault					
.1	1693	633	497	633	1693
.3	2215	865	699	865	2215
.5	3153	1273	1046	1273	3153
.7	5338	2215	1839	2215	5338
.9	16256	6898	5775	6898	16256
(c) Test Length for Detecting any Stuck-at Fault					
.1	7841	5025	4877	6352	15881
.3	2344	1558	1531	2031	5213
.5	3153	1273	1046	1273	3153
.7	5338	2215	1839	2215	5338
.9	16256	6898	5775	6898	16256

address decoder is internal to the memory. The memory outputs are not latched, and during memory write operations the outputs assume a fixed value. Only one address can be selected by each input vector. It is further assumed that the memory outputs are directly observable and that the circuitry driving the data-in lines, the address lines, and the read/write control line are independent.

This section describes how detection probabilities of faults in the combinational logic feeding either the data-in lines (called prelogic) or the address lines are affected by the presence of the RAM. Section 7.11.1 considers the propagation of detection probabilities of faults in the prelogic through the memory and Section 7.11.2 considers the propagation of faults in the address line logic through the memory. Since it is impossible to directly observe the data-in lines or the address lines, it is necessary to infer the presence of a fault from the reading of an occasional incorrect word on the memory output lines.

This section uses the same notation as in the previous section regarding the parameters involved. Namely, p_c is the signal probability of the read/write control line, as well as the probability of a write operation. Similarly, $1 - p_c$ is the probability of a read operation. The parameter r is the probability of accessing a word in the memory and is assumed to be equal for all addresses.

7.11.1 Propagation of Prelogic Faults

The prelogic has to be tested for stuck-at faults by applying random patterns to its primary inputs and observing the responses on the memory data-output lines, either directly or via a compressed signature. The memory, including the internal address decoder, is assumed to be fault-free.

Let q be the probability that an incorrect word appears at the data-in lines of the memory due to a prelogic fault. When the memory is actively writing, an incorrect word will be written to some address with probability q. The parameter q can be viewed as the detection probability of the fault as measured at the data-in lines. Given q, it is necessary to compute q', the probability that an incorrect word appears at the data-out lines. The parameter q' is the detection probability of the fault as measured at the data-output lines.

Figure 7.34 describes the Markov chain for detecting a prelogic fault through the embedded memory. The Markov chain displays the activities leading to the reading of an incorrect word from a specific address A in the memory. State S_0 corresponds to a correct word stored in address A. State S_1 corresponds to an incorrect word stored in address A. State S_2 is the detection state, namely, a state in which an incorrect word is read out from address A.

The transition probability between state S_0 and S_1 is the probability of writing an incorrect word into address A. For this transition to occur, the memory should be actively writing (probability p_c); the address selection lines should access address A (probability r); and the fault should be active at the data-in lines (probability q). Thus,

$$a = p_c r q$$

The transition probability between states S_1 and S_0, and between S_2 and S_0 is the probability of writing correctly into address A. For these transitions to occur, the memory must be actively writing; the address selection lines should access address A; and the fault should be inactive (probability $1 - q$). Thus,

$$b = p_c r (1 - q)$$

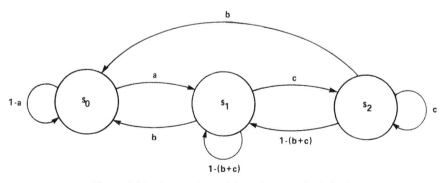

Figure 7.34. Markov chain of detecting a prelogic fault.

The transition probability between states S_1 and S_2 is the probability of reading out the error stored in address A. Thus,

$$c = (1 - p_c)r$$

Note that the probability of staying in state S_2 is also c. The remaining transition probabilities between states are derivable from a, b, and c, and are shown in Figure 7.34. The transition matrix for this Markov chain is

$$\mathbf{P} = \begin{bmatrix} 1 - a & a & 0 \\ b & 1 - (b + c) & c \\ b & 1 - (b + c) & c \end{bmatrix}$$

The objective is to compute the probability of being in state S_2 in steady state, namely the probability of being in this state after many random patterns have been applied. This probability constitutes the probability of reading an incorrect word from address A after the effect of the memory initialization has been eroded. In order to compute that, we may use the same technique discussed in Section 7.2.2 regarding stationary state distribution. We have to solve Equations (7.14) and (7.15); the solution for π_2 is the one we are interested in. The system of equations in this case is

$$\begin{bmatrix} -a & b & b \\ 0 & c & c - 1 \\ 1 & 1 & 1 \end{bmatrix} \begin{bmatrix} \pi_0 \\ \pi_1 \\ \pi_2 \end{bmatrix} = \begin{bmatrix} 0 \\ 0 \\ 1 \end{bmatrix}$$

The solution for π_2 is

$$\pi_2 = \frac{ac}{a + b} = (1 - p_c)rq$$

The probability π_2 is, in fact, the contribution to the output detection probability by a single address in the memory. In order to compute the output detection probability (in steady state) it is necessary to add the contribution from all addresses. Thus,

$$q' = \sum_{i=0}^{2^m - 1} \pi_2 = (1 - p_c)q \tag{7.84}$$

7.11.2 Propagation of Address Line Faults

Consider now the effect of a fault in the combinational logic driving the address decoder. Assume that the detection probability q of the fault at the decoder inputs is known. Then q is the probability that an incorrect address occurs at the decoder inputs. Since we assume that the memory itself as well as its internal decoder are fault-free, then every time the fault is present, exactly two addresses in the memory are affected. Consider a random vector which accesses word A in fault-free operation. When the fault is present the decoder sees a vector that represents a different address in memory, say address B. If the memory is actively writing, then the effect of the fault is to write to B the word which should have gone to A. If the memory is actively reading, then the effect of the fault is to read from address B as opposed to reading from the intended address A. Assume now that the memory is actively writing and the fault is present. Even though both A and B are affected, it does not necessarily follow that both will contain incorrect data. An error in A only occurs if the word which should have been written to A is different from the one stored in A. Similarly, B is in error only if the word erroneously written into it is different from the one expected at B. A similar situation occurs when the memory is actively reading and the fault is present. An output error is observed only when the data read from B is different from the word stored in A.

To reduce the amount of computation required to understand the effects of various address logic faults, we will make the assumption that the effect of any such fault is that which is most difficult to detect with random test patterns. This worst case fault is called a *single-mapping fault*. A single-mapping fault is a fault in the logic driving the address decoder inputs which, when active, causes all selected addresses to be mapped to a single *common address*.

A single-mapping fault is quite easy to visualize. Imagine that the terminal circuit on each of the m outputs of the address logic is a two-input AND gate. The output of each gate drives one input of the address decoder. One input of each gate carries the value of the address bit. The second inputs of each gate are tied together to form a global address deselect line. Imagine a fault upstream of this deselect line which, for some input vectors, causes the deselect line to go to a logical 0. In this situation the actual address accessed is the all-0's address, regardless of the address which should have been selected.

The use of a single-mapping fault concept results in a lower bound on the output detection probability of any kind of address logic fault. To see this consider a case where the memory is always writing (the read/write control line is set to a constant 1). Consider a fault which causes a stuck-at fault at one of the address decoder inputs, regardless of the random pattern being applied. For simplicity assume that this bit is the high order bit. The effect of this fault is to write only into the higher half of the memory. Every time the high order bit is different from its stuck value, two addresses are affected. These two addresses are called a *fault-coupled pair*. There are 2^{m-1} fault-coupled pairs of addresses. In each pair the last $m - 1$ bits are identical and only the first bits are different. Assume that the values on the address lines are equally likely in fault-free operation. Then, after a long period of time approximately half of the vectors will access the right addresses, and, therefore, half of the patterns will access their fault-coupled counterpart. Each time the fault-coupled counterpart is being written into due to the fault, both

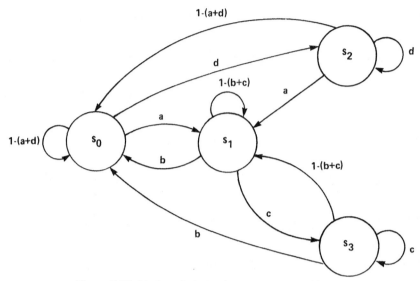

Figure 7.35. Markov chain for the noncommon address.

members of the fault-coupled pair are being affected. Since all 2^{m-1} fault-coupled pairs are different, the error density inside the memory will be quite high. On the other hand, a single-mapping fault under the same conditions (namely $q = 1/2$) will cause roughly half this density. While there are two addresses affected by each faulty write, one of the two is always the common address. For a large memory infested with a single-mapping fault, it approximately holds that only a single affected address occurs for each faulty write. Since the single-mapping fault produces the lowest error density inside the memory, it is the most difficult one to detect.

Figure 7.35 shows the Markov chain for detecting a fault from an arbitrary noncommon address, say address A. State S_0 corresponds to a state where the contents of A are correct. State S_1 corresponds to a state where the contents of A are incorrect. State S_2 corresponds to a state where a detection has occurred and the contents of A are still correct. State S_3 corresponds to a state where a detection has occurred and the contents of A are incorrect. The objective is to compute the sum of the probabilities of being in states S_2 and S_3 in steady state.

The one-step transition probability from state S_0 to state S_1 and that from state S_2 to state S_1 are identical and denoted by a in Figure 7.35. These transitions occur when the memory is actively writing (probability p_c); the address logic is selecting address A (probability r); the fault is active at the decoder inputs (probability q); and when there is a mismatch between the word currently stored in A and the word that ought to have been written into A, but because of the presence of the fault, was written into the common address. The probability of having a match between these two words is denoted μ, where

$$\mu = 2^{-n}$$

The probability of a mismatch is, therefore, $1 - \mu$. The transition probability a is therefore

$$a = p_c rq(1 - \mu)$$

The transition probability between states S_1 and S_0 and that between states S_3 and S_0 are identical and are denoted by b in Figure 7.35. For these transitions to occur the erroneous contents of A must be corrected. There are two ways in which the contents of A could be corrected. One way to correct the contents of A is by writing into address A while the fault is inactive. Another way of correcting the contents of A is by selecting address A while the memory is writing; have the fault strike, and, have a match between the word currently stored in A and the new word that ought to have gone to A, but because of the presence of the fault, went into the common address. The transition probability b is, therefore,

$$b = p_c r(1 - q) + p_c rq\mu = p_c r[1 - q(1 - \mu)]$$

The one-step transition probability between states S_1 and S_3, and the probability of staying in state S_3, is denoted by c in Figure 7.35. Again, there are two situations in which this transition can occur. One situation is to have the memory in read mode; have the address logic select address A; and have the fault dormant. The probability of this sequence of events is $(1 - p_c)r(1 - q)$. Another situation that will lead to these transitions is having the fault present while the memory is intending to read from address A. In this case the read will occur from the common address rather than from A. In order for detection to occur the memory must be actively reading and selecting address A; the fault should be active at the decoder inputs; the most recent write into the common address should have not been induced by an erroneous access to address A; and there should be a mismatch between the word read from the common address and the word that is currently stored in A. The reason why it is necessary to require that the most recent write into the common address be a result of something different from an erroneous access of address A is, that if indeed an erroneous access of address A led to the last write into the common address, then a subsequent erroneous read from address A will read the contents of the common address which is identical to the word that was expected to be in A in fault-free operation, and therefore no detection will occur. The probability that the common address was erroneously written into as a result of an erroneous access to address A is $p_c rq/[p_c(1 - r)q + p_c r]$ and the probability that the common address was written into as a result of a different reason is, therefore, $[(1 - 2r)q + r]/[(1 - r)q + r]$. The transition probability c is, therefore,

$$c = (1 - p_c)r(1 - q) + (1 - p_c)rq(1 - \mu)\frac{(1 - 2r)q + r}{(1 - r)q + r}$$

The transition probability between states S_0 and S_2, and the probability of staying in state S_2 is denoted by d in Figure 7.35. In order for these transitions to occur the memory must be reading from address A; the fault should be active at the decoder inputs, and there should be a mismatch between the contents of the common address and the contents of the word currently stored in A. It is also necessary to require that the contents in the common address not be due to a misdirected write to address A. Thus,

$$d = (1 - p_c)rq(1 - \mu)\frac{(1 - 2r)q + r}{(1 - r)q + r}$$

The remaining transition probabilities in Figure 7.35 are derivable from a, b, c, and d, and are shown in the figure. The transition matrix for this chain is

$$\mathbf{P} = \begin{bmatrix} 1 - (a + d) & a & d & 0 \\ b & 1 - (b + c) & 0 & c \\ 1 - (a + d) & a & d & 0 \\ b & 1 - (b + c) & 0 & c \end{bmatrix}$$

We are interested in $\pi_{12} + \pi_{13}$, where π_{12} and π_{13} are the steady state probabilities for S_2 and S_3. The solution of Equations (7.14) and (7.15) for this Markov chain yields

$$\pi_{12} + \pi_{13} = \frac{ac + bd}{a + b} = (1 - p_c)rq(1 - \mu)\left[1 - q\frac{(1 - 2r)q + r}{(1 - r)q + r}\right]$$

Consider now the possible errors that can occur when the memory attempts to read from the common address (notice that in the Markov chain of Figure 7.35 the memory always made an attempt to access the noncommon address). Figure 7.36 shows the Markov chain of detecting a fault when the memory is accessing the common address. State S_0 corresponds to a state where the contents of the common address C is correct. State S_1 corresponds to a state where the contents of C is incorrect. State S_2 corresponds to a state where a detection occurs and the contents of C is incorrect. Notice that in this chain, unlike in the previous chain, it is impossible to select the common address, read an error from it, and leave its contents correct. The reason for this is that when aiming at the common address, no matter whether the fault strikes or not, the actual word accessed is the common address. The only way to read an

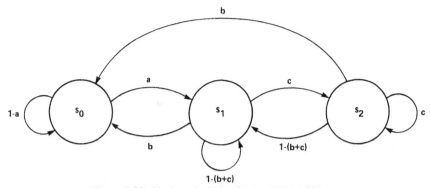

Figure 7.36. Markov chain for the common address.

error from the common address (while aiming at the common address) is by previously having a misdirected write into the common address and then reading from the common address.

The transition probability between states S_0 and S_1 is denoted by a in Figure 7.36. This transition occurs when the memory is attempting to write into one of the noncommon addresses while the fault strikes, and when there is a mismatch between the current content of C and the new word that has been misdirected to C. Thus,

$$a = p_c(1 - r)q(1 - \mu)$$

The transition between states S_1 and S_0 and that between S_2 and S_0 occurs when the erroneous content of C is corrected. There are two ways to correct the contents of C. One way is just by writing into C (probability $p_c r$). In this case, no matter whether the fault strikes or not, the word being accessed is C, and therefore its content is correct. Another way is to have the fault strike while the memory is attempting to write into one of the noncommon addresses and have a match between the current content of C and the word being misdirected into C. Thus,

$$b = p_c r + p_c(1 - r)q\mu = p_c[r + (1 - r)q\mu]$$

The transition between states S_1 and S_2 and the probability of staying in S_2 is denoted by c in Figure 7.36. The only condition necessary for these transitions to occur is to have the memory read from the common address. Notice that here, also, independent of whether the fault is active or not, the memory will be reading the content of the common address, and since this content is erroneous, an error will be detected. Thus,

$$c = (1 - p_c)r$$

The remaining transition probabilities of the Markov chain are derivable from a, b, and c, and are shown in Figure 7.36. The transition matrix for the Markov chain is

$$\mathbf{P} = \begin{bmatrix} 1 - a & a & 0 \\ b & 1 - (b + c) & c \\ b & 1 - (b + c) & c \end{bmatrix}$$

We are interested in π_{22}, the probability of being in state S_2 in steady state. Solution of Equations (7.14) and (7.15) yield

$$\pi_{22} = \frac{ac}{a + b} = \frac{(1 - p_c)r(1 - r)q(1 - \mu)}{r + (1 - r)q}$$

In order to compute the detection probability q' at the memory data-output lines, it is necessary to add the contribution of all the noncommon addresses to that of the common address. Thus,

$$q' = \sum_{i=0}^{2^m - 2} (\pi_{12} + \pi_{13}) + \pi_{22}$$

The final expression for the output detection probability in steady state is therefore

$$q' = (1 - p_c)(1 - r)q(1 - \mu)\left[2 - q\frac{r(1 - q)}{(1 - r)q + r} \right] \qquad (7.85)$$

7.12 RANDOM PATTERN TESTABILITY OF DELAY FAULTS

The maximum allowable path delay in a synchronous computer is determined by its clock rate. If the delay on a path of a manufactured network exceeds the time period between clocks, incorrect output values may be latched. The objective of delay testing is to ensure that the manufactured network operates correctly at the functional clock rate.

Given a desired clock rate, there are two different philosophies used to design the network paths within that rate. In a "worst case" design philosophy, the delay of a logic path is computed as the sum of the worst case delays of each gate in the path plus the worst case wiring delay, and each path must have a worst case delay no greater than the clock interval. However, since it infrequently happens that every gate in a path exhibits its worst case delay, these path lengths are quite conservative.

The second approach, a "statistical" design philosophy, recognizes that each gate in a path has a distribution of delays caused by manufacturing variations. The summation of these gate distributions results in a path delay distribution. Path lengths are chosen so that their statistical delay lies within the clock interval. When a statistical design philosophy is used, however, manufacturing variances can create networks whose gate delays are within specifications but whose path delays exceed the clock interval. In either design philosophy, delay testing has to detect the presence of excessive path delays caused by manufacturing defects. Additionally, if a statistical design is used, delay testing must detect the presence of slow paths.

When a logic circuit is manufactured, the actual gate delays may differ from their specifications. A gate whose propagation delay exceeds the specified worst case is quite possible. This behavior is referred to as a *gate delay defect*. Denote by DP_i the propagation delay associated with any latch-to-latch path i in the circuit. The clock rate of the machine determines the maximum acceptable path delay DP_{max}. Due to "slow down" delay faults, the actual DP_{max} of the manufactured circuit may exceed the DP_{max} predicted by the logic designer. Since the predicted DP_{max} was chosen based on the clocking rate of the machine, the faulty circuit may not operate correctly. A delay fault cannot be detected by a static test, because the logic function of the faulty gate is unaffected.

A gate delay defect does not necessarily cause a path delay longer than the clock period. The actual delay of a faulty gate can be greater than the predicted worst case value, but be compensated for by faster gates along the propagation path.

Delay testing is used to verify that the manufactured logic operates correctly within the constraints imposed by the system clocks. The objective is to ensure that the actual DP_{max} of a manufactured circuit is less than or equal to the clock period (or interval between clocks, if multiple clocks are used) by running delay tests at actual machine speed or faster. (It is prudent to run the tests at a clock speed which is slightly faster than the actual machine speed to account for possible differences between the operating conditions during test and in the application.) If any discrepancy from the expected test response is observed, a delay fault is said to have been detected. Thus, a path has a delay fault if the signal propagation time through the path is greater than the clock interval. We refer to this fault as a *path fault* [Smith (1985)].

7.12.1 Hazard-Free Delay Test

The network model to be used is shown in Figure 7.37. An input register rank consisting of n latches feeds an n-input combinational circuit whose k outputs in turn feed the k latches of the output register. The structure can be considered as one segment of an LSSD "single-latch design."

Under test, we assume that the combinational circuit is fed with random patterns at machine speed. The machine is clocked by a pair of nonoverlapping clocks C_1 and C_2 as shown in Figure 7.37. During test, the input register generates random patterns which are launched into the combinational logic by

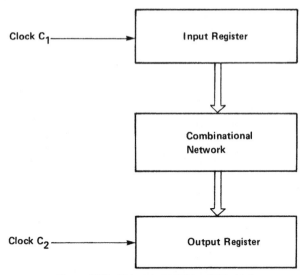

Figure 7.37. Network model for delay test.

clock C_1. The output register captures the responses from the combinational logic whenever triggered by clock C_2. To avoid having to interrupt the test after each C_2 clock for a check of the captured responses, assume that in test mode the output register is restructured into a multiple-input signature analyzer.

Any latch in the input register which changes from 0 to 1 or from 1 to 0 between two consecutive input patterns generates a transition at the input driven by that latch. We further assume that an input latch whose initial and final values are identical does not glitch between the two values. The input transition will propagate into the combinational logic, and may or may not reach the output register in time to be latched by clock C_2, depending first upon whether it is sensitized along a propagation path, and second upon whether or not the path delay along the sensitized path exceeds the interval between clocks. The time it takes a transition to propagate through the combinational logic is the sum of the gate delays along the path traversed by the transition. The values captured in the output register are the values existing at the combinational logic outputs when clock C_2 arrives.

In order to test for possible delay faults in the circuit it is necessary to propagate transitions along every path from inputs to outputs. A circuit is said to pass the delay test if all the necessary transitions arrive at the primary outputs before the application of clock C_2.

Consider the circuit of Figure 7.38(a). We are interested in testing whether or not the boldfaced path $x_2 G_1 G_4 G_6$ is slow. In order to guarantee that this path is free of delay faults, it is necessary to show that both a rising transition and a falling transition originating at input x_2 and propagating along the path

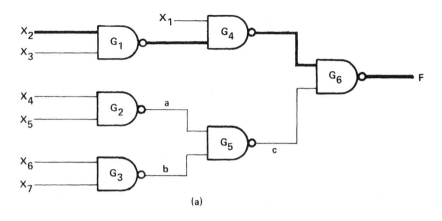

(a)

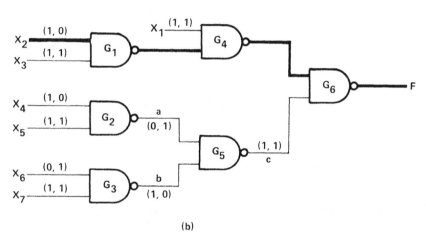

(b)

Figure 7.38. One possibility for a delay test.

can reach the output F before the arrival of clock C_2. For a transition generated at x_2 to propagate to the output F it is necessary to have the boldfaced path in a sensitized condition. This requires that inputs x_1, x_3 and line c, be at a value 1. Figure 7.38(b) shows a pair of vectors that satisfy these conditions. In order to test for a falling transition at input x_2 we apply the vector $v_1 = 1111101$ followed by the vector $v_2 = 1010111$. The pair of values (i, j) attached to a line in Figure 7.38(b) is the bit values that the vectors v_1 and v_2 assign to the line, respectively. Notice that all the requirements placed on lines x_1, x_3 and line c are satisfied.

The transition generated at input x_2 will, in fact, propagate to the output F, but this transition may include a glitch that under certain conditions will

Figure 7.39. An unintended spike during delay test.

invalidate the test. To see this, observe that both inputs to gate G_5 change values between the application of v_1 and v_2. If the delay of gate G_2 is shorter than that of G_3, the inputs to gate G_5 will change from $(a, b) = (0, 1)$ to $(a, b) = (1, 1)$ before reaching their final value $(a, b) = (1, 0)$. This will create a falling spike at line c. This momentary falling spike at line c may cause a rising spike at the output F as shown in Figure 7.39.

The timing of the rising spike at the output F depends on the gate delay values in the circuit. Assume for a moment that the boldfaced path is slow so that the final transition in Figure 7.39 is late, relative to clock C_2. If it just happens that clock C_2 arrives during the rising spike at the output F, the boldfaced path will be erroneously declared good.

Consider now the case of generating a rising transition at input x_2 using the vectors $v_1 = 1011101$ and $v_2 = 1110111$. Here also, line c experiences a hazard that may cause a momentary rising spike at the output F. In this case, however, the momentary spike can only occur after the transition generated at x_2 has reached the output F. If clock C_2 happens to arrive during this spike, an incorrect response will be observed at the output F. This incorrect response will cause the circuit to be declared faulty. This conclusion, however, is not an incorrect one. If this behavior really happens then the circuit is faulty, not because the boldfaced path is slow but because one of the paths passing through gate G_5 is slow. The only reason the spike can reach the output line during the application of clock C_2 is because there exists a path fault in the circuit. This path fault is not the one tested for by the delay test pair, but a different one. This phenomenon is true in general, namely, the existence of a hazard cannot cause a good circuit to be declared faulty, but it may cause a bad circuit to pass the test.

In order to avoid the problem of occasionally declaring a bad circuit as being good, it is necessary to perform a hazard-free delay test. A delay test is said to be *hazard-free* if no spikes can occur on the tested path during the application of the test regardless of gate delay values.

Figure 7.40 illustrates a hazard-free test of the boldfaced path from Figure 7.38. In order to assure that line c has a stable 1, it is necessary to insist that at least one of the inputs to gate G_5 be a stable 0. This can be accomplished by requiring that either line a or line b receive a value 0 for both vectors v_1 and v_2. Therefore, it is necessary to require that (v_1, v_2) assign values $a = (0, 0)$ and $b = (x, x)$, or vice versa, where x denotes a "don't care." Backtracing these requirements to the inputs reveals that line c can be made hazard-free

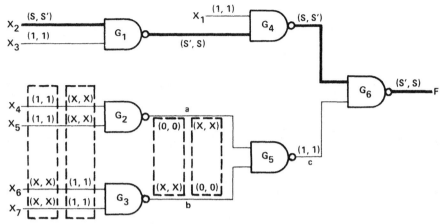

Figure 7.40. Hazard-free delay tests.

either by having inputs x_4 and x_5 at a value 1 for both vectors v_1 and v_2, or by having inputs x_6 and x_7 at a value 1 for both vectors v_1 and v_2. In order to propagate a rising transition along the boldfaced path, the value S in Figure 7.40 should be 0, and for the reverse polarity it should be 1.

As will be seen in Section 7.12.2, there are 62 pairs of vectors (v_1, v_2) that will perform a hazard-free delay test on the boldfaced path, 31 of which generate a rising transition at the primary input x_2, and the other 31 generate a falling transition at the same input. The pair $(v_1 = 1010111, v_2 = 1111011)$, for example, is a hazard-free test that launches a rising transition at the primary input x_2. This property of having the same number of tests for each polarity is true in general: Whenever a path can be tested with hazard-free pairs, the number of pairs that test for one polarity-type fault equals the number of pairs that test for the opposite polarity-type fault.

Not every path fault, however, can be tested with a hazard-free pair. An example of this is shown in Figure 7.41. The path $x_2 G_1 G_3 G_5$ cannot be tested for delay faults using hazard-free pairs.

7.12.2 Restricted Delay Test Pairs

This section sizes the number of random patterns necessary to detect a delay fault by taking a conservative approach. A path fault is considered tested only if it is detected by a hazard-free pair. Moreover, it is required here that the path fault be detected along a single propagation path to a primary output. These two restrictions will yield a test length figure which is higher than necessary. Thus, we consider a delay fault as being detected if it is tested by a restricted delay test pair whose definition follows.

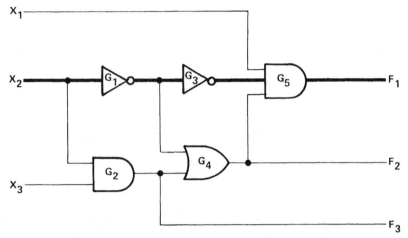

Figure 7.41. The boldfaced path cannot be tested with hazard-free pairs.

Definition 7.19. A *restricted delay test pair* (RDTP) is a hazard-free pair of vectors (v_1, v_2) such that if v_2 is applied after v_1 it provokes the path fault and propagates its effect along a single path to a primary output.

A restricted delay test pair is a special case of a delay test, and does not necessarily exist for every path. There may be paths whose delays can be measured only in conjunction with other paths, creating multiple path propagation. The analysis here excludes these cases, and therefore yields a conservative test length.

In this section it is assumed that the path under consideration can be tested with a restricted delay test pair.

Refer again to the circuit of Figure 7.38. The objective is to compute the probability that a pair of random patterns will constitute an RDTP that detects a fault along the boldfaced path. Without loss of generality assume that the RDTP has to launch a rising transition at the primary input x_2 in order to detect the fault. Similar arguments apply if the RDTP has to generate a falling transition at input x_2.

Let (v_1, v_2) be an RDTP that causes a rising transition at the primary input to the boldfaced path. Obviously, the inverse pair (v_2, v_1) is an RDTP that causes a falling transition at the primary input to the boldfaced path. Let q_1 be the probability that a random pattern will detect a stuck-at-1 fault on the primary input from which the boldfaced path emanates, by exactly sensitizing the fault along this single path. This probability is also the probability that the random vector will constitute the v_1 vector of the (v_1, v_2) pair. Thus, the probability that the first vector will be the v_1 vector is the probability that it either belongs to the set $A = \{101\text{xx}11 | \text{x} = 0, 1\}$ or that it belongs to $B =$

$\{10111\text{xx}|\text{x} = 0, 1\}$. Thus,

$$q_1 = \Pr(A) + \Pr(B) - \Pr(AB) = 2^{-5} + 2^{-5} - 2^{-7} = 7 \times 2^{-7}$$

Let q_2 be the probability that the second random pattern will be the v_2 vector of the (v_1, v_2) pair. The vector v_2 must either be a member of $C = \{111\text{xx}11|\text{x} = 0, 1\}$ or a member of $D = \{11111\text{xx}|\text{x} = 0, 1\}$. Moreover, the vector v_2 cannot be just any member of either C or D. The vector v_2 depends on what v_1 was. The vector v_2 must be such that the pair (v_1, v_2) will constitute a hazard-free pair. If the vector v_1 happened to be the vector 1011111, which is the intersection of the sets A and B, then v_2 can be any member of C or D. If v_1 was not the vector 1011111, but belonged to A, then v_2 can be any member of C. If v_1 was not the vector 1011111, but belonged to B, then v_2 can be any member of D. Thus,

$$
\begin{aligned}
q_2 &= \frac{\Pr\{(v_1, v_2) \text{ is an RDTP}\}}{q_1} \\
&= \frac{(7 \times 2^{-7}) \times (2^{-7}) + (2^{-5}) \times (3 \times 2^{-7}) + (2^{-5}) \times (3 \times 2^{-7})}{7 \times 2^{-7}} \\
&= \frac{31}{7} \times 2^{-7}
\end{aligned}
$$

Notice that q_2 is less than q_1, because of the insistence that the test be hazard-free. We define the *hazard factor* as q_2/q_1, denoted by ρ:

$$\rho = \frac{q_2}{q_1}, \qquad 0 < \rho \le 1$$

The boldfaced path of Figure 7.38 has a hazard factor of 0.632. The hazard factor differs for different paths and depends on circuit topology. Whenever hazards do not impose restrictions on v_2 the value of ρ is 1. The more restrictions that hazard-free testing imposes upon v_2, the smaller ρ is going to be.

The following theorem states the preceding result as a general property.

Theorem 7.15. *Given a path i that can be tested by a restricted delay test pair, the probability that two consecutive random test vectors will constitute an RDTP for a rising (or falling) transition is ρq_1^2.*

Proof. Directly from the definition of ρ.   Q.E.D.

Notice that for uniform random testing, because of the restriction to single-path sensitization, the probability of detecting a stuck-at-0 fault at an origin of a path is equal to the probability of detecting a stuck-at-1 fault at the origin of the same path. Thus the problem of detecting a slow-to-rise fault with RDTPs is completely symmetric to the problem of detecting slow-to-fall faults with RDTPs. Thus, q_1 is the probability of detection of either a stuck-at-0 or a stuck-at-1 fault (but not both) on the path i primary input by sensitization of the single path i. Notice this restriction on q_1: There may be many other tests which will detect stuck-at faults on the primary input of path i, but only those tests which propagate the effect of the fault along path i are counted in q_1.

7.12.3 Delay Test Length

It is interesting to compute the random pattern test length $T_i^{(1)}$ necessary to detect a delay fault on path i with RDTPs with an escape probability no larger than e_{th}.

Consider a test of length t. Let the pair (v_1, v_2) be an RDTP for a single polarity delay test on that path (single polarity meaning either a rising or a falling transition at the path input but not both).

Figure 7.42 describes the Markov chain for detecting a rising (or falling) path delay fault with RDTPs.

S_0 is the initial state.

S_1 is the state where the first vector of the pair, namely v_1, was the last vector applied to the circuit.

S_2 is the state where an RDTP has been generated. This state is an absorbing state since we are interested in the probability of generating at least one RDTP during the test sequence.

A transition from state S_0 to state S_1 occurs when the vector v_1 is applied. The probability of occurrence of this transition is q_1, the detection probability along path i of a stuck-at fault on the primary input of path i. When in state S_1, any application of a vector v_1 will leave the system in state S_1 (probability q_1). A transition from state S_1 to state S_2 occurs with the application of v_2 with probability ρq_1. This v_2 vector is logically hazard-free consistent with its

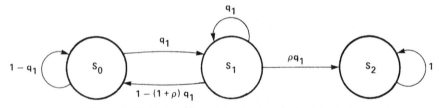

Figure 7.42. Markov chain for a single polarity delay test.

v_1 predecessor. The remaining transition probabilities between states are shown in Figure 7.42.

As a matter of fact the Markov chain of Figure 7.42 is a reduced chain of the actual behavior of a single-polarity delay test. To see this, refer again to the boldfaced path in Figure 7.38. As mentioned earlier, the vector v_2 depends on what v_1 was. In Figure 7.38 there were three different probabilities to get the second vector v_2, depending on what v_1 was. This could have been modeled by using three different states, S_{11}, S_{12}, and S_{13}, instead of state S_1 in Figure 7.42. However, this chain, with multiple S_1 states, is reducible to the one shown in Figure 7.42.

The transition matrix for the Markov chain of Figure 7.42 is

$$\mathbf{P} = \begin{bmatrix} 1 - q_1 & q_1 & 0 \\ 1 - (1 + \rho)q_1 & q_1 & \rho q_1 \\ 0 & 0 & 1 \end{bmatrix}$$

It is necessary to compute the state probability $s_2(t)$, which is the probability of being in state S_2 at test t and is also the cdp of the path fault with RDTPs. The boundary conditions for state probability $s_2(t)$ are

$$s_2(0) = s_2(1) = 0, \qquad s_2(2) = \rho q_1^2, \qquad s_2(3) = 2\rho q_1^2$$

The solution for $s_2(t)$ is given by

$$s_2(t) = 1 - \frac{1}{\gamma}\left[\left(\frac{1 + \gamma}{2}\right)^{t+1} - \left(\frac{1 - \gamma}{2}\right)^{t+1}\right] \tag{7.86}$$

where $\gamma = \sqrt{1 - 4\rho q_1^2}$, $q_1 < 1/2$.

The escape probability achieved after applying t random patterns is

$$\text{Pr}\{\text{escape}\} = 1 - s_2(t) = \frac{1}{\gamma}\left[\left(\frac{1 + \gamma}{2}\right)^{t+1} - \left(\frac{1 - \gamma}{2}\right)^{t+1}\right]$$

The first term of this equation is the dominant one in determining the escape probability, and also constitutes an upper bound for it,

$$\text{Pr}\{\text{escape}\} < \frac{1}{\gamma}\left(\frac{1 + \gamma}{2}\right)^{t+1} \leq e_{\text{th}} \tag{7.87}$$

An upper bound for the test length necessary to detect a single polarity delay fault along path i with RDTPs is therefore

$$T_i^{(1)} < \left\lceil \frac{\ln(\gamma e_{\text{th}})}{\ln\left(\dfrac{1 + \gamma}{2}\right)} \right\rceil - 1 \tag{7.88}$$

When $q_1 \ll 1/2$, the upper bound on the test length can be simplified by using the approximate relations

$$\gamma = \sqrt{1 - 4\rho q_1^2} \simeq 1 - 2\rho q_1^2$$

and

$$\gamma e_{th} \simeq e_{th}$$

yielding

$$T_i^{(1)} \simeq \left\lceil \frac{\ln e_{th}}{\ln\left(1 - \rho q_1^2\right)} \right\rceil \tag{7.89}$$

In a uniform random test, the worst case for $T_i^{(1)}$ occurs when there is only one vector that will detect the stuck-at-0 fault (and only one vector that will detect the stuck-at-1 fault) at the primary input attached to path i. In this case, assuming that the number of inputs to the circuit is large ($n = 5$ will suffice for this approximation), the worst case test length can be approximated by using the relation

$$\ln\left(1 - \rho q_1^2\right) \simeq -\rho q_1^2 = -4^{-n}$$

so that

$$T_i^{(1)} \simeq \left\lceil 4^n \ln(1/e_{th}) \right\rceil \tag{7.90}$$

The test length needed to test a path delay fault with escape probability no larger than 10^{-3} for paths with small q_1 is approximately

$$T_i^{(1)} \simeq \left\lceil \frac{6.9}{\rho q_1^2} \right\rceil \tag{7.91}$$

Example 7.18. Consider again the circuit of Figure 7.38. The length of a single polarity delay test that will detect path faults, along the boldfaced path, with escape probability no larger than 10^{-3} is

$$T_i^{(1)} \simeq \frac{6.9}{0.632 \times (7 \times 2^{-7})^2} = 3651 \quad \square$$

The random pattern single polarity delay test just described will detect either a slow-to-rise or a slow-to-fall fault on the blocks along path i, but not both. It is usual, however, to ensure that both slow-to-rise and slow-to-fall faults are tested since the absence of one does not preclude the presence of the other.

The random pattern test length $T_i^{(2)}$ required to generate both a rising and a falling RDTP for path i with an escape probability no larger than e_{th} is described in Savir and McAnney (1986).

An approximation to the test length $T_i^{(2)}$ necessary to detect both slow-to-rise and slow-to-fall faults along path i (also called a double polarity delay test) with RDTPs is quoted here without proof.

$$T_i^{(2)} \simeq \left| \frac{\ln\left(\dfrac{\gamma e_{th}}{1 + \gamma} \right)}{\ln\left(\dfrac{1 + \gamma}{2} \right)} \right| \qquad (7.92)$$

When $q_1 \ll 1/2$, the approximation to the test length in (7.92) can be simplified by using the approximate relations

$$\gamma = \sqrt{1 - 4\rho q_1^2} \simeq 1 - 2\rho q_1^2, \qquad \gamma e_{th} \simeq e_{th}, \qquad 1 + \gamma \simeq 2$$

Then

$$T_i^{(2)} \simeq \left| \frac{\ln\left(\dfrac{e_{th}}{2} \right)}{\ln\left(1 - \rho q_1^2 \right)} \right| \qquad (7.93)$$

It is interesting to compare this test length for a double polarity delay test with the corresponding single polarity delay test length given in (7.89). Taking the ratio of (7.93) to (7.89) gives

$$\frac{T_i^{(2)}}{T_i^{(1)}} \simeq 1 - \frac{\ln 2}{\ln e_{th}}$$

For an escape probability of 10^{-3}, the ratio is 1.1003, meaning that the double polarity delay test is roughly 10% longer than a single polarity delay test.

The test length needed to test for both a slow-to-rise and a slow-to-fall delay fault with an escape probability no larger than 10^{-3} for paths with small q_1 is approximately

$$T_i^{(2)} \simeq \left\lceil \frac{7.6}{\rho q_1^2} \right\rceil \qquad (7.94)$$

Example 7.19. Consider again the boldfaced path of Figure 7.38. The double polarity delay test length for this path is

$$T_i^{(2)} \simeq \frac{7.6}{0.632 \times \left(7 \times 2^{-7} \right)^2} \simeq 4021 \quad \square$$

7.12.4 Practical Considerations in Delay Testing

As noticed from Examples 7.18 and 7.19, the test length required to adequately test for path delay faults is quite large. Although the test length formulas are slightly conservative, they can still be used as a general guideline. There are several reasons why the formulas are conservative. The first reason is that we compute the test length using RDTPs where this restriction is not always necessary. A second reason for the conservativeness in the computed test length is the approximations done in deriving the formulas. These approximations are quite good, and only slightly affect the test length figure. The test length formulas derived here indicate that a substantial number of random patterns are needed to perform a thorough path delay test in a large network. Assume, for example, that only 1 min of testing is affordable. If the machine speed is 100 MHz, then only 6×10^9 random patterns can be applied within that time. Using Equation (7.94), and assuming a favorable case of $\rho = 1$, the threshold input stuck-at fault detection probability is $q_1 = 3.56 \times 10^{-5}$ for a double-polarity delay test. If any input stuck-at faults have a detection probability lower than this threshold, no adequate path delay fault will be performed by running the test for 1 min.

The path with the smallest ρq_1^2 is certainly the dominant factor in determining the test length. It is not necessarily true, however, that the path with the smallest ρq_1^2 will end up being the slowest path in the network. There may very well be paths with higher ρq_1^2's that may have higher propagation times between inputs and outputs. Since there are, in general, many paths between inputs and outputs, and since it is impracticable to consider all of them for path delay testing, the question naturally arises as to which ones to choose. This will affect the actual test length to be applied in practice.

A combinational network is free of delay faults of any size if there are no path faults. In choosing the list of paths that should be tested for path faults it is appropriate to take into consideration the nominal gate delays.

Definition 7.20. The *slack of a path* of a manufactured circuit is the time difference between the arrival of clock C_2 and the arrival time at the output latches of a transition propagating along this path. See Figure 7.43.

Notice that the slack of the path determines whether or not there is a path fault. A path with negative slack indicates a path delay fault. The slack of a path in the design may be different from the actual slack of the path in the manufactured circuit. A path, therefore, may have a positive slack in a design, yet have a negative slack in an actual machine due to a delay defect.

Definition 7.21. The *slack of a gate* in the circuit is the least slack of any path passing through this gate.

Consider the circuit of Figure 7.44 [Smith (1985)]. The nominal gate delay values are shown inside the gate blocks as multiples of a basic unit delay

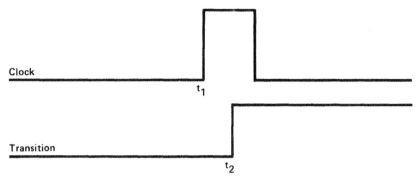

Figure 7.43. Slack $= t_1 - t_2$.

value. Gate G_3 participates in basically four different paths. In the good machine the slack of gate G_3 is the slack of path $G_2G_3G_5$, which amounts to $12 - (5 + 2 + 4) = 1$. An isolated delay defect at gate G_3 does not result in a gate delay fault unless the defect size exceeds 1 (gate delay exceeds 3). If, for example, there is a delay defect of 2 at gate G_3, it will result in a gate delay fault, and the slack of gate G_3 becomes -1. Since path $G_2G_3G_5$ is the only path with negative slack, this delay fault can only be detected by delay testing this path. If, on the other hand, there is a delay defect of size 7 (a gross delay fault) at gate G_3, the slack of G_3 becomes -6, and this delay fault can be detected by delay testing any path, because the slack of all four paths is negative.

This example indicates which paths should be considered for delay testing. If, for each gate, the path in the design with the worst (namely smallest) slack for that gate is tested, then the network is entirely tested for path faults under

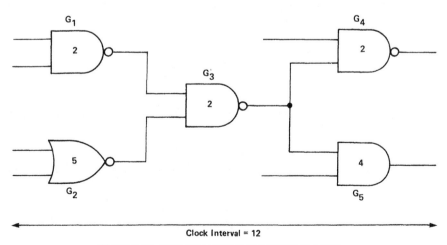

Clock Interval = 12

Figure 7.44. Gate delay defect size affects testability.

the assumption that there is no variation of delay values from machine to machine except for the occurrence of at most a single gate delay defect. Thus, the information on nominal gate delays may serve in identifying the paths with worst slacks. The stuck-at fault detection probability q_1 and the corresponding hazard factor ρ for these selected paths can then be computed or estimated. Then, the minimum ρq_1^2 for these paths can be used to determine the random pattern test length according to (7.89) or (7.93). This test length will be a more realistic figure than the one which is based on the global minimum of ρq_1^2.

7.13   TESTING FOR INTERMITTENT FAULTS

An intermittent fault is a fault that appears and vanishes in a random way. A transient fault, on the other hand, is a fault that strikes once and never reappears. A bad wire bonding may cause an intermittent fault. This wire may, in some instances, make the right connection, while in other instances it may either show up as an open or as a short circuit. Electromagnetic interference, on the other hand, may cause a transient fault in a digital system during operation. This electromagnetic interference may generate unwanted spikes throughout the system which may result in output errors. In some cases, these spikes may latch up in some memory elements and cause unrecoverable errors. In general, it is very difficult, or even impossible, to test for transient faults because for all practical purposes they occur only once. Intermittent faults, on the other hand, can be detected if the test is repeated enough times. This section discusses the testing problem of intermittent faults in combinational circuits by focusing on a random test approach.

7.13.1   Intermittent-Fault Parameters and Properties

It is assumed that the test designer has some statistical information on the behavior of the intermittent faults. More specifically, it is assumed that the test designer knows the following two sets of probabilities for the fault set $F = \{f_1, f_2, \ldots, f_m\}$:

$$\Pr\{\text{fault } f_i \text{ is present}|\text{circuit is faulty}\} = w_i$$

$$\Pr\{\text{fault } f_i \text{ is active}|\text{fault } f_i \text{ is present}\} = p_i, \qquad i = 1, 2, \ldots, m$$

Both w_i and p_i are assumed to be constant for all i, $i = 1, 2, \ldots, m$. The parameter w_i is called the *probability of presence* of fault f_i, and the parameter p_i is called the *probability of activity* of fault f_i.

The probability of activity of an intermittent fault can be interpreted as the fraction of the time for which the fault is active. If $p_i = .01$, it means that for any given test cycle there is a 1% chance that the fault f_i is active, given that f_i is known to exist in the circuit.

It is further assumed that only one out of m possible intermittent faults may be present in the circuit (the single fault assumption). Because of this assumption, the parameters $\{w_i\}$, $i = 1, 2, \ldots, m$, form a probability distribution:

$$\sum_{i=1}^{m} w_i = 1 \tag{7.95}$$

Also notice that $\{p_i\}$, $i = 1, 2, \ldots, m$, does not form a probability distribution.

Definition 7.22. The *error probability* q_i of a fault f_i is the conditional probability of an incorrect output under the application of a randomly chosen input vector, given that the fault f_i is active:

$$q_i = \Pr\{\text{input vector} \in AF_i\} \tag{7.96}$$

The following theorem describes the statistical behavior of an intermittent fault.

Theorem 7.16. *The error latency of an intermittent fault has a geometric distribution.*

Proof. In order to detect the fault, two events must occur simultaneously: The fault must be active and the applied input vector must be a test for that active fault. Hence

$$\Pr\{\text{detecting } f_i \text{ by a random input vector} | f_i \text{ is present}\}$$
$$= \Pr\{f_i \text{ is active} | f_i \text{ is present}\} \Pr\{\text{input vector} \in AF_i\}$$
$$= p_i q_i$$

Denote

$$b_i = p_i q_i \tag{7.97}$$

Now

$$\Pr\{EL_i = t | f_i \text{ is present}\}$$
$$= \Pr\{\text{no detection at step } j, 1 \leq j \leq t - 1, \text{ and detection at step } t\}$$

Therefore,

$$\Pr\{EL_i = t | f_i \text{ is present}\} = (1 - b_i)^{t-1} b_i \qquad \text{Q.E.D.} \tag{7.98}$$

The cdf of the error latency can be obtained from (7.98):

$$F_{EL_i}(t) = \Pr\{ EL_i \le t | f_i \text{ is present}\}$$

$$= \sum_{j=1}^{t} (1 - b_i)^{j-1} b_i$$

yielding

$$F_{EL_i}(t) = 1 - (1 - b_i)^t \qquad (7.99)$$

The expected value of the error latency (or the mean time to detect fault f_i) is

$$M_i = E(EL_i) = \sum_{t=1}^{\infty} t \Pr\{ EL_i = t | f_i \text{ is present}\}$$

hence

$$M_i = \sum_{t=1}^{\infty} t(1 - b_i)^{t-1} b_i = \frac{1}{b_i} \qquad (7.100)$$

The variance of the error latency can be found using the formula

$$\text{var}(EL_i) = E(EL_i^2) - E^2(EL_i)$$

hence

$$E(EL_i^2) = \sum_{t=1}^{\infty} t^2 (1 - b_i)^{t-1} b_i = \frac{2}{b_i^2} - \frac{1}{b_i}$$

or

$$\text{var}(EL_i) = \frac{1 - b_i}{b_i^2} \qquad (7.101)$$

The escape probability achieved after applying t random patterns is given by

$$\Pr\{\text{escape} | f_i \text{ is present}\} = \Pr\{ EL_i > t | f_i \text{ is present}\}$$

$$= 1 - F_{EL_i}(t) = (1 - b_i)^t$$

Letting e_{th} be the escape probability threshold, the least upper bound on the number of tests needed to achieve this confidence level is given by solving

$$(1 - b_i)^t \le e_{th}$$

yielding

$$T_i = \left\lceil \frac{\ln e_{th}}{\ln(1 - b_i)} \right\rceil \qquad (7.102)$$

7.13.2 Optimal Intermittent-Fault Detection Experiments

The problem is to find an optimal random intermittent-fault detection experiment for detecting all single faults. The criterion of optimality used here is maximizing the probability of fault detection. The objective, therefore, is to determine the optimal assignment of input probabilities that will maximize the probability of fault detection.

Definition 7.23. An *optimal assignment of input probabilities* is an input probability distribution which maximizes the probability of fault detection.

Clearly, the optimal assignment of input probabilities is not unique because different tests from the same affected subset might be used to detect a given fault. For example, in the case of only one fault with $AF = \{t_1, t_2\}$, all the distributions of the form $\Pr\{t_1\} = u$ and $\Pr\{t_2\} = 1 - u$, $0 \le u \le 1$, are valid optimal assignments. In these cases, we agree to use either $u = 0$, or $u = 1$ for the optimal distribution. The optimal assignment will, in general, depend on t, the experiment length.

Definition 7.24. A set S is called a *generating set* if it contains at least one element from each affected subset.

Definition 7.25. An *intermittent-fault detection experiment* for $F = \{f_1, f_2, \ldots, f_m\}$ is a string of t tests, which are all members of a generating set S. The length t of this string is called *the length of the experiment.*

Generating sets and intermittent-fault detection experiments are usually not unique. In general, different experiments yield different fault detection probabilities. Note that the shortest string formed using all members of any generating set is a permanent fault detection experiment of F. Such a string formed from the smallest possible generating set (of the smallest cardinality) is a minimal permanent fault detection experiment for F.

Definition 7.26. An *optimal generating set* S^{opt} is a generating set for which an optimal assignment of probabilities exists which assigns zero probability to every test not in S^{opt}.

The optimal generating set is not unique for the same reason that the optimal assignment is not unique. It is sufficient, therefore, to find just one optimal generating set.

An optimal intermittent-fault detection experiment is an application of vectors from an optimal generating set weighted with an optimal probability assignment.

Example 7.20. The following affected subsets, with their tests represented in decimal notation, were obtained for a combinational circuit consisting of

$F = \{f_1, f_2, f_3, f_4\}$:

$$AF_1 = \{1, 2, 7, 8\}$$
$$AF_2 = \{3, 7, 9, 12\}$$
$$AF_3 = \{2, 3, 11, 14\}$$
$$AF_4 = \{2, 4, 7, 14\}$$

Note that:

$S_1 = \{8, 9, 11, 12, 14\}$ is a generating set.

$S_2 = \{3, 12, 14\}$ is not a generating set, because AF_1 is not represented in it.

$S_3 = \{2, 3\}$ and $S_4 = \{3, 7\}$ are generating sets each of which constitutes a minimal permanent fault detection experiment for F.

$E_1 = 1, 2, 2, 3, 1, 9, 9, 7, 7$ is an intermittent-fault detection experiment of length 9 for F. $\square$

In order to carry out the optimization procedure, it is necessary to find an expression for the cdf of the error latency F_{EL} for the case of detection of single faults from a fault set F:

$$F_{EL}(t) = \Pr\{ EL \leq t | \text{circuit is faulty} \}$$

$$F_{EL}(t) = \sum_{i=1}^{m} \Pr\{ EL_i \leq t | f_i \text{ is present} \} \Pr\{ f_i \text{ is present} | \text{circuit is faulty} \}$$

Thus,

$$F_{EL}(t) = 1 - \sum_{i=1}^{m} w_i (1 - b_i)^t \qquad (7.103)$$

$F_{EL}(t)$ is the probability of detecting all the single faults in t input vectors. The optimal assignment of input probabilities is the input probability distribution that maximizes $F_{EL}(t)$. The optimal assignment of input probabilities in the general case can be found in Savir (1980a). The special case where the intermittent faults have disjoint affected subsets is treated here.

In a combinational circuit which has the property that distinct faults have disjoint affected subsets, it is only necessary to find the optimal assignment to the affected subsets. In this case, the optimal generating set requires only one element from each affected subset, and q_i is, therefore, the probability assigned to the test chosen from AF_i, $i = 1, 2, \ldots, m$.

Considering only the disjoint case implies that

$$\sum_{i=1}^{m} q_i = 1$$

The optimal assignment to the affected subsets is obtained by minimizing the function

$$Z = \sum_{i=1}^{m} w_i (1 - q_i p_i)^t + M \left(\sum_{i=1}^{m} q_i - 1 \right) \tag{7.104}$$

where M is the Lagrange multiplier. Hence

$$\frac{\partial Z}{\partial q_i} = -t w_i p_i (1 - q_i p_i)^{t-1} + M = 0, \qquad i = 1, 2, \ldots, m \tag{7.105}$$

$$\frac{\partial Z}{\partial M} = \sum_{i=1}^{m} q_i - 1 = 0 \tag{7.106}$$

Equation (7.105) can be rewritten as

$$t w_i p_i (1 - q_i p_i)^{t-1} = M$$

yielding

$$q_i = \frac{1}{p_i} \left[1 - \left(\frac{M}{t w_i p_i} \right)^{1/(t-1)} \right], \qquad i = 1, 2, \ldots, m \tag{7.107}$$

The value of the constant M can be obtained from the relation

$$\sum_{i=1}^{m} q_i = \sum_{i=1}^{m} \frac{1}{p_i} \left[1 - \left(\frac{M}{t w_i p_i} \right)^{1/(t-1)} \right] = 1$$

yielding

$$M = t \left[\frac{\sum_{i=1}^{m} \left(\frac{1}{p_i} \right) - 1}{\sum_{i=1}^{m} \frac{1}{p_i (w_i p_i)^{1/(t-1)}}} \right]^{t-1}, \qquad t > 1 \tag{7.108}$$

In most practical circuits the length of the experiment will be quite large. It is useful, therefore, to investigate the properties of the optimal assignment of probabilities for large t.

Definition 7.27. The *effective probability of activity* p_{ef} is defined to be

$$p_{\text{ef}} = \frac{1}{\sum_{i=1}^{m} \frac{1}{p_i}} \tag{7.109}$$

Theorem 7.17. *If the affected subsets are disjoint, then as the length of the experiment becomes large, the optimal assignment to the affected subsets approaches a limit distribution which is independent of w_i for all $i = 1, 2, \ldots, m$, and is given by*

$$\lim_{t \to \infty} q_i = \frac{p_{ef}}{p_i}, \qquad i = 1, 2, \ldots, m \qquad (7.110)$$

Proof. Substituting Equation (108) into Equation (107) yields

$$q_i = \frac{1}{p_i}\left[1 - \frac{1}{(w_i p_i)^{1/(t-1)}} \frac{\sum_{j=1}^{m}\left(\frac{1}{p_j}\right) - 1}{\sum_{j=1}^{m} \frac{1}{p_j (w_j p_j)^{1/(t-1)}}} \right]$$

Hence

$$\lim_{t \to \infty} q_i = \frac{1}{p_i}\left[1 - \frac{\sum_{j=1}^{m}\left(\frac{1}{p_j}\right) - 1}{\sum_{j=1}^{m} \frac{1}{p_j}} \right] = \frac{1}{p_i} \frac{1}{\sum_{j=1}^{m} \frac{1}{p_j}} \qquad \text{Q.E.D.}$$

Theorem 7.17 enables us to find the asymptotic properties of the optimal random testing procedure for the disjoint case. From Equations (7.108) and (7.109) the asymptotic behavior of the parameter M is found to be

$$M \simeq t(1 - p_{ef})^{t-1} \qquad (7.111)$$

The cdf of the error latency for the disjoint case, with optimal probability assignment to the affected subsets, becomes

$$F_{EL}^{(opt)}(t) = 1 - \sum_{i=1}^{m} w_i \left(\frac{M}{t w_i p_i} \right)^{t/(t-1)} \qquad (7.112)$$

The asymptotic expression for the cdf of the error latency can be obtained from (7.111) and (7.112):

$$F_{EL}^{(opt)}(t) \simeq 1 - \sum_{i=1}^{m} \frac{1}{p_i}(1 - p_{ef})^t = 1 - \frac{1}{p_{ef}}(1 - p_{ef})^t \qquad (7.113)$$

The meaning of the parameter p_{ef} can now be clarified. When the affected subsets are disjoint, testing for the existence of a fault from the fault set F is asymptotically equivalent to testing for a fault set containing a single fault with probability of activity p_{ef}. This is the reason why this parameter was called the effective probability of activity. The fact that the cardinality of the fault set F is virtually reduced is counterbalanced by the fact that the effective fault manifests itself less frequently, namely

$$p_{ef} \leq p_i, \quad \text{for all } i = 1, 2, \ldots, m$$

The escape probability achieved under optimal probability assignment to the affected subsets asymptotically approaches

$$\Pr\{\text{escape}|\text{circuit is faulty}\} = 1 - F_{EL}^{opt}(t) \simeq \frac{1}{p_{ef}}(1 - p_{ef})^t \quad (7.114)$$

Hence, the experiment length needed to gain an escape probability of e_{th} is given by

$$T \simeq \left\lceil \frac{\ln(p_{ef}e_{th})}{\ln(1 - p_{ef})} \right\rceil \quad (7.115)$$

Example 7.21. In a combinational circuit with $F = \{f_1, f_2, f_3\}$ and disjoint affected subsets, the fault parameters were estimated to be

$$w_1 = .1, \qquad w_2 = .3, \qquad w_3 = .6$$
$$p_1 = 10^{-5}, \qquad p_2 = 10^{-3}, \qquad p_3 = 10^{-4}$$

Find the optimal assignment to the affected subsets and the experiment length T needed to gain an escape probability of $e_{th} = 10^{-6}$.

SOLUTION. Because the p_i's are small we use the asymptotic formulas

$$p_{ef} = \frac{1}{\sum\limits_{i=1}^{3} \dfrac{1}{p_i}} = 9.009 \times 10^{-6}$$

$$q_1 = \frac{p_{ef}}{p_1} = .9009$$

$$q_2 = \frac{p_{ef}}{p_2} = 9.09 \times 10^{-3}$$

$$q_3 = \frac{p_{ef}}{p_3} = 9.09 \times 10^{-2}$$

$$T \simeq \frac{\ln(9.009 \times 10^{-12})}{\ln(1 - 9.009 \times 10^{-6})} = 2.82 \times 10^6$$

Notice that it was not necessary to use the information on the probabilities of presence w_i, $i = 1, 2, 3$, because for high test lengths the optimal assignment is fairly independent of these parameters.   □

Detection of intermittent faults in sequential circuits is discussed in Savir (1980c).

7.13.3   Practical Approach to Intermittent-Fault Detection

It is very hard, if not impossible, to estimate the intermittent-fault parameters $\{w_i | i = 1, 2, \ldots, m\}$ and $\{p_i | i = 1, 2, \ldots, m\}$. Thus, it is also impossible to compute the optimal assignment of input probabilities, and, therefore, the optimal random intermittent-fault detection experiment. A practical intermittent-fault detection strategy is to repeatedly apply the tests computed to detect permanent faults. This approach will not require any extra computational effort beyond what is usually spent on the task of computing tests to detect permanent faults. It will require, however, more testing time, but this is inevitable due to the nature of the intermittent faults.

It is also possible to use redundancy techniques to detect intermittent faults during normal operation. Techniques like parity checking and checksum methods may detect some of the intermittent faults and give some indication on their location in the circuit. Whenever an error is detected the last few operations can be repeated to determine if the error was due to an intermittent fault. If the retry is successful computation can resume, but if an error is detected again the fault is considered permanent and repair action is taken.

8

Built-In Test Structures

There is no one *best* built-in test structure. Built-in test is a collection of possibilities, the choice of which depends upon the application. Factors to consider include fault coverage required, the system overhead which is tolerable, the system performance and the performance impact of the built-in test technique, and the socket or test time which is allowable.

This chapter presents several different implementation approaches which have been suggested or used in the past, and discusses the hazards and rewards of each one. It concentrates almost exclusively upon random pattern or probabilistic built-in test structures. The chapter is divided into three parts: The first describes structures which build upon the concept of scan path; the second covers special topics in structures; and the third discusses structures which are constructed from off-the-shelf components.

8.1 SCAN-PATH STRUCTURES

Scan path refers to a disciplined design standard for all storage elements (other than memory arrays) which has the express purpose of making the stored values easy to control and easy to observe. With this facility the storage element becomes in effect both a primary input and primary output. Test input signals can be introduced or test results observed wherever one of the storage elements occurs in the logic circuit. Thus the test problem reduces to one for the combinational logic between the storage elements. Various scan-path-structured design techniques are described in Chapter 2 and are assumed to be known to the reader.

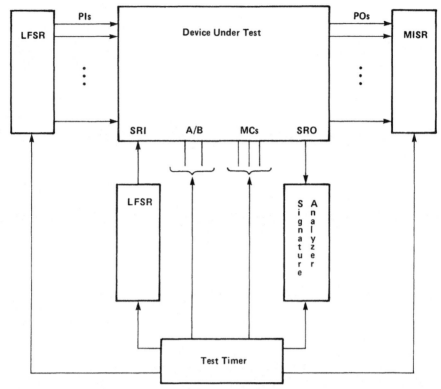

Figure 8.1. The random test socket.

8.1.1   The Random Test Socket

We begin the discussion of scan-path structures with the random test socket [Bardell and McAnney (1982)]. The random test socket, shown in Figure 8.1, lies just outside the domain of built-in test since all of the test hardware is external to the module or board under test. It is designed to perform a random pattern test on an LSSD-based device under test (DUT), either module, card, or board, which is inserted in the socket.

The I/O pins of the DUT are assigned to classes based on their function as follows: One pin is the input to the shift-register (SR) string and is marked SRI for shift-register input; similarly, one pin is marked SRO for shift-register output. (There may be more than one SR string on the DUT, in which case there will be one SRI and SRO pin for each string. In case of multiple strings, the socket circuitry shown attached to SRI and SRO in Figure 8.1 is replicated for each string.) Two pins are used for the A/B shifting clocks, and are marked A and B. At least two pins are system clocks and are marked MCs (for machine clocks). The remaining I/O pins are either input pins (PIs) or output pins (POs). (Pins which are common I/O or dual driver/receiver pins are being ignored.)

The aim is to drive the PIs of the DUT with random (or pseudorandom) patterns. The socket does this by connecting the PIs directly to individual stages of a linear feedback shift register (LFSR) whose feedback is such that it generates a maximal length SR sequence. With this feedback connection (and given enough time), the LFSR will generate (and apply to the PIs) all possible nonzero binary words, one word for each cycle of the LFSR shifting clocks.

The scan input to the LSSD shift-register string (SRI) is driven serially by another LFSR.

The primary outputs (POs) of the DUT feed a multiple-input signature register (MISR), and the scan-out port (SRO) of the DUT shift-register string feeds a signature analyzer.

A large number of tests are to be applied to the DUT. For each test the test timer circuitry must do the following:

Step 1: Load a random test vector into the SR string of the DUT by simultaneously clocking the SRI LFSR and the A/B shift clocks of the DUT. On each clock cycle, one new pseudorandom bit is generated by the LFSR and shifted into the SRI. Enough shift clock cycles are applied to fill the scan string completely.

Step 2: Apply one clock cycle to the PI LFSR to apply a new pseudorandom vector to the DUT PIs.

Step 3: Cycle through the machine clocks (MCs) to capture the responses to the random stimuli in the internal DUT latches.

Step 4: Unload the responses on the DUT POs to the MISR by applying one clock cycle to the MISR shift clocks.

Step 5: Unload the captured responses in the internal latches via the scan path to the signature analyzer by simultaneously clocking the DUT A/B clocks and the signature analyzer shift clocks.

Steps 1 and 5 of the process can be overlapped since it is possible to unload the SR string into the signature analyzer at the same time as the next pseudorandom pattern is loaded into the SRI. A pass/fail indication is obtained after the last test by comparing the binary values remaining in the MISR and the signature analyzer with the expected values.

One of the problems with the random test socket is a long test time (or socket time). The majority of the socket time is taken up by the loading and unloading of the SR string, and proportionately less by the actual tests. The random test socket could become a true built-in test by implementing all of the socket hardware on the DUT, but this would not help speed up the test since the limiting factor is the shifting frequency sustainable by the internal SRLs. The one approach that would help is reducing the number of SRLs per scan path by breaking up the shift register into several shorter strings. This is unattractive for the random test socket approach since it uses additional I/O port capacity on the DUT. As shall be seen later, however, partitioning the scan string is a useful method for built-in test in some implementations.

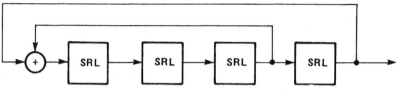

Figure 8.2. LFSR generating the sequence ··· 111100010011010 ··· .

Another concern with the random test socket is the linear correlation which exists between bits of the test data in the DUT shift-register string. This concern was described in detail in Chapter 3. The bit sequence being pushed into the SRI is generated by the LFSR feedback, a recurrence relationship that causes linear dependencies between the bits in the sequence when the number of latches in the DUT shift register is greater than the number of stages in the LFSR. For example, the four-stage LFSR of Figure 8.2 generates a 15-bit cyclic sequence of which one period is shown.

Suppose this LFSR is used to serially load a five-latch SR string. There are $2^5 = 32$ patterns that could be loaded into the string, but only 15 of the 32 will ever appear. Of the 17 missing patterns, 16 are those which have either 001, 010, 100, or 111 as the bits in latches 1, 4, and 5 of the string (numbering the latches from 1 to 5 starting with the SRI). These patterns cannot be generated by the LFSR because the generating recurrence requires the modulo 2 sum of latch bits 4 and 5 to be equal to bit 1. (The final missing pattern is 00000, and does not appear since it requires all 0's in the LFSR, a condition which causes the LFSR to remain all 0's.) This linear correlation between bits only occurs when the SR string is longer than the LFSR.

The concern with linear correlation is the possible exclusion of patterns which are necessary to detect certain faults. If for example a three-input OR gate were driven by shift-register latches 1, 4, and 5 of the structure just described, not one of the gate inputs would ever be tested for stuck-at-0. This a priori exclusion of certain patterns may be a significant detractor from the fault coverage of the random test socket.

There are several ways in which this linear correlation can be overcome. One of the simplest is just not to use every bit from the LFSR. Since the A/B shift clocks of the SR string are isolated from the clocks of the LFSR, we can shift the LFSR either once or twice for every clock cycle of the SR. If the choice of one or two clock cycles is done randomly, the procedure will break up the lock step of the generator correlation and the bit sequence loaded into the SR string of the DUT will be a "random" selection from the LFSR sequence.

A second approach is to ensure that the LFSR in the random test socket is longer (has more stages) than the span of any circuitry in the DUT. Span, in this sense, is the distance in SRL stages along the SR string between the most separated SRLs driving any logic cone of the circuitry. (A logic cone is identified by starting at each primary output and backtracing through the circuit until primary inputs are reached on all backtrace paths.) Alternatively, the span of the circuitry can be restricted in design to less than the fixed length of the socket LFSR. With either method there will be no correlation problems encountered since all logically independent cones will be driven from random PI values or random SRL values derived from within the LFSR. More detail on this subject can be found in Chapter 3.

8.1.2 Simultaneous Self-Test

Simultaneous self-test (SST) is true built-in test approach which significantly reduces test time over the random test socket [Bardell and McAnney (1985)]. SST requires the conversion of every shift-register latch (SRL) on the DUT into a self-test SRL. Figure 8.3 shows a self-test SRL designed for use in an LSSD "double-latch" design.

When the self-test SRL is not in test mode (with Test Mode at a logical 0), the SRL functions normally as a system latch using the system data (D) and system clock ($C1$ and $C2$) inputs, or as a shift-register element using the scan data input (I) and the shift A and shift B clocks. In test mode only the A and B clocks are used. (System clock $C1$ is held off.) When the A clock is on, the EXCLUSIVE OR (XOR) of the scan-path data (from the preceding $L2$ latch or from the SRI) and the system data is gated into the $L1$ latch. The $L2$ latch acts as a one-bit storage element, accepting data from $L1$ when the B clock is on.

Figure 8.4 shows the basic test mode structure for using SST in a double latch LSSD-based circuit. With Test Mode off, the scan path is first initialized by loading in a repeatable data pattern. Test Mode is then turned on. In Test Mode, all SRLs on the DUT are connected into one shift-register string with feedback that simultaneously performs both test pattern generation and re-

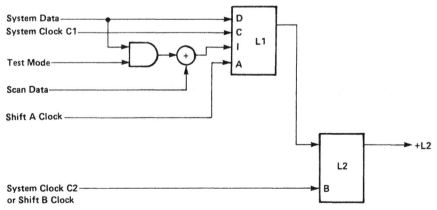

Figure 8.3. Double-latch design self-test SRL.

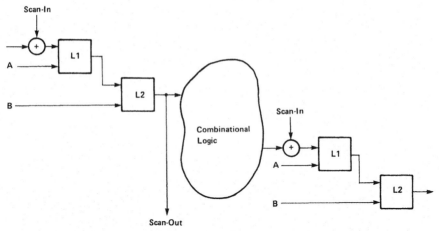

Figure 8.4. Simultaneous self-test in a double-latch LSSD design.

sponse data compression. When Test Mode is on, each A/B clock cycle is a test. The A clock captures the logic responses from the system data line through the XOR. The B clock changes the logic stimuli driving the downstream logic by gating the contents of $L1$ into the $L2$ latch. Thus, in test mode, the SRL is simultaneously collecting test results from the logic driving its system data port and supplying test values through its $L2$ outputs. At the end of the requisite number of tests, Test Mode is turned off and the contents of the scan path are read out as the signature to be compared with the desired value.

While SST performs a rather fast test (each A/B clock cycle is a new test), it has a few problems. For a manufacturing test of the DUT, the test procedure as described must be modified to enable testing of the I/O circuitry of the DUT (those circuits lying between the DUT PIs and the self-test SRLs and between the SRLs and the DUT POs). One interesting solution would be to wire the test socket so that the DUT POs are directly connected back to its PIs. If there are exactly as many PIs as POs, the socket wiring is one-to-one. In the general case, however, there will be a disparity; if more POs than PIs, several POs can be XORed together to drive a single PI; if more PIs than POs, each PO drives a PI directly and the extra PIs are driven by XORs of several POs. Another solution is to use an external (socket mounted) LFSR to drive the PIs and an external MISR to compress the POs, as was done with the random test socket. For system level built-in testing, there is no problem since the adjacent modules or cards will also be in self-test mode and will provide the stimulus and response compression in their own self-test SRLs.

A second concern is the fact that the signal paths used within the self-test SRLs are not the same paths used in the system design. Specifically, some form of additional test is necessary to test the system data and system clock inputs to the SRLs (refer to the SST circuit in Figure 8.3). These paths can be tested inferentially by cycling the machine clocks several times (10 or 20) just after initializing the scan path and just before turning Test Mode on. Stimulating the first machine clock $C1$ captures the logical values on the system data ports of those SRLs driven by $C1$. The changed $L1$ latch values are loaded into the $L2$ latches by machine clock $C2$ (also known as the shift B clock) and are immediately propagated through the logic to other system data ports to be captured in turn when their system clocks are stimulated. Continuing this process for 10 or 20 loops will test most, if not all, of the system data and system clock ports. At the end of the (fixed number of) loops, the $L1$ latches of the DUT are in a known and repeatable state and Test Mode can be turned on to start the test.

A third concern with SST has some good and some bad aspects. Let us again assume a double-latch design, and suppose that a fault indication appears on the input of a particular SRL. It is captured into the $L1$ latch of the SRL at A clock time. On the following B clock the error is shifted into $L2$ and immediately propagates through the logic driven by that latch, to be captured in possibly many downstream latches at the next shift A clock. The error is thus multiplied across the SRL boundary. (Note that this error multiplication may cause diagnostic problems later.) On the other hand, there is a particular chain of circumstances under which the captured error can vanish. Suppose that the error does not propagate from $L2$ to any downstream latches, because no sensitized paths exist from that particular $L2$ to any downstream $L1$s. Then the only indication of the error lies in one and only one $L2$. If, on the following A clock, the next SRL in line also has an error on its input, the XOR of the input error and the error from the preceding $L2$ will cancel. The fault indication has effectively been erased by this correlated error. If this type of error is unlikely, then there is no greater possibility of loss of fault information from the SST technique than there is in a MISR, and if error multiplication occurs in SST there is less.

Another concern is that in SST the modified SRL collects test results via the MISR input and simultaneously supplies stimuli to the logic driven by the SRLs. There is some concern about the randomness properties of these modified SRLs with respect to their suitability as sources of pseudorandom patterns. This concern is addressed in Section 8.1.6.

8.1.3 Self-Test Using MISR / Parallel SRSG

A scan-path structure called STUMPS, for self-test using a MISR and a parallel shift-register sequence generator, is a built-in test architecture for an LSSD-based multiple chip system, in which each field replaceable unit (FRU) has a large number of logic chips [Bardell and McAnney (1982); McAnney (1985)]. Consider first the assembly test of this replaceable unit. Each unit contains an added STUMPS test chip. The special test chip contains a parallel pseudorandom pattern source (a slight modification of the LFSR used to drive

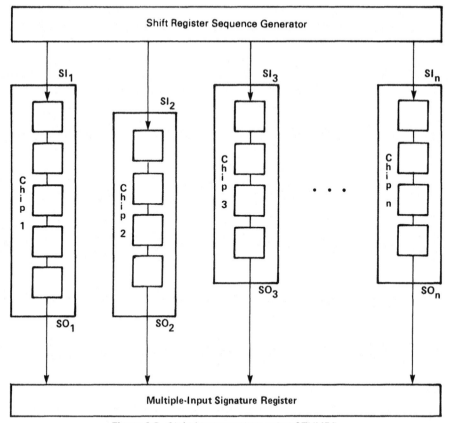

Figure 8.5. Global test structure using STUMPS.

the primary inputs of the random test socket previously described), called a shift-register sequence generator (SRSG), and a multiple input signature register (MISR). The test chip is wired only into the scan path of the DUT, and thus can be overlayed on an already completed scan-path design.

An illustration of the global structure of STUMPS is given in Figure 8.5. Assume that the DUT has n logic chips, and that each logic chip has a scan path containing a number of SRLs. The A/B shifting clocks of the SRLs and the shift clocks of both the random pattern source and the MISR are tied together. In the built-in testing mode, the scan-in port (SI) of each of the n logic chips is connected to a stage of the SRSG, the pseudorandom pattern source, and the scan-out port (SO) of each logic chip is wired to an individual stage of the MISR.

Using the A/B shift clocks, the SRLs are loaded with pseudorandom patterns from the source. The number of A/B clock cycles required is equal to the number of SRLs in the longest string on any logic chip. This will cause the shorter SR strings to overflow into the MISR, but will not affect the correctness of the final MISR signature.

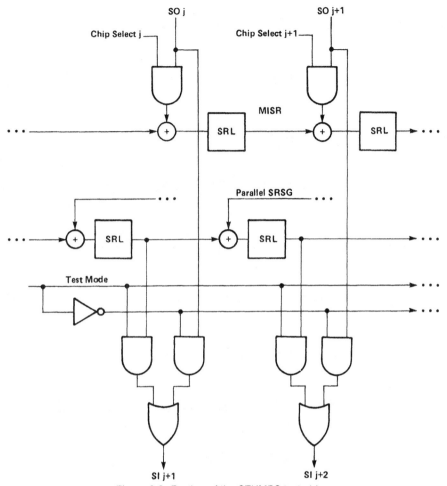

Figure 8.6. Portion of the STUMPS test chip.

After loading random patterns into the SRLs, the system clocks are cycled to capture the test results back into the SRLs. These results are then scanned out into the MISR, simultaneously loading the next random pattern set from the source. A test pass or fail indication is obtained after the last test by comparing the signature remaining in the MISR with the expected signature.

Figure 8.6 shows a portion of the STUMPS test chip. The test chip contains the pseudorandom signal generator, the data compression circuit, and the switching circuits which either connect the scan paths in parallel between different stages of the signal generator and the data compressor for built-in test or connect the scan paths together in series for other testing purposes.

Though only two stages are shown in Figure 8.6, all stages are identical. The output (SO) of the scan path of chip j is fed through an AND gate to a XOR circuit of the MISR and also to an AND circuit which couples the output through an OR gate to the scan input SI $j + 1$ of the scan path on the next chip. The OR gate is also driven by the pseudorandom pattern source through another AND gate.

The n-stage MISR also contains a feedback loop which connects the nth stage of the MISR with its first stage so that data in the output stage is wrapped around and fed through the input stage as data is stepped from stage to stage through the MISR. (If considered necessary, the MISR feedback may also implement a primitive polynomial.)

The pseudorandom generator SRSG contains an SRL for each logic chip on the DUT. These SRL stages are connected together into a linear feedback shift register to implement a primitive polynomial. The SRLs in both the MISR and the SRSG are scan-only SRLs (without the system data inputs).

In test mode, all Chip Select lines are held to a logical 1. For diagnostic purposes, various Chip Select inputs are left down while in test mode to deselect scan paths of certain chips from the MISR signature.

When not in test mode (Test Mode = 0), a normal scan path is structured from the SO of one logic chip to the SI of another.

There are a few concerns with STUMPS which must be considered to achieve a successful built-in test.

The random pattern source, the SRSG, illustrated in Figure 8.6 is oversimplified. During the discussion of the random test socket, we were concerned about the linear dependency which exists between bits in the LFSR-generated sequence, and the possible test degradation caused by the circuit structural dependency upon these correlated bits. This same correlation exists between bits in the array of SRLs formed by the STUMPS parallel channels when the output from each stage of the SRSG is simply fed directly to the scan path channels. One (rather expensive) solution is to build n independent LFSRs, each having more stages than there are SRLs in the longest channel, and use the output of each LFSR to drive the scan-in port of one of the SRL channels. A more attractive solution [Bardell and McAnney (1986)] is to load the individual channels from outputs of an XOR network (driven by the stages of a single LFSR). The network is such that its outputs provide shifted versions of the sequence generated by the LFSR. For more detail on linear and structural dependencies refer to Chapters 3 and 6.

For testing the DUT in isolation, the circuitry lying between the DUT primary inputs and the internal SRLs, and between the SRLs and the primary outputs, is not covered by STUMPS. The DUT socket must provide stimuli to its PIs and compress responses on its POs in a manner similar to that used in the random test socket. In the system environment, the functions performed by the circuits associated with the DUT socket can be performed by the system.

The DUT must be initialized to a repeatable (if not necessarily known) state after power-on and before turning Test Mode on to start the test. An acceptable procedure is to apply a constant logical 1 on the shift register scan-in pin and to run enough A/B clock cycles to load the string. The MISR and the SRSG are similarly initialized using their scan paths. The MISR may be initialized with any pattern (even all zeros), but the SRSG must be loaded with a nonzero pattern. The key to initialization is not in knowing the actual logical values stored in any latch, but rather in being able to repeat the identical initialization patterns each time.

It is not absolutely necessary to initialize the system SRLs as just described. The SRLs will be loaded with a repeatable state from the SRSG on the transmission of the first test pattern. However, if these SRLs are not initialized, the nonrepeatable state of the SRLs which is set at DUT power-on will be scanned into the MISR on the first test and will unpredictably affect the final MISR signature. This error could be avoided by turning off all Chip Select lines during the first test scan operation. The easier procedure is to initialize the system SRLs as previously described.

The number of SRLs on each logic chip of the DUT will vary and some logic chips may not have any. (A logic chip with no SRLs must contain pure combinational logic and is tested by stimuli provided by its surrounding chips). There is a test time advantage to be had by structuring the parallel SRL "channels" so that they each contain roughly the same number of SRLs. This may require that several logic chips are connected in series in one or more channels.

There is a considerable amount of global wiring required to connect the test chip with each of the logic chips. For 100 logic chips on the DUT, there are 200 global wiring nets connecting the chips to the STUMPS test chip. This is a disadvantage only if the DUT package is severely limited in wiring capacity.

STUMPS requires only one additional I/O pin, the Test Mode signal. Since the test chip is inserted into the scan path of the DUT, there is no adverse effect on critical paths of the system.

8.1.4   Boundary Scan with Scan-Path Random Pattern Test

One of the concerns about both the simultaneous self-test and the STUMPS architecture is diagnosis. There are differing diagnostic demands placed on the testing facility at different packaging levels. At the chip level, diagnosis to the failing net (or circuit) is required for resolution of zero yield situations at first silicon and later for yield improvement [Arzoumanian and Waicukauski (1981)]. At the subassembly level (multiple chips on a card or board), the diagnostics must identify a repair action since it is unlikely that the part is a throwaway item. At the system level, diagnosis must be to a field replaceable unit. A useful concept for diagnosis is called boundary scan, the inclusion of shift-register latches around the periphery of a packaging level [Eichelberger and Lindbloom (1983); Goel and McMahon (1982); Porter (1980); Zasio (1983)]. The most appropriate package level for boundary scan is the FRU, although it would also be appropriate and extremely useful for FRU repair if implemented at the VLSI chip level.

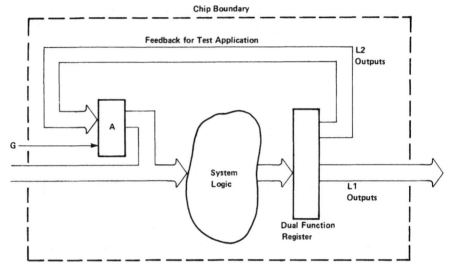

Figure 8.7. Half-series ECIPT with self-test.

Suppose we want to do a STUMPS-like built-in test at the FRU level. If all nonclock inputs and all outputs of the FRU are buffered by scannable latches, and the latches are incorporated into the STUMPS channels, a complete built-in test capability exists. The FRU tester becomes nothing but a powering and clock control socket. Testing at the system level just requires putting every FRU into self-test mode simultaneously. Notice that the fault-free FRU signature remains the same at the system level as at the FRU test level.

A structure called electronic chip in place test (ECIPT) was originally designed to ease the problems of test generation and diagnostic resolution for multichip FRUs by bounding each chip with shift-register latches [Goel and McMahon (1982)]. A version of ECIPT called half-series ECIPT (half-series because only the chip primary outputs are latched) can be used to construct a self-testing mechanism [Bottorff et al. (1983)] with structure as shown in Figure 8.7.

In Figure 8.7, if the dual function register were normal ECIPT SRLs and if the feedback path (for test application) were removed, the structure is just half-series ECIPT. Full ECIPT, in which both chip inputs and outputs are latched, forces network partitions of approximately one chip's worth of logic, and isolates the internal circuits of one chip from those of other chips. Thus both test generation and fault isolation are enhanced at the cost of additional SRL circuits. Half-series ECIPT reduces the SRL overhead by latching only the chip primary outputs, but at the cost of more complex test application software (to direct tests to the appropriate driving chip) and greater uncertainty regarding fault location. This combination of ECIPT with built-in test aids the diagnostic resolution of built-in test and removes the test generation burden from ECIPT.

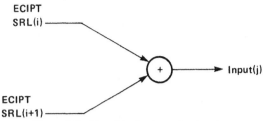

Figure 8.8. Circuit to feed surplus inputs.

In Figure 8.7, a single-latch LSSD network has had a bank of SRLs added, one SRL per output, to isolate the chip from its downstream circuitry. The $L1$ outputs of the buffering SRLs feed the chip output drivers (if any) to drive off-chip. The $L2$ outputs of the buffer SRLs are fed back (in test mode) to the nonclock primary inputs to the chip. The ideal is one buffer SRL driving one input, but typically there are more used inputs than outputs and an XOR circuit as shown in Figure 8.8 must be used to drive the surplus input pins.

The assumption is that the normal chip inputs can be disabled during test to allow the feedback path to drive the system logic. This can be done by placing all drivers of the FRU into the high impedance state during test, or by inserting a multiplexer between the chip inputs and the system logic.

The buffer register (dual function register of Figure 8.7) is designed to function as both a multiple-input signature register and as a pattern generator. The circuitry of the dual function register is shown in Figure 8.9. It has two control inputs, G_1 and G_2, with functions as follows:

$$G_1G_2 = \begin{cases} 10 & \text{for normal scan operation} \\ 01 & \text{for logic test} \\ 00 & \text{for test of system path through ECIPT} \\ 1x & \text{for normal system operation} \end{cases}$$

Testing is done in two parts.
Part 1:

Step 1: Set G_1G_2 to 10.

Step 2: Scan in the initial seed.

Step 3: Set G_1G_2 to 01. At this point the initial seed is in the $L2$ latches of the dual function register and is driving the chip inputs.

Step 4: Sequence the system clocks (if any) to the SRLs embedded in the system logic to propagate the signals through them.

Step 5: Apply one A/B shift clock cycle to the dual function register to compress the system logic output responses and to generate the next chip input pattern.

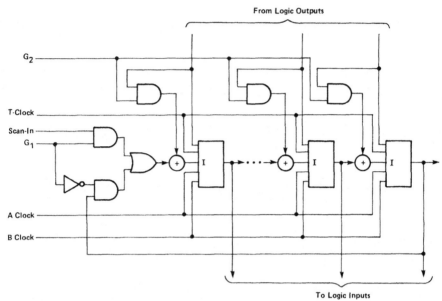

Figure 8.9. Dual function register.

Step 6: Go to Step 4 until enough patterns have been applied.

Step 7: Set G_1G_2 to 10.

Step 8: Scan out the final test signature.

Notice that during this test procedure the embedded system SRLs are not being scanned. A preliminary shift-register test is used (the flush-and-scan tests[1]) to ensure functionality of the scan strings.

As was the case with the simultaneous self-test method discussed earlier, the system path through the buffer SRLs (the dual function register) is not tested by the described procedure. The second part of the procedure should test these system paths.

Part 2:

Step 1: Set G_1G_2 to 10.

Step 2: Scan in the initial seed.

Step 3: Set G_1G_2 to 00.

---

[1]A flush test for string continuity holds both shift A and shift B clocks on so that the shift register acts as a delay line. A 1 applied to the scan-in port should appear (suitably modified to allow for possible inversions along the chain) at the scan-out port after a propagation delay down the chain. A scan test, on the other hand, tests for storage capability by applying an alternating pattern of 1's and 0's to the scan-in, using normal A and B clocks, and observing that the applied pattern appears at the scan-out.

Step 4: Sequence the system clocks to the embedded SRLs.

Step 5: Capture the logic outputs in the dual function register using the T clock.

Step 6: Using this pattern as the seed, cycle the dual function register several times (using the A/B clocks) to generate the next pattern.

Step 7: Return to Step 4 until enough patterns have been applied.

Step 8: Set $G_1G_2 = 10$.

Step 9: Scan out the signature for comparison with the expected good signature.

After completion of these two parts of the test procedure, all chips on the FRU have been tested but the wiring interconnections between chips have not. These interconnections can be tested, if desired, by enabling the normal chip inputs, allowing the buffer SRLs on the driving chips to feed the system logic, and repeating Part 1 of the procedure.

Of course, many other schemes for using boundary scan within a built-in test can be conceived.

8.1.5   Fast-Forward Test

STUMPS test consists of a sequence of A and B shift clock cycles followed by a cycling of the machine clocks. Since the number of A and B clock cycles is determined by the length of the longest STUMPS channel, it is apparent that the bulk of the socket time for STUMPS testing is consumed by cycling A and B clocks.

Obviously a method which would reduce the test application time over that of the basic STUMPS approach would be of significant value. Such a method is described here and is called fast-forward testing.

In fast-forward testing, the basic STUMPS test structure is modified to create two auxiliary parity checking networks. The first network takes the parity of the SRLs in each channel of the STUMPS structure on every test and loads that parity into the MISR. The second network takes the parity of the system data input to each SRL over all tests (or over a subset of all tests). Figure 8.10 shows the basic fast-forward STUMPS architecture. The SRL used in Figure 8.10 (called a SRL*) is shown in Figure 8.11. It is a modification of the standard LSSD SRL which was shown in Figure 2.7 (the modification being an additional clocked data input to the $L1$ latch).

Consider first the fast-forward SRL* shown in Figure 8.11. It is used to replace all SRLs in an LSSD single-latch system. During normal system operation the fast-forward clock $C2$ (F-F Clock) is always off. System data can be gated normally into $L1$ using the system clock $C1$. Normal scan operations use the scan-data input and the A and B shift clocks.

During fast-forward test, the SRL* computes the parity of the signals on its system data line. In fast-forward testing only the F-F clock and the B clock

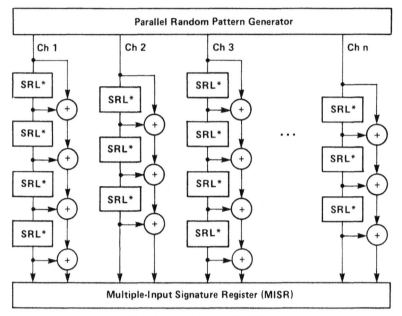

Figure 8.10. Fast-forward test STUMPS architecture

are used (in nonoverlapping mode), while the system clock $C1$ and the shift A clock are kept turned off. On the F-F clock pulse, the XOR of the system data line and the output of the $L2$ latch is gated into latch $L1$. On the succeeding B clock pulse, the value in $L1$ is latched into $L2$. Thus the SRL* value after the B clock is the "parity history" of its system data input since the start of test. The sequence of F-F/B clocks continues until the end of the fast-forward test sequence. At the end of the sequence, the value remaining in SRL* is the parity of the system data input values. This parity is then scanned into the MISR using normal STUMPS procedures (alternating shift A and shift B clocks).

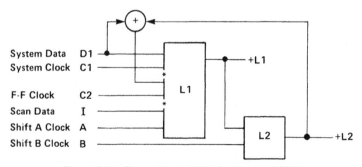

Figure 8.11. General form of the fast-forward SRL*.

Consider the fast-forward architecture shown in Figure 8.10. The SRL* is used instead of the normal SRL in the STUMPS channels. In addition, a chain of XORs is strung in parallel with each STUMPS channel and connected to the channel scan path as shown. The outputs of the XOR chains are connected to added inputs of the MISR, or alternatively may be used to replace the normal STUMPS inputs to the MISR during fast-forward testing. When connected as shown in Figure 8.10, the XOR chain computes the parity of the value out of the pattern generator and the values on the $L2$ latches of each SRL* in the channel on each test. This parity is loaded into the MISR on each test.

Two small additional modifications are required to the STUMPS pattern generator and MISR. During normal STUMPS testing, the A and B shift clocks drive the channel SRLs, the MISR, and the pattern generator in common. During fast-forward test, on the other hand, the A clocks of the pattern generator and the MISR are replaced by the F-F Clock. With this construction, each F-F clock/B clock cycle also shifts the pattern generator and the MISR.

Consider the procedure for testing with fast-forward. First the random pattern generator, the MISR, and all SRL*s are initialized to some repeatable pattern. (Normally the pattern generator must be initialized to a nonzero pattern, but the MISR and the SRL*s can have any initial pattern as long as it is repeatable.) Testing consists of alternating and nonoverlapping F-F clock and B clock pulses. Each pair of F-F and B clocks is a test. On the F-F clock pulse, each SRL* captures the parity of its system data line in its $L1$ latch, and since the F-F clock also drives the A clock of the MISR, the XOR chain outputs from each channel (the parity sum of the pattern generator channel input and the individual channel SRL* outputs) are captured by the MISR. The following B clock pulse shifts the pattern generator, shifts the MISR, and latches the parity value previously held in $L1$ into the $L2$ of each SRL*. This alternation of F-F clock and B clock continues until the end of the test sequence. The final step in the test procedure is to unload the SRL* parities into the MISR using normal A and B shift clocks. At the end of the test sequence the MISR contents are checked against the precomputed correct signature for a pass/fail decision.

Notice that for a single-latch designed system there must be as many F-F clock phases as there are different system clocks to avoid races. All SRL*s clocked by one particular system clock can also be clocked by one particular F-F clock phase. The individual F-F clock phases can be cycled in various sequences during fast-forward test to improve test coverage.

Consider now the fast-forward test time. Each cycle of F-F clock and B clock is a test. Make the reasonable assumption that the application frequency of the F-F and B clocks during fast-forward testing is the same as that of the A and B clocks during normal STUMPS testing. Then fast-forward testing improves the test time by a factor equal to the number of SRLs in the longest channel.

The test coverage of fast-forward testing is considered in the following:

1. The parallel random pattern generator is tested since on every test in the sequence the output of the generator is parity checked by the XOR chain as shown in Figure 8.10.
2. The MISR is tested since it is actively involved in computing the signature.
3. The SRL*s are tested since their parity is also checked by the XOR chain, and their scan path (and its A clock) is tested when the SRL* parity contents are scanned into the MISR as the last step in the test procedure.
4. The combinational logic between the SRL*s is tested using as stimuli the values in the $L1$ latches of the SRL*s. This value is the parity history of the values appearing at the SRL* system data input since the start of the test. It is not obvious just how "random" these values are, and it is at least conceivable that the parity in one or more SRL* remains fixed for several or perhaps many tests. In this case it will be necessary to occasionally substitute an A clock pulse for an F-F clock pulse. This will scramble the lock step parities by shifting a pattern generator bit into the topmost SRL* of each channel, shifting the parities one step down the chain, and loading the parity from the bottom SRL* into the MISR. Of course a test is lost on each substitution, so the method should only be used as necessary.
5. The system data and system clock inputs to the SRL* are not tested under the fast-forward procedure just described. This flaw can be remedied by sequentially stimulating the system clocks for a few cycles while holding the F-F clock off. Stimulating the first system clock $C1$ captures the logical values on the system data ports of those SRL*s driven by clock $C1$. The changed $L1$ values in these SRL*s are propagated through the logic to other system data ports to be captured in turn when their appropriate system clock is stimulated. Continuing this process for, say, 10 cycles of the system clocks will test most, if not all, of the system data and system clock lines. This test procedure can be inserted into the fast-forward test either at the start of test or at the end of test.

Consider the masking characteristics of the fast-forward double-parity data compression as compared to that of the normal STUMPS method. Masking occurs when a fault is detected by a test sequence but the evidence of the fault is lost by the compressor. A usual measure of the probability of masking is the ratio of the number of masking circuit response sequences to the total number of possible sequences. In the following analysis the masking characteristics of the MISR are ignored since they are common to both STUMPS and fast-forward.

Consider a simplified version of fast-forward which consists of a single STUMPS channel. Let the channel contain K SRL*s, and assume that N tests are to be applied. There are then NK response bits from the circuit, and 2^{NK} possible response sequences. Since one sequence represents the correct circuit response, there are $2^{NK} - 1$ incorrect sequences.

Now to compute the number of these sequences which will mask, it is necessary to prove the following theorem.

Theorem 8.1. *Masking can only occur under the double parity fast-forward scheme if each SRL* sees an even number of failing tests on its system data input and if every failing test fails an even number of system data lines.*

Proof.

1. Suppose a particular SRL* sees just one failing test. On that test the bit on its system data input either went from a correct 1 to an error 0 or vice versa. In either event, the parity computed by the SRL* will be wrong and the fail will be detected when the SRL* is unloaded into the MISR. Similarly any odd number of failing tests on a particular SRL* will be detected. However, an even number of system data input errors to a single SRL* will result in the correct parity. Hence masking can occur only if each SRL* sees an even number of failing tests $(0, 2, 4, 6, \dots)$.

2. Suppose that on the first failing test there was just one failing system data input. On the F-F clock, the fail incorrectly changes the parity history in the affected SRL*. At the following B clock, the SRL* output has incorrect parity and since it is the only incorrect parity, it is detected at the output of the XOR chain. Similarly if on the first failing test there were any odd number of system data input fails, the XOR chain will detect the error. Hence masking can occur on the first failing test only if that test fails an even number of system data lines. Let $F1$ be the set of SRL*s which failed on the first failing test. We have seen that masking occurs only if $F1$ has even cardinality. By the argument in paragraph 1 above there must be at least one more failing test or the failure will be detectable. Consider then the second failing test, and assume first that this second failing test also fails an even number of system data lines. There are three possibilities:

- The second failing test hits all SRL*s in $F1$ and no others. The second failing test will correct the parity of the SRL*s in $F1$ and will not be detected by the XOR chain.
- The second failing test hits some of the SRL*s in $F1$ and some other SRL*s. The hits on the SRL*s in $F1$ will correct their parity, and the hits on other SRL*s will leave them with incorrect parity. Regardless of the number of hits in $F1$ there will remain an even number of SRL*s with incorrect parity and the fails will not be detected by the XOR chain.

• The second failing test hits only SRL*s not in $F1$. The set $F1$ has even cardinality, and the set of SRL*s hit by the second test also has even cardinality. Hence the second test will not be detected by the XOR chain.

The only other situation is that the second failing test fails on an odd number of system data lines. Again there are three possibilities to consider:

• The second failing test hits only SRL*s in $F1$ and no others. The second failing test corrects the parity of an odd number of SRL*s in $F1$. Since $F1$ has even cardinality, the second test will leave an odd number of SRL*s with incorrect parity and will be detected by the XOR chain.
• The second failing test hits some of the SRL*s in $F1$ and some other SRL*s. Those hits on SRL*s in $F1$ correct their parity and the hits on other SRL*s leave them with incorrect parity. If an odd number in $F1$ are corrected, there must still be an odd number of $F1$ SRL*s with incorrect parity which, combined with an even number of non-$F1$ SRL*s with incorrect parity, will cause detection on the XOR chain. Similarly, if an even number in $F1$ are corrected there are still an even number in $F1$ uncorrected which, combined with the odd number of non-$F1$ errors, causes detection down the XOR chain.
• The second failing test hits only SRL*s not in $F1$. The set $F1$ has even cardinality, and the set of SRL*s hit by the second test has odd cardinality. Hence, the second test will be detected by the XOR chain.

These arguments are extended by recursion to the third and further failing tests. Hence for masking to occur it is necessary that every failing test fails an even number of system data lines. Therefore, masking will occur under fast-forward testing only if each SRL* sees an even number of failing tests on its system data input *and* if every failing test fails an even number of system data lines. Q.E.D.

To continue with the masking probability analysis, let N be the number of tests in the fast-forward test sequence for the single channel STUMPS structure, and let the channel contain K SRL* latches. Consider an N row by K column matrix as represented symbolically by Figure 8.12. Each row stands for one of the N tests and each column represents one SRL*. The entries in the matrix are 1 if the SRL* of that column has incorrect parity for the test of that row and 0 if it has correct parity, and the matrix is called an error matrix. There are 2^{NK} possible error matrices, each representing one possible response of the K output circuit to the N tests. Since one matrix contains all 0's (and thus represents the error-free circuit response), there are $2^{NK} - 1$ true error matrices.

Consider the meaning of the masking requirements in terms of such an error matrix.

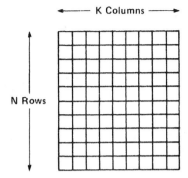

<--------- K Columns --------->

N Rows

Figure 8.12. $N \times K$ error matrix.

First, each SRL* must see an even number of failing tests (either $0, 2, 4, \ldots$) on its system data input. This requirement means that every column of each masking error matrix must have even parity (that is, there must be an even number of 1's in every column of the masking error matrix).

Second, every failing test must fail an even number of system data lines. This requirement means that every row of each masking error matrix must have even parity (that is, there must be an even number of 1's in every row of the masking error matrix, remembering that 0 is an even number).

Thus, of the $2^{NK} - 1$ true error matrices, only those with both even parity columns and even parity rows will mask. To determine the number of such masking error matrices, consider the $(N - 1) \times (K - 1)$ matrix of Figure 8.13(a). There are $(N - 1)(K - 1)$ entries in the matrix, each of which can be either 0 or 1. Construct each of the $2^{(N-1)(K-1)}$ possible versions of the matrix. For each of these versions construct a vector of length $(K - 1)$ such that when the vector is added as a row at the bottom of the version, it will cause even column parity on all of the $(K - 1)$ columns of the version, that is, each bit of the vector is the XOR of its corresponding column. Observe that each of the constructed row vectors is unique (that is that there is one and only one row

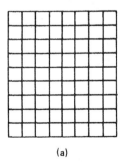

(a)

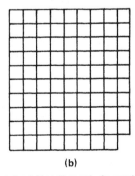

(b)

Figure 8.13. (a) $(N - 1) \times (K - 1)$ matrix. (b) Notched $N \times K$ matrix formed by adding row and column vectors.

vector which will cause even column parity on all columns for a particular version).

Similarly, it is possible to construct a vector of length $(N - 1)$ which, when added as a column to the right-hand side of each version, will cause even row parity on all rows of the version. Again the vector is constructed so that each bit is the XOR of its corresponding row. Each of the constructed column vectors is unique.

Take each of the $2^{(N-1)(K-1)}$ versions of the matrix and append to each version its constructed row and column vectors as shown in Figure 8.13(b). Now (except for a notch in the bottom right-hand corner) each version is an $N \times K$ matrix which satisfies the masking requirements; that is to say, that each version has even parity rows and columns. Since all possible $(N - 1) \times (K - 1)$ submatrices have been used, and since there is only one row and one column vector which can be added to satisfy row and column parity, there are no other (notched) $N \times K$ matrices which match the masking requirements.

Now consider the notch in Figure 8.13(b). It lies at the intersection of the appended row and column. To complete the $N \times K$ matrix, the notch must be filled with a value which satisfies even parity on both the appended row and the appended column. It was shown earlier that the values in the appended row were the XOR of their associated columns, and that the values in the appended column were the XOR of their associated rows. Obviously the XOR over the appended row is exactly the same as the XOR over the appended column since both represent an XOR over the entire original $(N - 1) \times (K - 1)$ matrix. The value to be inserted into the notch must satisfy even parity over both the appended row and the appended column, and is either 1 or 0 as the XOR over the appended row or column is 1 or 0, respectively.

Thus, starting with $2^{(N-1)(K-1)}$ submatrices, a complete and exhaustive set of the masking matrices of size $N \times K$ has been constructed. Notice that one of these matrices contains all 0's and represents the responses of the fault-free circuit. Hence, there are $2^{(N-1)(K-1)} - 1$ error matrices of size $N \times K$ which mask. Since there are a total of $2^{NK} - 1$ true error matrices, the probability of masking with fast-forward is

$$\Pr\{\text{masking}\} = \frac{2^{(N-1)(K-1)} - 1}{2^{NK} - 1} \simeq 2^{-(N+K)}.$$

The approximation given is valid when the product NK is reasonably large.

This is a rather remarkable result. Applying 200 fast-forward tests to a 25 output circuit gives a masking probability of less than 10^{-67}. Recall that this masking probability ignores the masking characteristics of the MISR which, however, is common to both STUMPS testing and fast-forward testing.

Thus, for all practical purposes STUMPS and fast-forward testing have virtually the same masking probability. The advantage of fast-forward testing is its low test time.

8.1.6 MISR / LFSR Modification to an LSSD SRL

In both the simultaneous self-test method and the ECIPT self-test method which were described earlier, a modified SRL is used to collect test responses and simultaneously to supply stimuli to the logic driven by the SRL. There is some concern about the randomness properties of these modified SRLs with respect to their suitability as sources of pseudorandom test patterns. Conceivably the logic structure driving the MISR input of such an SRL could cause extremely nonrandom patterns in it. It is possible, however, to modify an LSSD SRL to make it useful as a MISR and simultaneously as an LFSR, thus providing both the data compression attributes of the MISR and the well controlled pseudorandom patterns of the LFSR which are ideally suited to a simultaneous self-test. The modification is shown in Figure 8.14.

The assumed basic network is an LSSD single-latch design. The SRL in Figure 8.14 is modified by adding a data port driven by an XOR to the $L2$ latch (shown as the $L2^*$ latch in Figure 8.14) and adding an $L3$ latch driven by the $+L2^*$ output. The SRL functions normally as a system latch using the system data input along with the system clock, and as a shift-register element using the scan data input and the shift A and B clocks.

Just prior to entering self-test mode, the SRL $L1$ and $L2^*$ latches are initialized, using the normal scan data path, with an arbitrary but repeatable and nonzero pattern as the initial seed.

In self-test mode, the entire shift-register string on the unit under test is connected to form two feedback shift registers, both implementing a primitive polynomial, as shown in Figure 8.15.

The clock lines shown in Figure 8.14 are omitted in Figure 8.15 to preserve clarity, but should be understood to be driven in common to each SRL in the string. Similarly, the feedback connections are shown symbolically but are

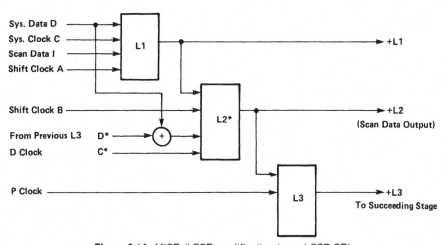

Figure 8.14. MISR/LFSR modification to an LSSD SRL.

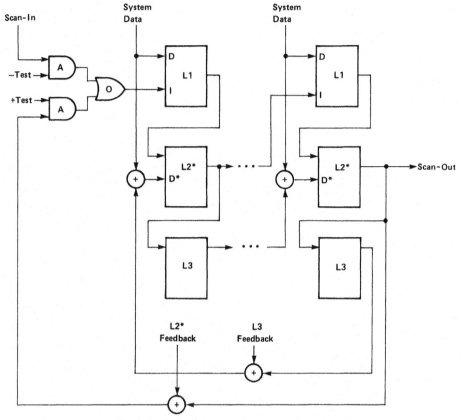

Figure 8.15. Dual feedback registers on the DUT.

understood to be such as to form a primitive polynomial. The two feedback shift registers of Figure 8.15 are running simultaneously, one as an LFSR and one as a MISR. Each $L1$ latch is an element in the LFSR and each $L3$ is an element in the MISR. The $L2^*$ latch is used in both the LFSR and the MISR to transfer data.

The clocking sequence in test mode is P, B, A, D as shown in Figure 8.16. At P clock time, the content of each $L2^*$ is transferred to its corresponding $L3$. The binary word now stored in the $L3$ latches is the current MISR state. At B clock time, the current LFSR state held in the $L1$ latches is shifted into the $L2^*$ latches. At A clock time, the $L2^*$ data is modified in its high-order bit by the feedback and is shifted into the $L1$ latches to form the new LFSR state.

At this point, the combinational logic of the unit under test is being stimulated by the LFSR values in the $L1$ latches and the responses of the logic to that stimulus are available at the system data lines of the SRLs. At D clock time, the XOR of the responses on the system data lines and the MISR state held in the $L3$ latches is captured in the $L2^*$ latches to form the new MISR state.

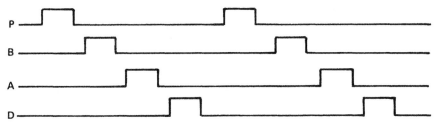

Figure 8.16. Timing diagram in test mode.

The clocking sequence of *P*, *B*, *A*, *D* continues to the end of the test. At the end of the test, the MISR state (held in the *L3* latches) can be shifted out of the unit under test for comparison with the expected signature. This shifting is by alternately pulsing the *D* and *P* clocks, and the state is observed at the scan data output. Notice that the MISR state scan-out is perturbed by the constant system data inputs to the XORs in series with the *L2** latches. The MISR state observed at the scan data output can later be corrected to the actual MISR state (if necessary) by using the system data values obtainable by correctly cycling the system clocks and off-loading the *L1* latches on the normal scan data path.

8.1.7   Brief Guidelines for Scan-Path Designs

A logical progression leading to complete built-in testing for designers using scan path could be

Step 1: Build pseudorandom pattern test generation, multiple input signature registers, and a test control facility into a low cost tester to reduce manufacturing test costs.

Step 2: Integrate pseudorandom pattern testing with boundary scan into the field replaceable unit to improve field maintenance testing and diagnostics and to reduce field costs.

Step 3: Integrate pseudorandom pattern testing with boundary scan into each VLSI chip for complete built-in testing and chip level diagnostics.

So far structures which support this progression have been discussed. Of course, at each level of the progression it is necessary to insure that the logic structures are random pattern testable. Methods of measuring random pattern testability of faults and mechanisms for modifying the logic structure around faults with unsatisfactory testability were discussed in Chapter 7.

8.2   SPECIAL TOPICS IN BUILT-IN TEST STRUCTURES

This section discusses special circuits which can be usefully used in built-in test structures and applications.

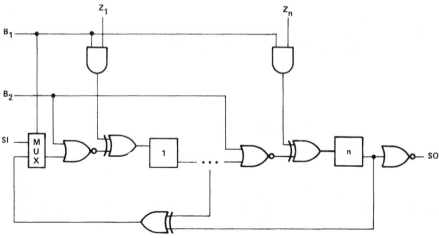

Figure 8.17. The BILBO register.

8.2.1 BILBO

Figure 8.17 shows the built-in logic block observer (BILBO) register
[Koenemann et al. (1979); Koenemann et al. (1980); Koenemann et al. (1983)].
As shown in Figure 8.17, the BILBO register constitutes a set of latches (for
simplicity assume D latches) where the inverted output of each latch (the $\overline{Q}$
output) is connected via a NOR and an XOR gate to the data input of the
succeeding latch. A multiplexer (MUX) allows either a serial input to the
latches in one mode of operation ($B_1 = 0$) or a connection of the XOR
feedback ($B_1 = 1$) in another mode of operation.

The BILBO register can be operated in a number of different modes. The
mode of operation is determined by the value of the control inputs B_1 and B_2.

When $B_1 = B_2 = 1$, the BILBO acts as a regular register with parallel
inputs Z_i and parallel outputs $\{Q_i, \overline{Q}_i\}$.

When $B_1 = B_2 = 0$, the BILBO acts a shift register with scan input SI and
scan output SO. In this mode of operation the EXCLUSIVE OR feedback is
disconnected from the first latch by the multiplexer. This mode of operation
may serve as a scan path if it is necessary to input external vectors to the
device. Notice that the vector stored in the latches is the bitwise complement
of the vector applied at the SI port.

When $B_1 = 1$ and $B_2 = 0$ the BILBO acts as a MISR or LFSR. In this case
the XOR feedback is active. If the Z_i's are outputs of the circuit under test,
then the BILBO acts as a MISR compressing the circuit responses. If,
however, $Z_i = 0$ for all i, then, provided there is a nonzero seed in the BILBO
register, it acts as an LFSR generating pseudorandom stimuli. Notice that in
this mode of operation the feedback function may be either an EXCLUSIVE
OR or EQUIVALENCE (the complement of EXCLUSIVE OR) depending on
the number of stages being fed back.

When $B_1 = 0$ and $B_2 = 1$ the BILBO register is reset to the all-zero state.

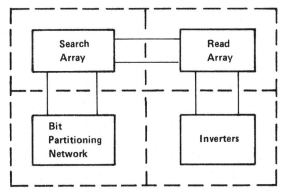

Figure 8.18. The partitioned blocks for self-test.

8.2.2 A PLA Self-Test Structure

Figure 8.18 describes the programmable logic array (PLA) planes and circuits that are to be tested separately in a built-in test [Daehn and Mucha (1981b)]. Assuming that both the search array and read array are realized by NOR planes, we observe that the following $k + 1$ test vectors detect all single stuck-at faults in a k-input NOR gate:

$$
\begin{array}{c}
000 \cdots 00 \\
100 \cdots 00 \\
010 \cdots 00 \\
\cdots \\
\cdots \\
\cdots \\
000 \cdots 01
\end{array}
$$

Assuming that shorts between input lines to a gate and between output lines or product term lines perform the logical AND function, they can be detected by the subset of vectors that has a single 1 in a given bit position and 0 elsewhere.

Similarly, a complete test set to detect faults in a two-bit partitioning network is

$$
\begin{array}{c}
000 \cdots 00 \\
100 \cdots 00 \\
110 \cdots 00 \\
011 \cdots 00 \\
\cdots \\
\cdots \\
\cdots \\
000 \cdots 01
\end{array}
$$

These deterministic patterns can be generated quite easily on chip.

Figure 8.19(a) shows a nonlinear feedback shift register that generates the patterns required to detect all single stuck-at and bridging faults in the NOR

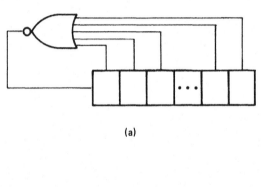

(a)

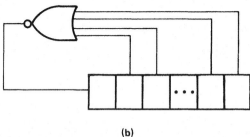

(b)

Figure 8.19. (a) Test pattern generator for the NOR planes. (b) Test pattern generator for the two-bit partitioning networks.

planes. The initial seed placed in the register is $000 \cdots 00$. Figure 8.19(b) shows the nonlinear feedback shift register that generates the test patterns required to test the two-bit partitioning networks. The initial seed placed in this register is also $000 \cdots 00$.

Figure 8.20 shows the PLA self-test architecture. The BILBOs in Figure 8.20 are made out of nonlinear feedback shift registers. To test the first NOR plane, test patterns are generated by BILBO 1 and responses are captured in BILBO 3. To test the second NOR plane, test patterns are generated by BILBO 2 and responses captured by BILBO 3. The output inverters and the bit partitioning network are tested by feeding back the primary outputs to the primary inputs and using BILBO 3 as a generator and BILBO 1 as a test response compressor.

8.2.3   The Bidirectional Double Latch

The bidirectional double latch is a modified $L1$-$L2$ latch commonly used in LSSD. As the name implies the bidirectional double latch has the capability of shifting information in two directions: left to right and right to left.

Figure 8.21 shows the NAND gate design of the bidirectional double latch [Savir (1986)]. Line SH is a control signal which determines the shift direction. When $SH = 0$ a left shift is performed, and when $SH = 1$ a right shift is

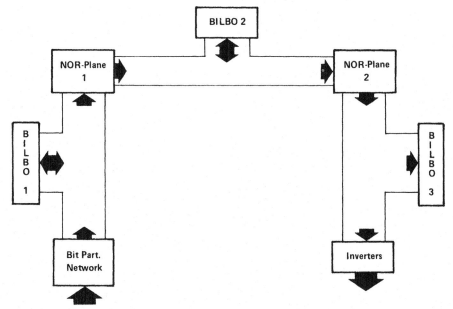

Figure 8.20. The PLA self-test architecture.

performed. $Q(j)$ is the output of the current stage; $Q(j-1)$ and $Q(j+1)$ are the outputs of the previous and next stage, respectively.

Figure 8.22 describes a bidirectional LFSR composed of bidirectional double latches. Line SH is a control signal determining the LFSR shift direction. When $SH = 0$, the LFSR shifts left using the lower feedback connections and when $SH = 1$, it shifts right using the upper feedback

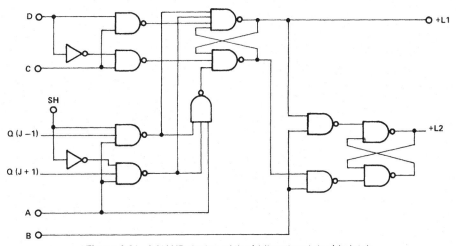

Figure 8.21. A NAND design of the bidirectional double latch.

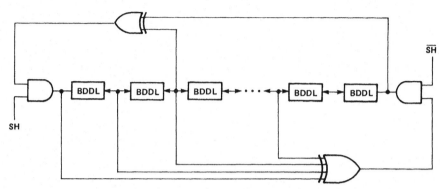

Figure 8.22. A bidirectional LFSR.

connections. Thus, the bidirectional LFSR can serve as a pseudorandom number generator with two different feedback polynomials. It is important to note that if the shift lines of the bidirectional LFSR are fed with random values, it will reduce the linear dependencies between the output bits due to the use of the bidirectional shift capability. In a similar fashion bidirectional double latches can be connected to form a bidirectional MISR with two different primitive polynomials.

Stuck faults in a shift-register string built out of bidirectional double latches (BDDLs) can be easily diagnosed. This task is quite difficult in a shift register with a unidirectional scan. If the string is built out of BDDLs, a defective latch can be identified by simply applying constant values, opposite to the stuck value, in one direction, until the output coming out at the other end has a fixed value (the latch stuck value), and then repeating the scan in the opposite direction. Then, by counting how many shifts it takes for the shift-register output to be the constant stuck value one can determine the position of the faulty latch.

8.3 BUILT-IN TEST STRUCTURES FROM OFF-THE-SHELF HARDWARE

This section describes built-in test possibilities offered by commercially available off-the-shelf hardware.

8.3.1 6000 and 20,000 Gate Semicustom CMOS Array Chip

The 6000 gate array chip [Resnick (1983)] has an on-chip maintenance system (OCMS) which occupies about 12% of the chip area and requires only 4 dedicated pins. OCMS, which is fixed for all options, utilizes the BILBO approach in the self-test mode.

The chip is two-layer metal CMOS technology and has the following features:

- 3953 cells of 6 transistors each available to the user. This is equivalent to 5930 two-input gates, at $1\frac{1}{2}$ gates per cell. (Because of the array structure it is extremely difficult to utilize all of the gates. Functions should be designed to utilize 80–85% of the gates. At this utilization 100% automatic wire routing should be easily achievable.)
- A clock bus driven by a clock buffer.
- 172 pin package with pins allocated in the following way:

    66 inputs,

    88 bidirectional lines with tristate capability that may be used either as inputs or outputs,

    1 clock,

    1 hold off to control timing between chips,

    4 for maintenance and test,

    12 for power.

- All inputs can be configured individually to either CMOS or TTL compatibility.
- A class of macros is available for logic design. Macros include shifters, decoders, and others.
- An on-chip oscillator to monitor AC performance.
- A built-in maintenance system based on the BILBO concept.

Figure 8.23 shows the architecture of the chip.

The input shift register can serve either as a data source or a data sink. Attached to the input register is an operand generator that may produce pseudorandom operands. The input register can be loaded in either a serial or a parallel fashion. The output register can also serve as either a source or a sink. Attached to it is a collection of XOR gates that allows it to operate as a MISR or LFSR. The control (function) register is divided into two eight-bit sections, one that is controlled by the system and the other by the user. The system-controlled section determines the particular function to be performed by the input or output register (such as putting the register into checksum mode). The user-controlled section is used by the logic designer. A test strobe control signal (TS signal) controls the registers operation. When TS is low, the control register is connected serially to the two other registers, and can be shifted together. When TS is high the clock of the control register is disabled, preserving its contents.

In self-test mode the input register is configured into an LFSR generating pseudorandom patterns, and the output register is configured into a MISR recording the gate array outputs. The external maintenance control system has to supply the pseudorandom number seed, the test length, and the expected

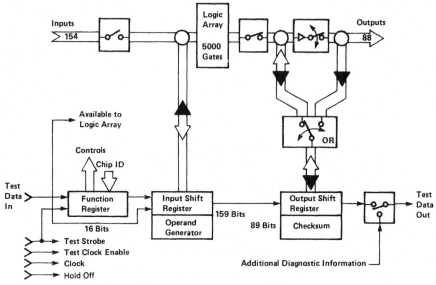

Figure 8.23. The architecture of the 6K gate array.

signature to be found in the output register, computed beforehand by simulation.

The chip also has an interconnect test mode, where operands are activated on the chips and recorded in the input registers of the interconnected chips. From the results of this test it is possible to identify opens and shorts.

It is possible to connect more than one such chip on a board. To do this the serial data-output pin of one chip is connected to the serial data-input pin of the next chip, creating a longer scan path. If the number of interconnected chips is excessive, it is possible to form several maintenance loops in order to reduce the access time to a particular chip on the scan path.

The chip is marketed by National Semiconductor Corporation as the SCX6260 CMOS gate array.

The 20,000 gate array chip is a derivative of the 6000 gate array chip and has similar capabilities [Lake (1986)]. The gate array product is licensed to Honeywell for commercial sale and is referred to as HC20000 (HC20K).

The HC20K is a CMOS gate array with density of 20,000 NAND gates. The array contains 12,065 internal logic cells of six transistors each, arranged in a matrix structure. This equates to 18,097 internal two-input NAND gates. The built-in evaluation and self-test (BEST) network is incorporated into the chip, and requires 2000 gates of internal logic. The HC20K has 284 I/O pins which are divided to perform several functions: 40 pins for power and ground, 4 pins for the BEST system, 1 pin for clock, 1 pin for hold off function, and 238 data I/O pins. The four pins used by the BEST system are test clock enable (TCE), test strobe (TS), test data in (TDI), and test data out (TDO).

The BEST system comprises a 24-bit control register, a 242-bit input register with operand generation capability, and a 148-bit output register with checksum capability. These three components are arranged into a serial shift-register configuration. TCE gates the system clock to the maintenance registers, while TS engages the maintenance function. When TS is low, the maintenance registers are separated from the array logic so that data may be serially shifted through the maintenance registers with no effect on system operation. When TS is high, a function code is frozen in the control register, and the contents are gated to the array control nodes. TDI is used to shift data serially to the first bit of the control or input registers in order to provide the input register with an initial seed value and to define the function in the control register. TDO serially shifts data from the last bit of the output register to the outside world, allowing the designer to examine the output register contents.

The input register serves as either a data source or destination for nodes between the input buffers and the internal gates of the logic array. The register may be loaded either in parallel from the input buffers or serially through the TDI pin. The input register can also be reconfigured into an LFSR by feeding back some of its stages via an XOR gate to the first stage. Given a user-defined seed value, the LFSR will generate a unique set of patterns with a cycle length of 10^{35}.

The output register can be reconfigured into a MISR with feedback implementing a pure cycling register.

The 24-bit control register is partitioned into a 10-bit system portion and a 14-bit user portion. The individual bits in the system portion each control a distinct function in the BEST system.

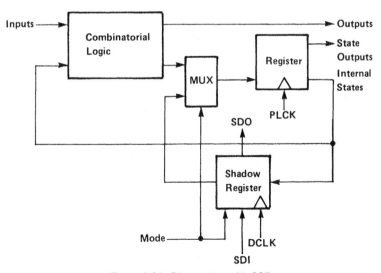

Figure 8.24. Diagnostics with SSR.

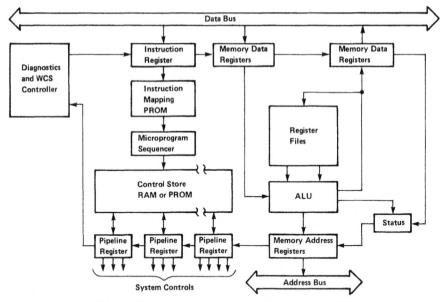

Figure 8.25. A typical computer architecture with SSRs.

The HC20K provides 42 internal control nodes for use at the designer's discretion. These control nodes may be connected within the design net list to increase the design's testability. The control nodes may be divided into two separate classes: Nodes that are outputs of the BEST logic and nodes that are inputs to the BEST logic. Thus the internal control nodes ease the controllability and observability of internal signals within the chip.

8.3.2 AMD Am29818 Serial Shadow Register (SSR™)[2] Device

This is the first off-the-shelf device that has a built-in scan path. This device and comparable ones (like SN54/74S818) enable companies with no semiconductor manufacturing facility to produce scan-path designs to alleviate the sequential test problem.

The SSR provides sufficient observability and controllability to practically turn the sequential test task into a combinational one. This is done by providing means to both control and sample the states of the sequential circuit.

The SSR diagnostics shown in Figure 8.24 use a multiplexer on the input of each state flip-flop to select between the combinational logic signal and a duplicate or shadow flip-flop signal. The shadow register can be loaded serially from the SDI (serial data input) to achieve the controllability feature. The

---

[2] SSR is a trademark of Advanced Micro Devices Inc.

contents of the shadow register can then be copied into the internal state register by controlling the multiplexer and clocking with PLCK (pipeline clock). Internal state information can be observed by loading the shadow register with the contents of the state register and shifting out via the SDO output (serial data output).

Figure 8.25 shows a typical computer architecture with SSRs for system testing and diagnostics. All the state registers have been connected to form a serial scan path. If the length of the scan path is too large, it is possible to break it down to several scan paths, paying the price of additional I/O for scan-in and scan-out.

9

Limitations and Other Concerns of Random Pattern Testing

This chapter deals with some of the difficulties encountered when incorporating built-in random pattern test into a digital system. Although the treatment here is not extensive, it is intended to alert the user to some of the important things to consider during the planning and design phase of a project. While no one of these problems is fatal, each must be dealt with in the context of the particular network design. Factors of technology, packaging, the intended application of the design, design time, and desired manufactured quality are all important in dealing with the concerns discussed here. Clearly, different design projects will deal with these concerns in different fashions, but each must address them.

9.1 INDETERMINATE STATES

In any digital system, there are combinations of data and control signals that can cause the state of a storage element to be unpredictable. An example of this indeterminate or "X" state is an SR flip-flop where both the set and reset signals are present. Unless specially designed to be set dominant (or reset dominant) the resulting state of the flip-flop is not predictable from the signals present. This is a simple case of a much more prevalent problem. In the design of a digital system, often only a small fraction of the possible states are actually valid for proper machine operation. The normal design criteria is that for every valid signal combination, the state of the memory elements is predictable.

However, when pseudorandom patterns are used as test stimuli, signal combinations that are not valid for proper machine operation can appear in the network. If special precautions are not taken during the design phase, these invalid signal combinations can cause X states that can be captured in the memory elements of the network (latches, flip-flops, and RAMs). Since built-in testing compresses the contents of the various storage elements at each cycle into a signature, even one indeterminate state may give rise to a nonunique signature. In order for there to be a single valid signature for built-in test, no X state can be allowed to propagate to and be captured by any storage element. This is the basis for one of the most important design ground rules for built-in test; each signal must be latched (stored) at a determined clock cycle, independent of the data. An example might be the carry output from an adder. If the add operation takes five clock cycles and the carry is available somewhere between cycles 2 and 5, depending on the operands, it must not be latched until cycle 5 in order that a valid signature be formed.

There are several ways in which indeterminate states can arise from RAMs used in conjunction with logic in digital networks. First consider uninitialized or improperly initialized RAMs. If the contents of an improperly initialized word are propagated to an observation point such that the signature is indeterminate, the test result is corrupted. Thus special care must be taken to have every cell of every RAM in a defined state at the beginning of a built-in test protocol.

Another problem can exist in a multiport RAM, where multiple write ports can access a single word. Unless the dominance of the write operations is established in the hardware design, an X state can exist during random pattern testing. A simple rule that port A always overwrites what port B has written if they both write at the same address in the same cycle is sufficient to prevent X states from occurring in a multiport RAM. Without such a rule, it is not possible to predict the state of the word in the RAM from the signals present during the write cycle. Subtle differences in the arrival times of the write signals could corrupt the test signature.

Yet another source of indeterminate states arises from invalid (unused) op-codes appearing on an instruction bus. The design must be extended to consider how the network will respond to *all* combinations of bits in the op-code field, since random patterns will generally present all such combinations to the network during test.

In order to have a valid signature for built-in test, all cases of indeterminate or X states must be eliminated. The hardware must be thoroughly examined to eliminate these situations and test protocols must be designed with proper initialization of all storage elements.

9.2 SIMULTANEOUS SWITCHING

The power supply system of any digital network is designed to support a maximum current surge and still maintain an acceptable voltage level. This design point usually places a limit on the number of chip output (off-chip) drivers that can be allowed to switch at one time. While the bulk power system can easily handle the surge, the distribution to the chip or module may have sufficient series inductance to impose a real limit. When an emitter coupled logic (ECL) driver launches a transition along a 50 Ω transmission line, a current change of 16 mA occurs (800 mV across 50 Ω). Since this transition can occur in less than a nanosecond, a collection of such drivers presents a severe challenge to the power supply design. Similarly, a CMOS driver that drives a 5 V transition in 10 ns with a 50 pF load creates a 25 mA current transient. A thorough discussion of the problem is given in Katopis (1985).

Too many drivers switching simultaneously causes the power supply voltage to dip momentarily. If the disturbance is severe enough,[1] one or more of the internal storage elements may change state (flip) as a result of the transient. The designer is usually provided with design rules that govern how many outputs can switch simultaneously during machine operation. These rules are implemented by considering the valid machine states and determining the switching activity for various operands. This is an admittedly complex analysis and various engineering approximations are often used to facilitate the design. Nevertheless, during random pattern testing, it is usually not possible to control the switching activity at the chip outputs. For this reason, care must be taken during the design phase to assure that the power supply system is robust enough to prevent switching transients from causing one or more internal memory elements (latches or flip-flops) to change state unpredictably and thus corrupt the test signature.

9.3 DRIVERS

There are some special problems associated with random pattern testing that are caused by the drivers that are commonly used to communicate between chips and modules in digital networks. These issues are discussed in the following subsections.

9.3.1 Tri-State Drivers

Tri-state drivers are designed to communicate with buses such that only one driver at a time is forcing a value to the bus (signal orthogonality). This discipline is not present during random pattern test and the driver situation must be reexamined to allow built-in test with random patterns. The drivers must be current limited or designed to be "burn-out proof." In other words, they must be able to supply enough current to force a condition on a normal (no contention) bus in the presence of a similar forcing current in the opposite

---

[1]Simple calculations show that a 100 mA transient in 1 ns through a 10 nH series inductor generates a 1 V transient.

direction from another driver. This first condition is necessary but not suffi-cient, since if the opposing forcing currents are equal, the bus may take on a voltage value in the middle between 0 and 1 and an X state will be generated. In order to prevent these indeterminate states, the bus must be terminated so that it assumes a valid logic state under these conditions. This design will also cover the case when all drivers on the bus are set to the high impedance state. Clearly the termination will resolve that potential X state also.

9.3.2 Bidirectional Driver–Receivers

Bidirectional driver–receivers, sometimes referred to as common I/O (input/output) circuits, present no fundamental problems if they are terminated as in Section 9.3.1 so that a defined value is present when the driver is in its high impedance state. The other considerations for parallel connections are the same as discussed for tri-state drivers. These bidirectional driver–receivers are usually used at a package boundary. When the packaged network is tested with built-in random pattern tests, an external circuit is usually connected to each of these bidirectional circuits at the package boundary. This external circuit may be part of a tester that is supporting the built-in test protocol or it may be another part of the system to which the package belongs. In either case, if the external circuit is passive, that is, it is only a receiver, no special constraints on the bidirectional circuit are neces-sary. However, if the external circuit is active (a driver) and the test protocol requires that the bidirectional circuit in the package be driven by the external circuit, special controls must be included in the design.

One method of controlling these bidirectional circuits is to have a control circuit that forces the driver portion of the circuit to assume the high impedance state. Under this condition the circuit is forced to be a receiver and the external circuit can drive it without any interference from the part under test. If such a control circuit is incorporated in every bidirectional circuit on the package and a single driver inhibit line is brought to the package boundary, the proper test protocols can be applied. When it is desired that the external circuit be active, the driver inhibit input is applied such that it forces the bidirectional circuits to be receivers. When the driver inhibit signal is not present, the bidirectional circuits are drivers and the external circuit must act as a receiver. Thus the extra control line resolves any conflicts that might arise at the bidirectional circuits at package boundaries.

9.4 TEST EFFECTIVENESS PROBLEMS

The one constant in various approaches to the testing of digital networks is the need for a clear measure of the effectiveness of the tests used. Without such a measure, no estimate of quality can be made. There are measures for the effectiveness of random pattern tests. Since there are basic underlying assumptions involved in most of them that limit their applicability, a brief discussion of them follows.

9.4.1 Delay Testing

Delay testing with pseudorandom patterns was thoroughly discussed in Section 7.12. Formulas for test length are derived there. The key feature of a delay test, sometimes incorrectly called an AC test, is a timed change between two test patterns. The test length formulas are derived assuming that two independent test patterns are chosen at random and applied in sequence to the network. A necessary condition for applying these tests is an accurately timed interval between the two patterns. This interval can be supplied by the network itself or, more often, by some form of external support equipment such as a tester. However, a timed interval, while necessary, is by no means sufficient to guarantee that a delay test is performed.

In the actual network under test, unless all of the control conditions along the path are directly controllable at the package boundary, there may be structure that precludes the existence of the sequential pair of patterns required to test a particular path. This arises from the need to either scan-in both patterns or derive the second pattern as a result of the signal traversing some amount of logic within the network. The constraints due to scan derive from the need to scan-in the second pattern only one shift behind the first (with the proper change at the path input) while holding the path controlling conditions unchanged at other points. The problem associated with traversing the network is similar. When the second pattern is derived as a result of a random pattern traversing a portion of the network prior to arriving at the clocked point (typically a scan latch), there may be structure in the logic of the portion of the network traversed that precludes obtaining the desired second pattern. This pattern must propagate the change along the path as well as support the desired control conditions.

In order to have any guarantee of test effectiveness, the actual pattern pairs that are used in the built-in test protocol must be analyzed. In the case of testing for delay faults, simulation is probably the best method of analysis. Only by analyzing the exact pattern pairs can the effects of structure discussed here be evaluated.

9.4.2 Random Pattern Resistance

The effectiveness of random patterns in detecting stuck faults was extensively covered in Chapter 7. The analysis techniques available to the designer include the cutting algorithm and fault simulation. When an instance of random pattern (RP) resistance is encountered, the designer may want to modify the logic (Section 7.7). This is often a major effort. Alternatively, the test protocol may have to be changed to one which includes weighted random patterns (Section 7.6). This may also have impacts on the design. Since the cutting algorithm performs a statistical analysis and many fault simulation techniques use surrogate patterns, the designer may ask if change is really necessary. For these reasons and to assure a valid test effectiveness measure, a fault simulation using the actual patterns against the RP resistant faults should be done.

Even with exact information, there will indeed be instances of RP resistance. Often the remedies are painful, either in design time or in performance impacts. The proper trades between test effectiveness, logic modification, and test protocols is a significant engineering decision. It should be provided for in the initial planning of any project incorporating built-in test.

9.5 CIRCUIT RELIABILITY CONSIDERATIONS

When circuitry is added to a chip for built-in test purposes, the question of decreased reliability is sometimes raised. This is a question that deserves a serious answer. First of all, the amount of circuitry that is added for test must be determined. In many implementations of built-in test, less than 10% extra circuits are added to perform the built-in test function. In some cases, the overhead is as low as 2%. Thus the question is what impact on circuit reliability does 10% added circuitry have?

Network A, which is 1.1 times as large as network B, will have a failure rate 1.1 times as high if the per circuit failure rates are the same. This makes the mean time between failures (MTBF) of A 0.909 times that of B, the smaller network. Therefore, if 10% extra circuitry is added for built-in test, the MTBF of the network is reduced by 9%, everything else staying constant. (For the 2% overhead case, MTBF is reduced only 2%.) This may not be significant in the application, but often it is. The benefits of reduced time to diagnose and repair (MTTR) must outweigh the reduced MTBF.

Consider the following example: Network B has an MTBF of 1000 hr and an MTTR of 2 hr. Over a 10,000 hr period (slightly more than 13 months of full time operation), 10 failures are experienced and 20 hr are spent in repair (and network B is out of service). Now consider network A, which has built-in test with a circuit overhead of 10%. Its MTBF is only 909 hr, but its MTTR is only 1 hr. This reduced MTTR is the result of the built-in test in network A. In the same 10,000 hr period, 11 failures are experienced, but only 11 hr of down time are experienced—a reduction of 45%. In most situations, a 45% reduction in down time will more than pay for a 10% increase in failure rate.

9.6 PARAMETRIC TESTING

As part of conventional approaches to testing integrated circuits, considerable DC parametric testing is performed. Typical tests measure input currents and driver output current capability. These parameters are measured to assure proper device operation under maximum load conditions. This type of test is not covered by the built-in tests discussed in this book. The normal manner in which this problem is handled at higher levels of assembly (modules, printed circuit boards, and system subassemblies) is to perform logic tests at various power supply settings away from nominal. In a multiple voltage system, the sets of power supply voltages at their extremes are commonly called corners. For example $V1_{max}$, $V2_{max}$, $V3_{min}$ is a power supply corner. Usually some analysis will determine which corner results in minimum output drive current capability. The test may be conducted at this corner to give some level of assurance that the driver outputs are sufficient.

For integrated circuits with built-in test, a similar approach may be sufficient in the parametric area. If the voltage corners are found where input current drain is the maximum and where output drive is the minimum, the tests can be conducted with the power supplies set at these corners.

The determination of which corner to use should not be based solely on the analysis and characterization of properly functioning "good product." Failing and marginal parts must be characterized also. Often, certain faults and failure modes can be accentuated by testing at a voltage corner that is not obvious from the analysis of defect-free integrated circuits. This need to analyze both good and bad product is not unique to built-in test, but is a method of problem discovery and elimination common in the semiconductor industry.

10

Test System Requirements for Built-In Test

Any electronic assembly with built-in test requires some interaction with the outside environment to support the test function. This may merely be power and a "start test" signal. In more complex situations, initialization sequences and special clocking may be required. The means by which these outside support and control functions are provided to the part under test is given the generic title of a "tester." The tester may be used to invoke various levels of built-in test, some which rely mainly on the test structure built into the part and others that rely more on the tester's intelligent manipulation of the part under test. This extended function of the tester prompts one to think of it as a system including its computer and control programs. In the following, the functions of the test system will be explored.

Most built-in test implementations are designed to provide a go/no-go indication at a package or assembly boundary. This identification of a field or line replaceable unit is a key attribute of built-in test, but there is a need for more precise diagnosis if repair is to be accomplished on the failed unit. This repair may take several forms. Usually a unit that is removed in the field is sent to a repair depot. At this stage, the system or use environment must be simulated and additional diagnosis performed to determine if the cause for failure is within the capabilities of the depot to repair or if the unit must be returned to the manufacturer. Here the tester must provide an accurate simulation of the environment in which the unit was reported faulty. Throughout the discussion of failure diagnosis, there is the very real problem of "false alarms" or no defect found at retest. While important, its complexity is beyond the scope of the present discussion.

At the manufacturing facility, in addition to repairing failing units that have been returned from the field, the repair of defects that are introduced during initial manufacture may be crucial to the economic production of the assemblies. Here equipment is necessary to enable diagnostic procedures to isolate a fault to a component which can be replaced.

Even at the semiconductor manufacturing plant, diagnosis to a repair action is important. In the introduction of a new part into production, the problem of zero yield at first silicon requires the ability to diagnose faults within the VLSI chip to a precision of a few circuits. Often there are subtle design or mask flaws that give common-mode failures at this early stage of a product's life. Only by isolating the failure to a very small area can the trouble be clearly identified and corrected. Here again some external equipment (a tester) is required to support and augment the test structure built into the part.

In the following, the characteristics of the test equipment that are needed to support diagnosis of failures in components and assemblies that incorporate built-in test will be discussed. The support requirements of other functions, such as parametric testing and delay testing, will also be examined. A key feature of any automatic test equipment is the software required to support it and the software tools required by the test engineer. These will be described in general terms.

10.1   TEST SYSTEM SUPPORT OF DIAGNOSIS

Three modes of diagnosis that are applicable to digital networks with built-in test will be discussed. They are

1. boundary scan techniques,
2. probe-based techniques, and
3. simulation-based techniques.

They may not all be appropriate for a given situation, but they show how to approach the diagnostic problem.

10.1.1   Boundary Scan Diagnosis

When boundary scan techniques are built into the part under test, relatively little is required of the test system. Each unit—board, module, chip, or subsection of a chip—has boundary scan latches that provide stimuli and collect responses. As an example, consider a circuit board with a number of packages each containing one VLSI chip. Each of the chips contains a boundary scan implementation. At the board edge, the ability to initialize, provide or control clock sequences, and read the signature of each chip is provided.

The test and diagnosis process follows this scenario:

Step 1: The test system performs a power and ground test to insure no power hazards. Then the part under test is powered.

Step 2: The test system tests the boundary scan rings.

Step 3: The test system initializes the part under test:
- sets all random-access memory to known states,
- sets all shift-register latches to known states, and
- places proper seeds in pattern generators and MISRs.

Step 4: The test system provides sufficient clocking sequences to complete the test.

Step 5: The test syst scans out the signatures of each chip individually and compares them to the stored correct values. Because of boundary scan, these signatures are the same as the signatures of the unassembled individual chips.

Step 6: Upon a miscompare, the failing chip is identified for subsequent repair. If diagnosis to a smaller domain than that isolated by the smallest boundary scan domain (the chip in the example) is needed, a fault simulation-based diagnostic protocol is required.

Step 7: The test system performs a test of interchip wiring on the printed circuit board.

To summarize, in this test and diagnosis procedure for a boundary scan part, the test system had to

1. initialize the part under test,
2. apply clocks to the part under test,
3. scan-out signatures from the part under test, and
4. compare signatures.

The advantages (relative to diagnosis) of boundary scan are

1. the same chip signatures at all packaging levels,
2. simple diagnosis to a falling chip, and
3. easy chip-to-chip wiring (board wiring) test and relatively simple diagnosis.

10.1.2 Probe-Based Diagnostics

Turning now to probe-based diagnostic techniques, assume that the tester has a probe that can collect signatures of the responses at interior probe points in the network. The probe is under the control of the tester either directly, in the form of perhaps a robot arm, or indirectly, by displaying coordinates that a human operator will probe. An example of the points to be probed could be the terminals of dual in-line packages mounted on a printed circuit board. Another could be interior points on a ceramic multichip carrier.

For an example, consider a printed circuit board with input and output registers configured as BILBOs. The network contains RAMs and flip-flops, but these can be initialized from the card edge. Any global feedback nets come to the card edge so that they can be disabled during test. Assume that the BILBOs are functioning correctly. (Special test and diagnostic procedures are used to assure that the BILBOs are functioning properly.) The test system has access to the signature dictionary and a map of the logic structure. The test and diagnosis proceeds as follows:

Step 1: The part under test is powered up. Care is taken to avoid catastrophic failure from power to ground shorts.

Step 2: The test system initializes the part under test:
- sets all random-access memory to known states,
- sets all flip-flops to known states,
- places proper seeds in BILBOs, and
- the portion of the tester that interfaces directly with primary I/O of the part under test is initialized.

Step 3: The test system applies sufficient clocking sequences to complete the test.

Step 4: The test system scans out the signature in the output BILBO which acted as a MISR and compares it with the known good signature stored in the signature dictionary. The signatures formed in the test system by the responses of the primary outputs are also compared with the signature dictionary.

Step 5: Upon a miscompare, the diagnostic procedure begins. If a primary output has an incorrect signature, the backtrace procedure skips to Step 8. If the BILBO signature is bad, the test system brings the part to the initial state and positions the signature probe at the input of the BILBO, formed out of the output register, to gather signatures.

Step 6: The test system again applies sufficient clocking sequences to complete the test.

Step 7: The signatures formed on the inputs to the BILBO are compared with their counterparts in the signature dictionary to determine which line(s) capture(s) the effects of a fault. These lines are candidates for backtrace. The packages identified as providing incorrect inputs to the BILBO are designated failure capturing packages (FCPs).

Step 8: The test system again brings the part to the initial state and positions the signature probe at the inputs of one of the FCPs.

Step 9: The test system again applies sufficient clocking sequences to complete the test. If the input signature of an FCP compares correctly with the reference in the dictionary, the fault is contained within the FCP and this portion of the diagnosis is complete. However, if the input signature of an FCP is bad, the test system must backtrace each faulty input.

- The test system initializes the part and positions the probe at the package on backtrace from a faulty FCP input.
- The test system again applies sufficient clocking sequences to complete the test. As before, if a package has a faulty output and a correct input, it contains a fault. If this is not the case, the backtrace continues until the fault is located.[1]

Step 10: After resolving one FCP, the test system repeats the procedure for all FCPs identified by incorrect inputs to the BILBO. In this way, a repair action is constructed.

There are several problems associated with a diagnostic backtrace that the program controlling the test system must deal with. One of these is an open circuit at an input to a circuit package. The driving circuit has the correct signal, but the receiving circuit sees a faulty signal. The backtrace technique must provide a method of resolving this ambiguity. Another problem occurs when an output fans out to several receiving circuits. A fault in one branch of the fan-out net may cause an error in another branch of the net. If this situation is not handled properly, the diagnostic program may incorrectly identify the driving circuit package as the FCP instead of pointing to the input fault on the other receiving package. To provide effective diagnostics, the test system and its control programs must be able to cope with problems such as these.

To summarize the operations the test system had to perform to support this test and diagnosis procedure based on probing interior nets, the test system had to

1. initialize the part under test,
2. apply clocks to the part under test,
3. apply stimuli to primary inputs of the part under test,
4. collect responses from primary outputs of the part under test and scan out the BILBO,
5. position a probe on the part under test,
6. perform logic backtrace, and
7. compare signatures.

---

[1]If the backtrace passes through another FCP and if this FCP's inputs are faulty, the FCP is marked as diagnosed as good. However if the inputs are correct then the fault is in this FCP.

10.1.3   Simulation-Based Diagnostics

Finally, simulation-based diagnosis is used when more information about the fault is needed than is obtained by the boundary scan or probe techniques just described. The basic idea is to obtain a failing response to a known stimuli and use fault simulation to determine the fault equivalence class(es) that can cause the observed response. Usually several different stimulus/faulty response pairs are needed to resolve the fault to a single fault equivalence class.

The key difference between fault simulation-based diagnosis and both boundary scan and probe diagnosis is that the failing response to a single test vector is required by simulation, while boundary scan and probing use a faulty signature for the entire test sequence which may be 10^4–10^6 test vectors. Getting those stimulus/faulty response pairs is a task that the test system must do to support fault simulation-based diagnosis.

Consider the previous boundary scan example. Suppose it is desired to diagnose within the faulty chip to a fault equivalence class. This would typically be in pursuit of some subtle common-mode fault introduced at the semiconductor fabrication process. The primary added data requirement is a signature dictionary of intermediate signatures. Suppose the complete test sequence is 10^5 patterns long. The signature dictionary for this part will have intermediate signatures for intervals of 100 patterns. The logic model of the part under test is also available to the test system.

The diagnostic procedure would be as follows.

Step 1: Upon determination of a failing boundary scan test, the test system reinitializes the part under test to the beginning of the test sequence:
- sets all random access memory to known states,
- sets all shift registers latches to known states, and
- places proper seeds in pattern generators and MISRs.

Step 2: The test system would apply the following to each 100 test group until a failing signature is observed:
- Apply sufficient clock sequences to apply 100 test vectors to the part under test.
- Scan out the signature from the part under test.
- Compare the signature from the part under test with its counterpart in the signature dictionary.
- If the part displays the correct signature, the test system must restore the state of the part under test to what it was at the completion of the last 100 tests before the signature was scanned out. The next 100 test group is initiated.
- If the signature of the part under test differs from the corresponding signature in the dictionary, at least one failing response occurred in the most recent 100 test sequence. At this point, the test system must reset the state of the part under test to the starting point of this most recent 100

test sequence. The sequence is repeated, but instead of compressing the responses in the on-chip MISR, each response is scanned out and saved in the test system. The test system must now restore the state of the part under test to that of a good part at the end of the most recent 100 test sequence. The next 100 test sequence is initiated.

Step 3: When sufficient failing 100 test sequences have been collected, the test system is finished with the physical part under test. The remainder of the diagnosis can be done off-line using known deterministic diagnostic techniques.

The test system has now collected one or more groups of responses to 100 tests, each group containing at least one failing response. The logic model of the part under test is available to the test system. Since each test group can be associated with a specific number of clock sequences from the initialization of the part, the test system can calculate the stimulus provided by the on-chip pattern generator. Alternatively these could be calculated in advance and stored in the test system's database. Each of the stimuli used is applied to a good machine model of the network to determine the correct response. These correct responses are compared with the observed responses to determine which of the observed patterns are faulty. In this way, the stimulus/faulty response pair(s) in the 100 test group is (are) determined.

In analyzing a stimulus/failing response pair to determine the fault equivalence classes that can be present, first a reduced model of the part under test is built. The model consists of the logic cone on backtrace from each failing SRL/PO in the stimulus/response pair under analysis. In this way, the fault simulation step of the analysis can be more efficient than if a model of the entire part were used. The stimulus is then fault simulated to determine which faults in the reduced model can be sensitized by the stimulus and observed at the outputs that have been identified. The group of faults thus identified must contain the fault that is present and causing the faulty response. Another stimulus/response pair is analyzed and its group of faults is intersected with those of the first pair. Under the assumption of a single fault, the fault that is sought is in the intersection of the two groups. This procedure can be continued until a single fault equivalence class is identified.

To summarize the test system functions required to support this fault simulation-based diagnostic procedure for a boundary scan part, the test system had to

1. initialize the part under test,
2. apply clocks to the part under test,
3. scan out signatures from the part under test,
4. compare signatures,
5. restore the part under test to a specific intermediate state, and
6. scan out and store uncompressed responses.

One consideration that must be included in any diagnostic procedure utilizing response compression is the possibility that a failing response may mask to a good response signature. In this case, incorrect diagnosis may result in an incorrect repair action. Thus, if a repair action does not correct a failure, and the same repair action is indicated, a diagnostic procedure must provide for the possibility of such masking.

10.2   TEST SYSTEM SUPPORT OF PATH DELAY TEST

Since built-in test implementations are configured from the same circuitry as the networks that they are designed to test, the tests can be applied at functional or use speeds. Thus path delay testing or the verification of path traversal adequacy can be implemented in suitable built-in test structures. In Section 7.12 path delay was treated in detail, resulting in formulas relating test length with coverage or quality. Here only a summary of path delay fundamentals will be presented.

In the circuit shown in Figure 10.1, an initial pattern [0100] is applied to the circuit. To measure the delay along the upper path 1-5-7, a transition pattern [1100], is applied and the delay between the time of application of the transition at node 1 and the arrival of the transition at node 7 is the path delay for path 1-5-7.

As an example of path delay measurement in an LSSD environment, consider the circuit shown in Figure 10.2. The clock timing is shown in Figure 10.3.

This LSSD double-latch design is just one of many that could be considered. The clock shown is "active high," meaning that the latches accept data only when the clock is high. During normal operation of the circuit, the data from the combinational portions of the network is captured in the L1 latches by the capture clock (system or latch clock). That is, the data present at the

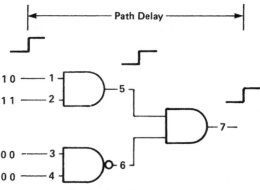

Figure 10.1. Path delay test.

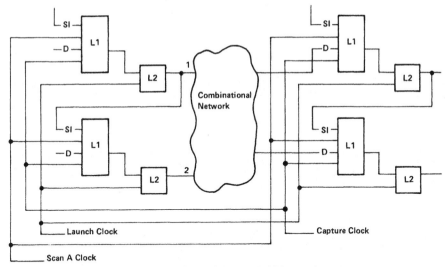

Figure 10.2. Double-latch LSSD circuit.

data (D) inputs has propagated through the combinational network and is present at the input ports of the $L1$ latches at the time the capture clock goes active. The latch output assumes this data value during the clock period and is stabilized when the capture clock goes down.

To start the next cycle, the launch clock (the normal shift B or trigger clock) is applied. When the launch clock rises, the data in the $L1$ latches, which is sitting at the input of the $L2$ latches, is propagated through the $L2$ latches and launched into the combinational network toward the next $L1$ latch. Thus the critical timing interval (path length) is from rising transition 1 (launch clock leading edge) to falling transition 2 (capture clock falling edge).

In scan mode, the launch clock is used as the scan B clock so the normal scan clock sequence is scan A, launch, scan A, launch, ..., thus transferring the data into the $L1$ and then from $L1$ to $L2$ and on to the next $L1$ in the chain.

In path delay test, one or more launch/capture cycles must be simulated. As discussed in Section 7.12, in order for a path delay test to be performed,

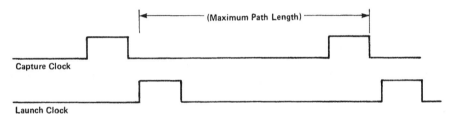

Figure 10.3. Clock timing relationships.

the two patterns must form a delay test pair, that is, cause at least one transition that propagates from input ($L2$) to output ($L1$). A necessary but not sufficient condition for a path delay test is an accurately timed pair of patterns. As discussed in Chapter 9, only a careful analysis of the network under test can determine if a path delay test is actually performed by the pattern pairs that the test system initiates. There are two protocols that can be developed to provide a test of path delay for the example circuit. Protocols A and B establish the timing relationships for such delay test pairs, but the logic structure determines whether the values of the two patterns actually constitute a delay test pair. Because the constraints placed on the patterns differ between protocol A and protocol B, both may be used to obtain a richer set of test pairs.

Protocol A starts with the loading of the $L1/L2$ latches with a scan-in sequence scan A, launch, scan A, ..., launch, scan A while providing a data stream to the input port of the various shift-register chains of the device under test. By ending the scan clocking sequence with an A clock, the data in the $L1$ latch has not been transferred to its $L2$ latch; thus, different values can exist in $L1$ and $L2$ in each $L1/L2$ latch pair. The values in the $L2$'s provide an initial pattern to the combinational network because the $L2$ outputs are stable and propagate through the network to $L1$ inputs. If, after the network has stabilized, a single launch clock is applied and an appropriate time later, a single capture clock is applied, a timed pair of patterns have been applied to the combinational network. If the network structure is such that a path delay test can be performed by protocol A, such a test is applied by this sequence of clocks. When the launch clock is applied, the values that are in the $L1$ latches are gated through the $L2$'s to the network to provide a transition pattern to the logic. The successful capture of the network's response to the transition pattern (provided that, indeed, a transition pattern was present) in the $L1$'s provides an indication of path traversal adequacy. The clocking sequence for protocol A is shown in Figure 10.4.

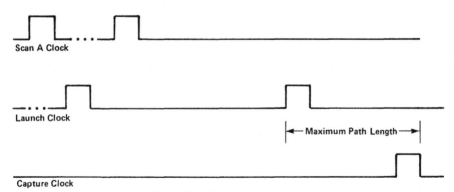

Figure 10.4. Path delay protocol A.

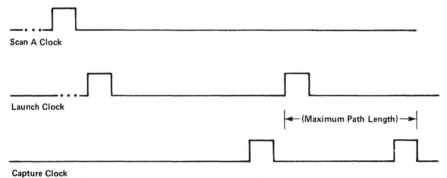

Figure 10.5. Path delay protocol B.

The timing for Protocol B is shown in Figure 10.5. In this case, the scan-in sequence is terminated normally with a launch clock. This results in the same data being in both elements of an $L1/L2$ latch pair. After the network has stabilized, the initial pattern is present throughout the network, and a capture clock is issued to move the data at the inputs of the $L1$'s to the inputs of the $L2$'s. Next, a launch clock pulse followed by an appropriately timed capture clock performs the path delay timing.

Different constraints on possible initialization/transition pattern pairs occur for protocol A than for protocol B. In protocol A, both the initial and transition pattern are scanned in. This causes some fixed relationships that reduce the possible combinations of initial and transition patterns. Referring to Figure 10.2, the initial value at input 1 is the transition value for input 2. In protocol B, while the initial pattern is scanned in, the transition pattern is the result of another initial pattern having traversed a combinational network and having been modified by that transit. As mentioned, a detailed discussion of the requirements of path delay test pairs is given in Section 7.12. However, from the preceding example it is clear that a richer set of pattern pairs exists when both protocols are used.

Turning now to the test system requirements to support path delay testing, it is clear that flexible clocking is the prime requirement. To provide a sense of the timing accuracy required, a hypothetical system that runs at 50 MHz will be discussed. That means that the clocks shown in Figure 10.3 have a period of 20 ns. As rule of thumb, the pulse width is 25% of the period (5 ns in this case). The edge placement accuracy should be 1.0 ns or 5% of the period for efficient testing in production volume. These relations more or less scale with the basic system speed and can be used to determine the timing requirements of a test system.

Thus to support path delay testing, the test system must provide clocking facilities that

1. scan in data and end with either an A clock or a B (launch) clock on demand,

2. issue single or pairs of clock pulses with well defined and timed leading and trailing edges, and

3. issue pulses on different lines (launch and capture) that have accurate timing between them.

10.3 THE IDEAL TESTER

The multivalued relation between diagnostics, the tester requirements, and the design rules has been explored. Use of a probe diagnostic means that the design must provide access to the peripheral signals of each chip and the tester must allow the access, support probe electronics, have storage for all package net signatures, and provide stimulus for package input pins and response collection for package output pins. Boundary scan, on the other hand, means greater design involvement in the scan latches and the control of various test modes, while tester design is simpler since it need not provide stimuli or collect responses. Simulation requires the packaging designer to minimize false alarms due to equivalence-class spanning and requires the tester to support failing test identification and either on-line or off-line simulation capacity.

Assume a package containing more than a single chip but less than a complete system. The application is at a manufacturing house and pass/fail testing plus diagnosis of all failures to a repair action (which may be a chip replacement) is needed. The built-in test uses pseudorandom pattern stimuli and signature analysis compression on responses.

The tester for this package would need the following capabilities:

1. ability to power and cool the package;
2. ability to reset package storage elements:
   - random-access memories and
   - system latches;
3. ability to command the package:
   - select test or use mode,
   - do a single pass test sequence,
   - continuous loop test (do again),
   - single step,
   - reset—go to start,
   - go to test X,
   - backstep, and
   - stop on test X;
4. ability to monitor the package:
   - collect PO signatures,
   - unload package signatures,

- make pass/fail decisions (compare signatures), and
- count tests;
5. ability to select test pattern:
   - external from random generators and
   - internal from built-in test;
6. ability to select test sequence
   - up to K different sequences;
7. ability to clock the package:
   - system clocks,
   - scan ring shift clocks, and
   - array write/read clocks;
8. ability to set package stability states
   - up to J different states.

It was noted earlier (Chapter 7) that logic design rules and perhaps logic modification techniques are necessary to restrict the random pattern test length to a reasonable value. This value is also a tester constraint since it impacts tester throughput. The design used to control or test embedded random-access memories obviously must be reflected into tester specifications. Orthogonality limits (Chapter 9) which are not resolved at the technology or design level must also be accommodated by the tester.

10.4   TEST SYSTEM SUPPORT OF FIELD TESTING

Field testing is the testing of the electronic assembly in use conditions in a maintenance or function verification mode. In these circumstances, the part under test is either an entire system or a portion of such a system that is embedded in the larger system. In either event, there has to be a mode of system operation that permits the built-in self-test to be effective. This function is usually performed by an auxiliary processor called a maintenance processor or support controller. During the period that the built-in test is to be invoked, system control is passed to the maintenance processor. The maintenance processor must then:

1. Initialize all memory elements in the portion of the system to be tested. Load seeds or starting values into on-board generators and response compressors.
2. Set the necessary stability states so that the portion of the system to be tested is properly conditioned and isolated from other parts of the system if required.
3. Provide the necessary clocking signals to run the built-in test.

4. Provide a means to read the compressed response of the portion of the system under test and determine if it is correct.

5. If the response is correct, the maintenance processor passes control back to the main system control program. If there is an improper response, the maintenance processor enters into its fault recovery strategy. This may take many forms, depending on the sophistication of the system. Actions that can be elements of such a strategy include abort, reconfigure, retry, and call in additional resources.

The maintenance processor may be built into the total system in the case of large fixed installations such as a mainframe computer or a shipboard signal processor. In the case of airborne avionic systems or portable equipment, the maintenance processor may be a portable unit that the line maintenance technician uses to control the execution of the built-in test. In any event, the processing requirements on the maintenance processor are modest. It must be able to take system control and set the proper states in the system under test; this may be through scan-path techniques or other maintenance facilities built into the system. There is a need for considerable data for the maintenance processor, since it must supply the proper values to all memory elements in the portion of the system under test. There may be several such parts that are tested separately, each requiring its unique set of initial and final values. Several megabytes of disk storage are typically required for the maintenance processor.

10.5 SOFTWARE FOR TESTER SUPPORT

Software systems for tester support provide data for several different purposes. These include

1. product data to the tester,
2. results data from the tester,
3. status data about the product and/or the tester, and
4. engineering workstation support.

Clearly there is a software system that controls the tester execution locally. This is the tester resident control system. However, in any development or production installation there is also an establishment data system. This establishment system may be merely a local area network (LAN) of test systems or it may be a much larger computer integrated manufacturing system (CIMS). In either case, it is important to consider how the data flows to and from the test system. One of the primary considerations may arise from the need for engineering workstation support. Many engineering workstations require access to a substantial database. A key system design issue is where this database should reside, at the tester or in the establishment data system. Engineering workstation activity at the test system includes test program debug, product characterization, and product debug using specially developed test procedures. If the database that this activity must access is large, it probably is not economical to have that database at the tester. However, if the workstation is attached to the establishment system to facilitate database access, the real-time interactions with the test system may be seriously impaired. One compromise is to have the database on the establishment side and the workstation attached to the test system. In this case, the test system must have enough local storage to contain a copy of that portion of the establishment database needed for the current task at the workstation.

Once the database issues are resolved, the remainder of the software for tester support can be specified. In the case of providing data to the tester, these needs are for test data (test programs for various products) and product descriptions (logic models) if diagnosis is to be done on a routine basis. If test programs come from the establishment data system, they can either be brought over in tester executable format or in establishment format. In the latter case, a compiler is needed in the test system to translate from establishment format to tester executable format.

The results of the tests must also be handled by the test system. Clearly, results processing for go/no-go decisions should be done locally by the tester. Any additional data needed for diagnosis should also be gathered under local control. After the test is complete, some form of the results most probably will need to be sent to the establishment data system. Most production systems have a product history database in which data of this type resides.

Other data that is needed in a production environment is status data about both the product and the test system. The shop floor control portion of the establishment data system requires information about the completion of test for each lot of product. The maintenance system needs data of test system performance such as uptime/downtime and calibration status. From these data, mean time to failure (MTBF), mean time to repair (MTTR), and stability can be calculated.

In summary, the software system requirements to support test systems that work with products with built-in test are not qualitatively different from those that support products without built-in test. The detail implementations may be different, but the variations from one establishment to another are probably larger and more significant than the particular requirements imposed by built-in test.

A Primitive Polynomial for Every Degree through 300

The polynomials listed have the fewest number of terms for a primitive polynomial of each degree. Only the exponents are shown in the table. For example, the entry

$$8: \quad 6 \quad 5 \quad 1 \quad 0$$

represents the polynomial

$$x^8 + x^6 + x^5 + x + 1$$

There are no primitive trinomials (having three terms) of degree 8 so the polynomial listed has five terms. In fact there are no primitive trinomials of degree $8n$ for any n [Golomb (1982)].

The first 168 entries in the table are reproduced from Stahnke (1973).

Exponents of Terms of Primitive Binary Polynomials

1:	0				13:	4	3	1	0	25:	3	0			
2:	1	0			14:	12	11	1	0	26:	8	7	1	0	
3:	1	0			15:	1	0			27:	8	7	1	0	
4:	1	0			16:	5	3	2	0	28:	3	0			
5:	2	0			17:	3	0			29:	2	0			
6:	1	0			18:	7	0			30:	16	15	1	0	
7:	1	0			19:	6	5	1	0	31:	3	0			
8:	6	5	1	0	20:	3	0			32:	28	27	1	0	
9:	4	0			21:	2	0			33:	13	0			
10:	3	0			22:	1	0			34:	15	14	1	0	
11:	2	0			23:	5	0			35:	2	0			
12:	7	4	3	0	24:	4	3	1	0	36:	11	0			

37:	12	10	2	0
38:	6	5	1	0
39:	4	0		
40:	21	19	2	0
41:	3	0		
42:	23	22	1	0
43:	6	5	1	0
44:	27	26	1	0
45:	4	3	1	0
46:	21	20	1	0
47:	5	0		
48:	28	27	1	0
49:	9	0		
50:	27	26	1	0
51:	16	15	1	0
52:	3	0		
53:	16	15	1	0
54:	37	36	1	0
55:	24	0		
56:	22	21	1	0
57:	7	0		
58:	19	0		
59:	22	21	1	0
60:	1	0		
61:	16	15	1	0
62:	57	56	1	0
63:	1	0		
64:	4	3	1	0
65:	18	0		
66:	10	9	1	0
67:	10	9	1	0
68:	9	0		
69:	29	27	2	0
70:	16	15	1	0
71:	6	0		
72:	53	47	6	0
73:	25	0		
74:	16	15	1	0
75:	11	10	1	0
76:	36	35	1	0
77:	31	30	1	0
78:	20	19	1	0
79:	9	0		
80:	38	37	1	0
81:	4	0		
82:	38	35	3	0
83:	46	45	1	0
84:	13	0		
85:	28	27	1	0
86:	13	12	1	0
87:	13	0		
88:	72	71	1	0
89:	38	0		
90:	19	18	1	0
91:	84	83	1	0
92:	13	12	1	0
93:	2	0		
94:	21	0		
95:	11	0		
96:	49	47	2	0
97:	6	0		
98:	11	0		
99:	47	45	2	0
100:	37	0		
101:	7	6	1	0
102:	77	76	1	0
103:	9	0		
104:	11	10	1	0
105:	16	0		
106:	15	0		
107:	65	63	2	0
108:	31	0		
109:	7	6	1	0
110:	13	12	1	0
111:	10	0		
112:	45	43	2	0
113:	9	0		
114:	82	81	1	0
115:	15	14	1	0
116:	71	70	1	0
117:	20	18	2	0
118:	33	0		
119:	8	0		
120:	118	111	7	0
121:	18	0		
122:	60	59	1	0
123:	2	0		
124:	37	0		
125:	108	107	1	0
126:	37	36	1	0
127:	1	0		
128:	29	27	2	0
129:	5	0		
130:	3	0		
131:	48	47	1	0
132:	29	0		
133:	52	51	1	0
134:	57	0		
135:	11	0		
136:	126	125	1	0
137:	21	0		
138:	8	7	1	0
139:	8	5	3	0
140:	29	0		
141:	32	31	1	0
142:	21	0		
143:	21	20	1	0
144:	70	69	1	0
145:	52	0		
146:	60	59	1	0
147:	38	37	1	0
148:	27	0		
149:	110	109	1	0
150:	53	0		
151:	3	0		
152:	66	65	1	0
153:	1	0		
154:	129	127	2	0
155:	32	31	1	0
156:	116	115	1	0
157:	27	26	1	0
158:	27	26	1	0
159:	31	0		
160:	19	18	1	0
161:	18	0		
162:	88	87	1	0
163:	60	59	1	0
164:	14	13	1	0
165:	31	30	1	0
166:	39	38	1	0
167:	6	0		
168:	17	15	2	0
169:	34	0		
170:	23	0		
171:	42	3	1	0
172:	7	0		
173:	10	2	1	0
174:	13	0		
175:	6	0		
176:	43	2	1	0
177:	8	0		
178:	87	0		
179:	4	2	1	0
180:	52	2	1	0
181:	89	2	1	0
182:	121	2	1	0
183:	56	0		
184:	41	3	1	0
185:	24	0		
186:	53	2	1	0
187:	20	2	1	0
188:	186	2	1	0
189:	49	2	1	0
190:	47	2	1	0
191:	9	0		
192:	112	3	1	0

193:	15	0			229:	21	2	1	0	265:	42	0		
194:	87	0			230:	25	2	1	0	266:	47	0		
195:	37	2	1	0	231:	26	0			267:	29	2	1	0
196:	101	2	1	0	232:	23	2	1	0	268:	25	0		
197:	21	2	1	0	233:	74	0			269:	117	2	1	0
198:	65	0			234:	31	0			270:	53	0		
199:	34	0			235:	45	2	1	0	271:	58	0		
200:	163	2	1	0	236:	5	0			272:	56	3	1	0
201:	14	0			237:	163	2	1	0	273:	23	0		
202:	55	0			238:	5	2	1	0	274:	67	0		
203:	45	2	1	0	239:	36	0			275:	28	2	1	0
204:	86	2	1	0	240:	49	3	1	0	276:	28	2	1	0
205:	21	2	1	0	241:	70	0			277:	254	5	1	0
206:	147	2	1	0	242:	81	4	1	0	278:	5	0		
207:	43	0			243:	17	2	1	0	279:	5	0		
208:	83	2	1	0	244:	96	2	1	0	280:	146	3	1	0
209:	6	0			245:	37	2	1	0	281:	93	0		
210:	31	2	1	0	246:	11	2	1	0	282:	35	0		
211:	165	2	1	0	247:	82	0			283:	200	2	1	0
212:	105	0			248:	243	2	1	0	284:	119	0		
213:	62	2	1	0	249:	86	0			285:	77	2	1	0
214:	87	2	1	0	250:	103	0			286:	69	0		
215:	23	0			251:	45	2	1	0	287:	71	0		
216:	107	2	1	0	252:	67	0			288:	11	10	1	0
217:	45	0			253:	33	2	1	0	289:	21	0		
218:	11	0			254:	7	2	1	0	290:	5	3	2	0
219:	65	2	1	0	255:	52	0			291:	76	2	1	0
220:	53	3	1	0	256:	16	3	1	0	292:	97	0		
221:	18	2	1	0	257:	12	0			293:	154	3	1	0
222:	73	2	1	0	258:	83	0			294:	61	0		
223:	33	0			259:	254	2	1	0	295:	48	0		
224:	159	2	1	0	260:	74	3	1	0	296:	11	9	4	0
225:	32	0			261:	74	2	1	0	297:	5	0		
226:	57	2	1	0	262:	252	2	1	0	298:	78	2	1	0
227:	21	2	1	0	263:	93	0			299:	21	2	1	0
228:	58	2	1	0	264:	169	2	1	0	300:	7	0		

References

Aboulhamid, M. E., and E. Cerny (1983), "A Class of Test Generators for Built-In Testing," *IEEE Trans. Comput.*, **C-32**(10), 957–959.

Abraham, J. A. (1981), "Functional Level Test Generation for Complex Digital Systems," 1981 International Test Conference, Philadelphia, PA, October, pp. 461–462.

Agarwal, V. K. (1983), "Increasing Effectiveness of Built-In Testing by Output Data Modification," 13th Annual Fault-Tolerant Computing Symposium, Milan, Italy, June, pp. 227–233.

Agarwal, V. K., and E. Cerny (1981), "Store and Generate Built-In-Testing Approach," 11th Annual Fault-Tolerant Computing Symposium, Portland, ME, June, pp. 35–40.

Agrawal, V. D., and M. R. Mercer (1982), "Testability Measures—What Do They Tell Us?," 1982 International Test Conference, Philadelphia, PA, November, pp. 391–396.

Ando, H. (1980), "Testing VLSI with Random Access Scan," Digest of Papers, COMPCON 80, February, pp. 50–52.

Arzoumanian, Y., and J. Waicukauski (1981), "Fault Diagnosis in an LSSD Environment," 1981 International Test Conference, Philadelphia, PA, October, pp. 86–88.

Bardell, P. H., and W. H. McAnney (1981), "A View from the Trenches: Production Testing of a Family of VLSI Multichip Modules," 11th Annual Fault-Tolerant Computing Symposium, Portland, ME, June, pp. 281–283.

Bardell, P. H., and W. H. McAnney (1982), "Self-Testing of Multichip Logic Modules," 1982 International Test Conference, Philadelphia, PA, November, pp. 200–204.

Bardell, P. H., and W. H. McAnney (1985), Simultaneous Self-Testing System, U.S. Patent 4,513,418, April 23.

Bardell, P. H., and W. H. McAnney (1986), "Pseudorandom Arrays for Built-In Tests," *IEEE Trans. Comput.*, **C-35**(7), 653–658.

Bardell, P. H., and T. H. Spencer (1984), "A Class of Shift-Register Sequence Generators: Hurd Generators Applied to Built-In Test," IBM Corporation Technical Report TR 00.3300, September, IBM, Poughkeepsie, NY.

Barzilai, Z., J. Savir, G. Markowsky, and M. G. Smith (1981), "The Weighted Syndrome Sums Approach to VLSI Testing," *IEEE Trans. Comput.*, **C-30**(12), 996–1000.

Barzilai, Z., D. Coppersmith, and A. L. Rosenberg (1983), "Exhaustive Generation of Bit Patterns with Applications to VLSI Self-Testing," *IEEE Trans. Comput.*, **C-32**(2), 190–194.

Beauchamp, K. G. (1975), *Walsh Functions and Their Applications*, Academic Press, NY.

Bennetts, R. G., C. M. Maunder, and G. D. Robinson (1981), "CAMELOT: A Computer-Aided Measure for Logic Testability," *IEE Proc. E*, **128**, 177–189.

Berg, W. C., and R. D. Hess (1982), "COMET: A Testability Analysis and Design Modification Package," 1982 International Test Conference, Philadelphia, PA, November, pp. 364–378.

Bhattacharya, B. B., and B. Gupta (1983), "Anomalous Effect of a Stuck-At Fault in a Combinational Logic Circuit," *Proc. IEEE*, **71**(6), 779–780.

Bhavsar, D. K. (1985), "Concatenable Polydividers: Bit-Sliced LFSR Chips for Board Self-Test," 1985 International Test Conference, Philadelphia, PA, November, pp. 88–93.

Bhavsar, D. K., and R. W. Heckelman (1981), "Self-Testing by Polynomial Division," 1981 International Test Conference, Philadelphia, PA, October, pp. 208–216.

Bhavsar, D. K., and B. Krishnamurthy (1984), "Can We Eliminate Fault Escape in Self Testing by Polynomial Division (Signature Analysis)?," 1984 International Test Conference, Philadelphia, PA, October, pp. 134–139.

Birkhoff, G., and S. MacLane (1953), *A Survey of Modern Algebra*, rev. ed., Macmillan, NY.

Bottorff, P. S., R. E. France, N. H. Garges, and E. J. Orosz (1977), "Test Generation for Large Logic Networks," 14th IEEE Design Automation Conference, June, pp. 479–485.

Bottorff, P. S., S. DasGupta, R. G. Walther, and T. W. Williams (1983), "Self-Testing Scheme Using Shift Register Latches," *IBM Tech. Disclosure Bull.*, **25**(10), 4958–4960.

Bozorgui-Nesbat, S., and E. J. McCluskey (1980), "Structured Design for Testability to Eliminate Test Pattern Generation," 10th Annual Fault-Tolerant Computing Symposium, Kyoto, Japan, October, pp. 158–163.

Brahme, D., and J. A. Abraham (1984), "Functional Testing of Microprocessors," *IEEE Trans. Comput.*, **C-33**(6), 475–485.

Brglez, F., P. Pownall, and R. Hum (1984), "Applications of Testability Analysis: From ATPG to Critical Delay Path Tracing," 1984 International Test Conference, Philadelphia, PA, October, pp. 705–712.

Brillhart, J., D. H. Lehmer, J. L. Selfridge, B. Tuckerman, and S. S. Wagstaff, Jr. (1983), *Factorization of $b^n \pm 1, b = 2, 3, 5, 6, 7, 10, 11, 12$, up to High Powers*, American Mathematical Society, Providence, RI.

Buehler, M. G., and M. W. Sievers (1982), "Off-Line, Built-In Test Techniques for VLSI Circuits," *Computer*, **15**(6), 69–82.

Carter, J. L. (1982), "The Theory of Signature Testing for VLSI," 14th ACM Symposium on the Theory of Computing, San Francisco, CA, May, pp. 66–76.

Carter, W. C. (1982a), "The Ubiquitous Parity Bit," 12th Annual Fault-Tolerant Computing Symposium, Santa Monica, CA, June, pp. 289–296.

Carter, W. C. (1982b), "Signature Testing with Guaranteed Bounds for Fault Coverage," 1982 International Test Conference, Philadelphia, PA, November, pp. 75–82.

Chen, C. L. (1986), "Linear Dependencies in Linear Feedback Shift Registers," *IEEE Trans. Comput.*, **C-35**(12), 1086–1088.

Chen, T.-H., and M. A. Breuer (1985), "Automatic Design for Testability via Testability Measures," *IEEE Trans. Computer-Aided Design*, **CAD-4**(1), 3–11.

Curtin, J. J., and J. A. Waicukauski (1983), "Multi-Chip Module Test and Diagnostic Methodology," *IBM J. Res. Develop.*, **27**(1), 27–34.

Daehn, W., and J. Mucha (1981a), "Hardware Test Pattern Generation for Built-In Testing," 1981 International Test Conference, Philadelphia, PA, October, pp. 110–113.

Daehn, W., and J. Mucha (1981b), "A Hardware Approach to Self-Testing of Large Programmable Logic Arrays," *IEEE Trans. Comput.*, **C-30**(11), 829–833.

David, R. (1978), "Feedback Shift Register Testing," 8th Annual Fault-Tolerant Computing Symposium, Toulouse, France, June, pp. 103–107.

David, R. (1980), "Testing by Feedback Shift Register," *IEEE Trans. Comput.*, **C-29**(7), 668–673.

David, R. (1984), "Signature Analysis of Multi-Output Circuits," 14th Annual Fault-Tolerant Computing Symposium, Kissimmee, FL, June, pp. 366–371.

de Bruijn, N. G. (1946), "A Combinatorial Problem," *Nederl. Akad. Wetensch. Proc. Ser. A*, **49**(Part 2), 758–764. Also *Indag. Math.*, **8**, 461–467.

Dervisoglu, B. L. (1984), "VLSI Self-Testing Using Exhaustive Bit Patterns," Technical Report 685, MIT Lincoln Laboratory, May, Lexington, MA (ESD-TR-84-019).

Eichelberger, E. B., and E. Lindbloom (1983), "Random-Pattern Coverage Enhancement and Diagnosis for LSSD Logic Self-Test," *IBM J. Res. Develop.*, **27**(3), 265–272.

Eichelberger, E. B., and T. W. Williams (1978), "A Logic Design Structure for LSI Testability," *J. Design Automat. Fault-Tolerant Comput.*, **2**(2), 165–178.

Eichelberger, E. B., T. W. Williams, E. J. Muehldorf, and R. G. Walther (1978), "A Logic Design Structure for Testing Internal Arrays," 3rd USA–JAPAN Computer Conference, San Francisco, CA, October, pp. 266–272.

Eldred, R. D. (1959), "Test Routines Based on Symbolic Logical Statements," *J. Assoc. Comput. Mach.*, **6**(1), 33–36.

Frohwerk, R. A. (1977), "Signature Analysis: A New Digital Field Service Method," *Hewlett-Packard J.*, **28**(9), 2–8.

Fujiwara, H., and T. Shimono (1983), "On the Acceleration of Test Generation Algorithms," 13th Annual Fault-Tolerant Computing Symposium, Milan, Italy, June, pp. 98–105.

Funatsu, S., N. Wakatsuki, and T. Arima (1975), "Test Generation Systems in Japan," 12th Design Automation Symposium, June, pp. 114–122.

Goel, P. (1980), "Test Generation Costs Analysis and Projections," 17th Annual IEEE Design Automation Conference, June, pp. 77–84.

Goel, P. (1981), "An Implicit Enumeration Algorithm to Generate Tests for Combinational Logic Circuits," *IEEE Trans. Comput.*, **C-30**(3), 215–222.

Goel, P., and M. T. McMahon (1982), "Electronic Chip-In-Place Test," 1982 International Test Conference, Philadelphia, PA, November, 83–90.

Goel, P., and B. C. Rosales (1981), "PODEM-X: An Automatic Test Generation System for VLSI Logic Structures," 18th IEEE Design Automation Conference, June, pp. 260–268.

Goldstein, L. H. (1979), "Controllability/Observability Analysis of Digital Circuits," *IEEE Trans. Circuits Systems*, **CAS-26**(9), 685–693.

Goldstein, L. H., and E. L. Thigpen (1980), "SCOAP: Sandia Controllability/Observability Analysis Program," 17th ACM/IEEE Design Automation Conference, Minneapolis, MN, June, pp. 190–196.

Golomb, S. W. (1982), *Shift Register Sequences*, rev. ed., Aegean Park Press, Laguna Hills, CA.

Good, I. J. (1946), "Normal Recurring Decimals," *J. London Math. Soc.*, **21**(Part 3), 167–172.

Grason, J. (1979), "TMEAS, A Testability Measurement Program," 16th Design Automation Conference, San Diego, CA, June, pp. 156–161.

Gutfreund, K. (1983), "Integrating the Approaches to Structured Design for Testability," *VLSI Design*, **4**(6), 34–37, 40–42.

Hassan, S. Z., and E. J. McCluskey (1984), "Increased Fault Coverage through Multiple Signatures," 14th Annual Fault-Tolerant Computing Symposium, Kissimmee, FL, June, pp. 354–359.

Hassan, S. Z., D. J. Lu, and E. J. McCluskey (1983), "Parallel Signature Analyzers— Detection Capability and Extensions," 26th IEEE Computer Society International Conference, COMPCON Spring '83, San Francisco, CA, February–March, pp. 440–445.

Hayes, J. P. (1975a), "Detection of Pattern-Sensitive Faults in Random-Access Memories," *IEEE Trans. Comput.*, **C-24**(2), 150–157.

Hayes, J. P. (1975b), "Testing Logic Circuits by Transition Counting," 5th Annual Fault-Tolerant Computing Symposium, Paris, France, June, pp. 215–219.

Hayes, J. P. (1976a), "Transition Count Testing of Combinational Logic Circuits," *IEEE Trans. Comput.*, **C-25**(6), 613–620.

Hayes, J. P. (1976b), "Check Sum Methods for Test Data Compression," *J. Design Automat. Fault-Tolerant Comput.*, **1**(1), 3–17.

Hayes, J. P., and A. D. Friedman (1974), "Test Point Placement to Simplify Fault Detection," *IEEE Trans. Comput.*, **C-23**(7), 727–735.

Hsiao, M. Y. (1969), "Generating *PN* Sequences in Parallel," 3rd Annual Princeton Conference on Information Sciences and Systems, March, pp. 397–401.

Hsiao, M. Y., A. M. Patel, and D. K. Pradhan (1977), "Store Address Generator with On-Line Fault-Detection Capability," *IEEE Trans. Comput.*, **C-26**(11), 1144–1147.

Hsiao, T.-C., and S. C. Seth (1984), "An Analysis of the Use of Rademacher–Walsh Spectrum in Compact Testing," *IEEE Trans. Comput.*, **C-33**(10), 934–937.

Hurd, W. J. (1974), "Efficient Generation of Statistically Good Pseudonoise by Linearly Interconnected Shift Registers," *IEEE Trans. Comput.*, **C-23**(2), 146–152.

Ibarra, O. H., and S. K. Sahni (1975), "Polynomially Complete Fault Detection Problems," *IEEE Trans. Comput.*, **C-24**(3), 242–249.

Illman, R. J. (1985), "Self-Tested Data Flow Logic: A New Approach," *IEEE Design Test Comp.*, **2**(2), 50–58.

Jain, S. K., and V. D. Agrawal (1985), "Statistical Fault Analysis," *IEEE Design Test Comp.*, **2**(1), 38–44.

Katopis, G. A. (1985), "Delta-I Noise Specification for a High-Performance Computing Machine," *Proc. IEEE*, **73**(9), 1405–1415.

Kjeldsen, K., and E. Andresen (1980), "Some Randomness Properties of Cascaded Sequences," *IEEE Trans. Inform. Theory*, **IT-26**(2), 227–232.

Koenemann, B., J. Mucha, and G. Zwiehoff (1979), "Built-In Logic Block Observation Techniques," 1979 Test Conference, Cherry Hill, NJ, October, pp. 37–41.

Koenemann, B., J. Mucha, and G. Zwiehoff (1980), "Built-In Test for Complex Digital Integrated Circuits," *IEEE J. Solid-State Circuits*, **SC-15**(3), 315–318.

Koenemann, B., J. Mucha, and G. Zwiehoff (1983), Logic Module for Integrated Digital Circuits, U.S. Patent 4,377,757, March 22.

Koren, I. (1979), "Analysis of the Signal Reliability Measure and an Evaluation Procedure," *IEEE Trans. Comput.*, **C-28**(3), 244–249.

Lake, R. (1986), "A Fast 20K Gate Array with On-Chip Test System," *VLSI System Design*, **7**(6), 46–47, 50–51, 54–55.

Lempel, A., and M. Cohn (1985), "Design of Universal Test Sequences for VLSI," *IEEE Trans. Inform. Theory*, **IT-31**(1), 10–17.

Losq, J. (1978), "Efficiency of Random Compact Testing," *IEEE Trans. Comput.*, **C-27**(6), 516–525.

MacWilliams, F. J., and N. J. A. Sloane (1976), "Pseudo-Random Sequences and Arrays," *Proc. IEEE*, **64**(12), 1715–1729.

Markowsky, G. (1987), "Bounding Signal Probabilities in Combinational Circuits," *IEEE Trans. Comput.*, to appear.

McAnney, W. H. (1985), Parallel Path Self-Testing System, U.S. Patent 4,503,537, March 5.

McAnney, W. H., and J. Savir (1986), "Built-In Checking of the Correct Self-Test Signature," 1986 International Test Conference, Washington, DC, September, pp. 54–58.

McAnney, W. H., P. H. Bardell, and V. P. Gupta (1984), "Random Testing for Stuck-At Storage Cells in an Embedded Memory," 1984 International Test Conference, Philadelphia, PA, October, pp. 157–166.

Nomura, T., H. Miyakawa, H. Imai, and A. Fukada (1971), "Some Properties of the $\gamma\beta$-Plane and Its Extension to Three-Dimensional Space," *Electron. Comm. Japan*, **54-A**(8), 27–34.

Parker, K. P. (1976), "Compact Testing: Testing with Compressed Data," 6th Annual Fault-Tolerant Computing Symposium, Pittsburgh, PA, June, pp. 93–98.

Parker, K. P., and E. J. McCluskey (1975), "Probabilistic Treatment of General Combinational Networks," *IEEE Trans. Comput.*, **C-24**(6), 668–670.

Peterson, W. W., and E. J. Weldon, Jr. (1972), *Error-Correcting Codes*, 2nd ed., MIT Press, Cambridge, MA.

Porter, E. H. (1980), "Testability Considerations in a VLSI Design Automation System," 1980 Test Conference, Philadelphia, PA, November, pp. 26–28.

Ratiu, I. M. (1983), "VICTOR: Global Redundancy Identification and Test Generation for VLSI Circuits," Memorandum No. UCB/ERL M83/27, May, Electronics Research Laboratory, College of Engineering, University of California, Berkeley, CA.

Ratiu, I. M., A. Sangiovanni-Vincentelli, and D. O. Pederson (1982), "VICTOR: A Fast VLSI Testability Analysis Program," 1982 International Test Conference, Philadelphia, PA, November, pp. 397–401.

Reed, I. S., and R. M. Stewart (1962), "Note on the Existence of Perfect Maps," *IRE Trans. Inform. Theory*, **8**, 10–12.

Resnick, D. R. (1983), "Testability and Maintainability with a New 6K Gate Array," *VLSI Design*, **4**(2), 34–38.

Robinson, G. D. (1983), "HITEST—Intelligent Test Generation," 1983 International Test Conference, Philadelphia, PA, October, pp. 311–323.

Roth, J. P. (1966), "Diagnosis of Automata Failures: A Calculus and a Method," *IBM J. Res. Develop.*, **10**(4), 278–291.

Sainte-Marie, C. F. (1894), "Solution to Question No. 48," *L'Intermediare des Mathematiciens*, **1**(6), 107–110.

Savir, J. (1980a), "Testing for Single Intermittent Failures in Combinational Circuits by Maximizing the Probability of Fault Detection," *IEEE Trans. Comput.*, **C-29**(5), 410–416.

Savir, J. (1980b), "Syndrome-Testable Design of Combinational Circuits," *IEEE Trans. Comput.*, **C-29**(6), 442–451; correction **C-29**(11), 1012–1013.

Savir, J. (1980c), "Detection of Single Intermittent Faults in Sequential Circuits," *IEEE Trans. Comput.*, **C-29**(7), 673–678.

Savir, J. (1983), "Good Controllability and Observability Do Not Guarantee Good Testability," *IEEE Trans. Comput.*, **C-32**(12), 1198–1200.

Savir, J. (1986), "The Bidirectional Double Latch (BDDL)," *IEEE Trans. Comput.*, **C-35**(1), 65–66.

Savir, J., and P. H. Bardell (1984), "On Random Pattern Test Length," *IEEE Trans. Comput.*, **C-33**(6), 467–474.

Savir, J., and W. H. McAnney (1986), "Random Pattern Testability of Delay Faults," 1986 International Test Conference, Washington, DC, September, pp. 263–273.

Savir, J., G. S. Ditlow, and P. H. Bardell (1984), "Random Pattern Testability," *IEEE Trans. Comput.*, **C-33**(1), 79–90.

Segers, M. T. M. (1981), "A Self-Test Method for Digital Circuits," 1981 International Test Conference, Philadelphia, PA, October, pp. 79–85.

Seth, S. C., L. Pan, and V. D. Agrawal (1985), "PREDICT—Probabilistic Estimation of Digital Circuit Testability," 15th Annual Fault-Tolerant Computing Symposium, Ann Arbor, MI, June, pp. 220–225.

Shanks, J. L. (1969), "Computation of the Fast Walsh–Fourier Transform," *IEEE Trans. Comput.*, **C-18**(5), 457–459.

Shedletsky, J. J., and E. J. McCluskey (1975), "The Error Latency of a Fault in a Combinational Digital Circuit," 5th Annual Fault-Tolerant Computing Symposium, Paris, France, June, pp. 210–214.

Shedletsky, J. J., and E. J. McCluskey (1976), "The Error Latency of a Fault in a Sequential Digital Circuit," *IEEE Trans. Comput.*, **C-25**(6), 655–659.

Smith, G. L. (1985), "Model for Delay Faults Based upon Paths," 1985 International Test Conference, Philadelphia, PA, November, pp. 342–349.

Smith, J. E. (1980), "Measures of the Effectiveness of Fault Signature Analysis," *IEEE Trans. Comput.*, **C-29**(6), 510–514.

Sridhar, T., D. S. Ho, T. J. Powell, and S. M. Thatte (1982), "Analysis and Simulation of Parallel Signature Analyzers," 1982 International Test Conference, Philadelphia, PA, November, pp. 656–661.

Stahnke, W. (1973), "Primitive Binary Polynomials," *Math. Comp.*, **27**(124), 977–980.

Stephenson, J. E., and J. Grason (1976), "A Testability Measure for Register Transfer Level Digital Circuits," 6th Annual Fault-Tolerant Computing Symposium, Pittsburgh, PA, June, pp. 101–107.

Stewart, J. H. (1977), "Future Testing of Large LSI Circuit Cards," 1977 Semiconductor Test Symposium, Cherry Hill, NJ, October, pp. 6–15.

Stewart, J. H. (1978), "Application of Scan/Set for Error Detection and Diagnostics," 1978 Semiconductor Test Conference, Cherry Hill, NJ, October–November, pp. 152–158.

Susskind, A. K. (1983), "Testing by Verifying Walsh Coefficients," *IEEE Trans. Comput.*, **C-32**(2), 198–201.

Tang, D. T., and C. L. Chen (1983), "Logic Test Pattern Generation Using Linear Codes," 13th Annual Fault-Tolerant Computing Symposium, Milan, Italy, June, pp. 222–226.

Tretter, S. A. (1974), "Properties of PN^2 Sequences," *IEEE Trans. Inform. Theory*, **IT-20**(2), 295–297.

Tzidon, A., I. Berger, and M. Yoeli (1978), "A Practical Approach to Fault Detection in Combinational Networks," *IEEE Trans. Comput.*, **C-27**(10), 968–971.

Van der Waerden, B. L. (1949), *Modern Algebra*, vol. I and vol. II (1950), Ungar, NY.

Wadsack, R. L. (1978), "Fault Modeling and Logic Simulation of CMOS and MOS Integrated Circuits," *Bell System Tech. J.*, **57**(5), 1449–1474.

Waicukauski, J. A., E. Lindbloom, B. Rosen, and V. Iyengar (1986), "Transition Fault Simulation by Parallel Pattern Single Fault Propagation," 1986 International Test Conference, Washington, DC, September, pp. 542–549.

Writer, P. L. (1975), "Design for Testability," IEEE Automatic Support System Symposium for Advanced Maintainability, Westbury, NY, October, pp. 84–87.

Zasio, J. J. (1983), "Shifting Away from Probes for Wafer Test," 26th IEEE Computer Society International Conference, COMPCON Spring '83, San Francisco, CA, February–March, pp. 395–398.

Zasio, J. J. (1985), "Non Stuck Fault Testing of CMOS VLSI," 30th IEEE Computer Society International Conference, COMPCOM Spring '85, San Francisco, CA, February, pp. 388–391.

Zorian, Y., and V. K. Agarwal (1984), "Higher Certainty of Error Coverage by Output Data Modification," 1984 International Test Conference, Philadelphia, PA, October, pp. 140–147.

Index

CPSIA information can be obtained at www.ICGtesting.com
Printed in the USA
BVOW09*1752091214

378177BV00004B/13/A